SONY

OTHER BOOKS BY JOHN NATHAN

Mishima, A Biography

AS TRANSLATOR

Teach Us to Outgrow Our Madness
Kenzaburo Oe

A Personal Matter
Kenzaburo Oe

The Sailor Who Fell from Grace with the Sea
Yukio Mishima

SONY

THE PRIVATE LIFE

John Nathan

HOUGHTON MIFFLIN COMPANY

Boston New York 1999

For information about permission
to reproduce selections from this book, write to Permissions,
Houghton Mifflin Company,
215 Park Avenue South, New York, New York 10003.

Library of Congress Cataloging-in-Publication Data

Nathan, John, date.
Sony : the private life / John Nathan.
p. cm.
Includes index.
ISBN 0-395-89327-5
1. Soni Kabushiki Kaisha — History. 2. Electronic
industries — Japan — History. I. Title.
HD9696.A3J367623 1999
338.7'6213'0952 — dc21 99-29810 CIP

Printed in the United States of America

QUM 10 9 8 7 6 5 4 3 2 1

The author gratefully acknowledges permission to quote from:
David Sanger in the *New York Times*, September 28, 1989; October 4,
1989; and February 18, 1990. Copyright 1989, 1990, by the New York
Times Co. Reprinted by permission. Michael Cieply and Alan Citron in
the *Los Angeles Times*, April 15, 1990. Copyright 1990, Los Angeles
Times. Reprinted by permission. Jon Landau in *Billboard*, August 24,
1990. Copyright 1990 BPI Communications Inc. Used with
permission from Billboard Magazine.

For my wife, Diane,

and our children, Emily and Toby

CONTENTS

Acknowledgments ix

Preface xi

1 The Founding Fathers: In Search of a Postwar Dream 1

2 Ibuka the Muse: From "Talking Paper" to Trinitron 25

3 Akio Morita: Discovering America 51

4 Morita the Dazzler: The Man Behind the Mask 68

5 Sony's First American: Lessons in Logic from Harvey Schein 93

6 Maestro Ohga: The Art of Profit 116

7 Extending the Family: The Rise of Mickey Schulhof 158

8 One for Chairman Akio: The Columbia Pictures Acquisition 180

9 Hollywood Continued: Tailspin 220

10 Ohga and Schulhof: A Tale of Love and Hubris 241

11 Idei the Heretic: Empire's End 280

Index 327

ACKNOWLEDGMENTS

I could not have written this book without the cooperation of the Sony Corporation, which allowed me unlimited access to its executives in Japan and the United States, and I wish to thank Sony for its openness. In particular, I am grateful to the individuals who granted me lengthy interviews and to whom I returned frequently over the course of two years as new questions arose. Readers will recognize many of the names that follow as principal contributors to the company's growth from its founding in the fall of 1945 to the present day:

Kunitake Ando, Teruo Aoki, Sato Ashizawa, John Briesch, H. Paul Burak, John N. Calley, Toshitada Doi, Glen Doman, Michel Galiana-Mingot, Peter Guber, Tsunao Hashimoto, Kozo Hiramatsu, Tamotsu Iba, Makoto Ibuka, Nobuyuki Idei, Shigeru Inagaki, Ken Iwaki, Yoshiyuki Kaneda, Nobutoshi Kihara, Makoto Kikuchi, Henry A. Kissinger, Junichi "Steve" Kodera, Yasuhiro Kuga, Hiroko Kurata, Otto Graf Lambsdorff, Yang Hun Lee, Teruo "Ted" Masaki, Frank Mascarich, Senri Miyaoka, Junya Morihara, Masayoshi "Mike" Morimoto, Minoru Morio, Hideo "Joe" Morita, Kazuaki Morita, Masaaki Morita, Masao Morita, Yoshiko Morita, Masahiko Morizono, Thomas Mottola, Masakazu Namiki, Yoshiyuki "Yuki" Nozoe, Norio Ohga, Mitsuru Ohki, Kozo Ohsone, Hiroko Onoyama, Peter G. Peterson, Tom Pollock, Irving Sagor, Harvey Schein, Jack Schmuckli, Michael P. Schulhof, Charles A. Steinberg, Howard Stringer, Kenji Tamiya,

Yoshio "Terry" Terasawa, Mario Tokoro, Teruhisa "Terry" Tokunaka, Katsuaki Tsurushima, Hajime "Jimmy" Unoki, Neil Vander Dussen, Toshiyuki Yamada, Hiroshi Yamaguchi, Tadashi Yamamoto, Carl Yankowski, Walter Yetnikoff, Susumu Yoshida, Noboru Yoshii.

I would like to register a special salute of admiration and respect for Akira Higuchi, the most inspiring and delightful octogenerian I have had the honor to meet. Mr. Higuchi was among the original seven engineers who followed Masaru Ibuka to Tokyo in October 1945 to found the company that would become Sony. He was the first factory manager, and, subsequently as head of personnel, the gatekeeper who employed many of the executives who lead Sony today. Currently the only member of the Sony family with the official title of "adviser," he drives himself to work by 8:00 A.M. and spends his day receiving visitors and organizing his papers. In the center of his large office is a globe covered with close to one hundred tiny paper flags that mark the mountains he has scaled. In the winter of 1998, just after his eighty-fifth birthday, he phoned me to say hello from Lake Tahoe, where, as head of the Mountaineering Club, he had led a group of Sony skiers. Our afternoons together, sipping green tea as he recalled the earliest years of Sony history in astonishing detail, are memories I shall always treasure.

Throughout the writing of the book, I relied on Sony's offices of Corporate Communications in Tokyo and New York for logistical support and primary source materials. My thanks in particular to Mitsuru Ohki for opening the doors to me, to Aldo Liguori and Ms. Maki Kawanishi in the International PR Group for going out of their way to respond to my often troublesome requests, and to Ms. Mika Ishida, my constant companion during the interminable and stressful rounds of interviewing in Tokyo. Needless to say, I alone am responsible for my interpretations and for any errors that may appear.

PREFACE

In the spring of 1995, the Sony Corporation commissioned me to produce a documentary video for its fiftieth anniversary. My theme was defined nebulously as "the Sony founders' spirit." Sony's cofounders, Masaru Ibuka and Akio Morita, had suffered strokes within a year of each other in 1992 and 1993 which had deprived them both of speech, and I attempted to convey a composite portrait of these extraordinary men by editing together reminiscences from their associates and friends in Japan and the United States.

I had been hired to create an anniversary confection, and my on-camera questions were designed to evoke bright memories, which were abundant. Sony is one of the business wonders of the modern world. The story of its birth in the rubble of postwar Tokyo and product-by-product growth to market leadership in consumer electronics is a lucent paradigm of Japan's economic resurgence. Other Japanese companies founded in the early postwar years have also flourished; Honda Motors, Kyoto Ceramics, and Canon Camera come to mind. But none has managed to achieve Sony's global status: in 1998, according to a Harris poll, Sony overtook Coca-Cola and General Electric to become the brand name best known and most highly esteemed by the American consumer.

But phenomenal success was only part of the story. Listening to firsthand accounts of the company's four decades in the United States,

I sensed difficulty and conflict on a fundamental cultural level. I began to perceive that the interior drama of Sony's growth modeled the Japanese intellectual and social experience in the postwar period as the country struggled to establish and maintain authentic individuality in its relationship to the West. By the time I had finished the film, I was certain that examining that struggle inside Sony at close range, illuminating the personal relationships at the heart of the organization, would open a window of understanding on the definitive challenge of Japan's coming of age in the twentieth century: reconciling two disparate and often irreconcilable views of reality and approaches to coping with it, one deeply rooted in native tradition, the other intractably foreign.

In September 1996, I went to see Mitsuru Ohki, then head of Corporate Communications, and told him that I wanted to write a book. His first question was how much I would expect to be paid for such a project. I explained that the intimate history I had in mind would have to be written independently of the company to be taken seriously. In fact, to approach a reputable publisher, I would need a letter from Sony guaranteeing me access to anyone I wished to see and relinquishing rights to review and approval. Within a month, I had a letter in English which closely followed a draft I had proposed. As anyone who has attempted to penetrate a global business organization will agree, particularly in Japan, it granted me extraordinary freedom:

> November 14, 1996
> Dear Professor Nathan:
>
> We are pleased to learn that you will be writing a book about Sony Corporation, its history and operations in the United States, including Hollywood. With the depth of your understanding of Sony, we feel that you are in a favorable position to write such a book. We are prepared to cooperate with your efforts to research this book by arranging all requested interviews and providing relevant documents and photographs from our archives.
>
> We understand that your book is a project entirely independent of Sony and that we have no editorial control whatsoever.
>
> We look forward to a book that we hope will contribute to improved

understanding of not only Sony, but of Japanese business society in general.

Signed, Mitsuru Ohki
Senior General Manager
Corporate Communications
Sony Corporation

Sony kept its word. Interviews were arranged, often more than once, with everyone I asked to meet. Between June 1997 and December 1998, on seven different trips to Tokyo, I conducted 115 interviews in Japanese with all the key figures in the Sony story who are alive today. In New York and Los Angeles, I interviewed other Sony executives as well as current and former associates of the company who have played important roles in Sony's international history.

With people outside the organization, I met privately. For interviews arranged by Sony with Sony employees, there was a procedure. I was asked to submit "sample questions" in advance, and was always accompanied by one and sometimes by two staffers from Corporate Communications, who took notes and made their own tape recordings of the conversations to ensure that I misquoted no one. As the volume of taped interviews grew to hundreds of hours, I warned my young companions that I was quoting bits and pieces from everywhere, and jokingly inquired how they intended even to locate the original passages in Japanese. Each chapter, they assured me with mock seriousness, would be assigned to a different department member.

In the beginning, I was concerned that the presence of observers in the room would inhibit candor; I even imagined that their purpose was to prevent sensitive information from being revealed. I soon learned that I had no cause to worry. To be sure, my probing into delicate matters was often unavailing—there is no language better suited to obfuscation than Japanese. At other times, I was told things that have never been disclosed inside or outside the organization. At these revelations—literally breathtaking moments that occurred with surprising frequency—I often observed my companions sitting stiffly at the far end of the room, struggling, as I was, to maintain composure. Lest they be blamed for the sensitive material that appears in these pages,

I should add that I was interviewing the legends of Sony history, in the Japanese phrase, "people above the clouds"—interrupting them to steer the conversation elsewhere was unimaginable. As I probed more deeply into the company's personal history, I grew apprehensive, looking for signs that the organization was becoming alarmed, half expecting an ultimatum to submit the manuscript before I had finished my research. None was delivered.

With Masaru Ibuka and Akio Morita incapacitated by illness, Norio Ohga, the current chairman, is the only person who knows the critical details of the Sony story from its earliest postwar days to the present, and Ohga generously accommodated my repeated requests for his time. These sessions were unpredictable. Sometimes he would subject me to set speeches I had heard before, about the value of the Sony brand or Akio Morita's "nose," his merchant's intuition about potential products. At other times, he would lead me on fascinating detours, comparing cockpit design in Lear and Falcon jets or recalling details of an after-hours meal with Leonard Bernstein at a Tokyo restaurant, which he had been obliged to book for the evening because the maestro insisted on signing autographs before he was ready to dine.

But he was just as likely to speak from the heart, as if I were an old friend, about the very things that mattered most to me, like his relationship to "Mickey" Schulhof, the American he considered his kid brother, or his loneliness as he agonized over choosing his successor. Ohga is a formal man; in all the hours I spent with him, he never removed his jacket or loosened his tie, but I did notice (and learned to recognize it as a signal) that he would slip his feet out of his patent leather shoes when he was preparing to take me into his confidence.

Before meetings with Ohga's chosen successor, Sony president Nobuyuki Idei, I invariably had a knot in my stomach. Idei is by nature an edgy person, and his chronic jet lag and dyspepsia tend to make him cranky. Whenever I asked to see him, half a dozen times in the course of a year, he agreed, but when his mood was sour he would shift uncomfortably in his chair, glance frequently at his watch, and sigh and groan at my questions. At such times, I longed to be at home with my family in Santa Barbara. Other days, when he was feeling better, he shared his heterodox view of Sony history and his unsentimental approach to the American side of the business with frankness that amazed me. Although he referred more than once to nondisclosure agreements and asked me to use caution, he never declared anything he told me out of bounds.

I cannot account for the license I was given or for the willingness of Sony's top management over time to let down their guard with me. Japan has been described as "a black hole of the universe in terms of communications, receiving signals but not emitting them." In the official domain, this point is hard to argue, but individuals are a different matter. Judging from my own experience over the course of forty years, in their own language and under the right circumstances, the Japanese are immoderately fond of what they call "frank talk." I like to think that Norio Ohga and others at Sony were responding to my attentiveness and to what was clearly my determination to understand them.

A different partial explanation, which I advance somewhat uncomfortably, is that some people may have spoken to me as openly as they did because they assumed mistakenly that Sony would be reviewing my manuscript. There were grounds for a misunderstanding. When I began the first round of interviews in September 1997, I read the memo being sent to executives introducing me and was unsettled to discover that Corporate Communications for reasons of its own had positioned the book as a Sony project:

> *"Regarding the research for a new 'Sony Hardcover' to be written by John Nathan":*
>
> Since the publication in 1986 of [Akio Morita's memoir,] *Made in Japan*, Sony has undergone profound change as a result of rapid growth, the expansion of international operations, and diversification to meet the needs of the Digital Age. It is our view at Corporate Communications that a new book in English is needed to promote understanding of the new Sony in countries around the world. . . . Accordingly, we have asked University of California Professor of Japanese Cultural Studies John Nathan to write such a book, and are proceeding with plans for publication by Houghton Mifflin Company in the fall of 1999. . . . Your cooperation will be appreciated."

Although I emphasized at the beginning of every interview that I had not been employed by Sony and was conducting my research independently, I must acknowledge the possibility that no explanation coming from an outsider could have corrected the impression created by an official document.

In detailing the openness I was vouchsafed at Sony, I don't wish to imply that I have exposed the organization—nothing could be further from my intention—or that I came away with answers to all my important questions. On the contrary, I was often obliged to abandon an inquiry when it became clear that no one who knew the answer was prepared to disclose it. More than once, as in *Rashomon* or a Pirandello play, I encountered multiple, mutually contradictory, versions of "the truth"—the wisdom of the postmodernist assertion that history is a phantom spun from rhetoric had never seemed so certain.

Nevertheless, this book is an attempt to write history, and while I have engaged in speculation, I have felt the historian's responsibility not to fabricate. In particular, I have resisted the siren call to imagine and reconstruct conversations that I was not present to overhear. Every line of dialogue in quotation marks is indeed a direct quotation. As I conducted the majority of my interviews in the Japanese language, I have been obliged to translate from my tape recordings, and while translation is never entirely adequate, I daresay my renderings convey both the significance and the tone of the original. In cases where Japanese was inlaid with English, as often happened, I have italicized the English words used by the speaker.

Montecito, California
May 1, 1999

SONY

1

THE FOUNDING FATHERS: IN PURSUIT OF A POSTWAR DREAM

At the center of the postwar social organism called Sony Corporation stands one of business history's most productive and intriguing relationships. For over forty years, Masaru Ibuka and Akio Morita grew Sony together, from adjoining offices, reveling in each other's company. Their personal secretaries, women who devoted themselves to their well-being for dozens of years and who remain at work maintaining their deserted offices today, every book and electronic gadget in place, like to remember them facing each other on the rug, playing with a prototype that Ibuka had snatched from the hands of one his engineers and carried upstairs gleefully to show Morita. Sometimes, if one of them had just returned from a trip to America, the focus of their pleasure would be a shopping bag full of actual toys from F.A.O. Schwarz. Ibuka's son, Makoto, recalls somewhat ruefully the mechanical toys his father bought for him on trips abroad; the gifts were handed over only after having been taken apart and reassembled, or partially reassembled, in the office of the president. Morita's second son, Masao, remembers that a visit to the founders' offices was like stepping inside a toy box. His father always purchased two of any toy or mechanical gimmick—an electric potato peeler, for example—one for himself and one for Ibuka. Ibuka had a lifelong passion for electric trains and was president of the Japan Association of Microtrains; for a period of years, a narrow-gauge track was installed along the walls of his office. Morita collected mechanical organs, music boxes, and player pianos. His favorite toy was a remote-controlled helium balloon.

When they were both at work, Ibuka and Morita took lunch together, sometimes inviting someone else to join them in the company dining room. Each knew the other's office as well as his own: Ibuka was an inveterate tinkerer, and as he sat at his desk, fixing a watch or a radio, he would call out to his secretary, "Go into Morita's office and see if there isn't a set of small screwdrivers in the third drawer of his desk."

If Ibuka and Morita had serious disagreements over the years, they resolved them privately. Regarding company policy they spoke with one voice; no one in or outside Sony ever heard either of them criticize the other. Nor has anyone who has ever seen them together failed to remark on the exclusive bond that seemed to unite them in a mysterious way. According to Makoto Ibuka: "They were closer than lovers, even Mrs. Morita felt that. They were bound together by a tie so tight it was more like love than friendship. The connection was so deep that not even their wives could break into it when they were together. Even now, when they're both sick, when Mr. Morita comes to visit my father or my father pays Mr. Morita a visit, they sit together in silence, holding hands, the tears running down their cheeks, and they're communicating without words. That's the kind of friendship they always shared."

Morita's two sons echo Makoto Ibuka's feelings about the bond between their fathers. Hideo Morita, the elder, remembers them at the dining table in his house in Tokyo: "They would sit there, talking to each other, and we would listen but we had no idea what they were saying. Each one seemed to be talking his own story, different from the other's. It was like gibberish to us, but they were understanding each other, and interrupting them for any reason was forbidden!" To Morita's younger brother, Kazuaki, as to many others in and outside the family, the intimacy of the relationship was beyond understanding: "It was truly strange. Men usually get along really well for three or maybe five years and then there's some kind of an argument. But they managed for such a long time, right up against each other, matching perfectly in their work lives and their personal lives. They were incredibly lucky to have found each other. I never saw anything like that combination!"

There were moments of discord, typically the product of tension between Ibuka's impulsiveness and naïveté and Morita's business pragmatism. Most of the anecdotes are trivial, remembered only because they were rare: Ibuka on vacation commandeers a company car in Honolulu and Morita reprimands him; Ibuka decides to sponsor an acquaintance and guarantees a loan from one of Sony's banks, angering Morita. The same variety of conflict was a recurrent theme in the management of the business, as when Ibuka nearly drove the company to ruin in the early 1960s by refusing to abandon the disastrous Chromatron technology for color television.

But even then, when Morita was desperately anxious to cut losses, there was never any question that he would attempt to block Ibuka: for Akio Morita, interfering with a plan conceived by his senior partner, obstructing his dream or vision, or in any way disappointing whatever desire he chose to entertain, however childish or irrational, would always be unthinkable. On the contrary, applying his magical persuasiveness to help Ibuka's visions come to life was in the nature of the relationship.

Morita also considered himself responsible for protecting Ibuka, whom he described as a "pure and simple soul," from those who would exploit his guilelessness. Until he was in his seventies, Ibuka customarily left the office alone when he went out on business. One day at Tokyo Station he went into the public toilet and was approached by a stranger who recognized him and wanted to say hello. When Morita heard, he was sufficiently alarmed to order that a male assistant accompany Ibuka wherever he went.

Morita had a giant, driving appetite for personal success and recognition, but in his relationship to Ibuka, he was clearly able to achieve a degree of selflessness. Hideo Morita sees in his father's devotion to the older man a longing for a big brother. He spoke to me in English: "My father was born the eldest son in a big old family and raised as a prince. All eldest sons in that position need a big brother. I know because I was similar. He had to act as head of family ever since childhood, as if he can do anything, but of course he can't, his life is full of contradiction.

He needed someone he could rely on mentally. Not for decisions or advice; he had people to advise him. Ibuka was not a businessman; he was a great straight dreamer. My father loved him for this: my father loved him for the way he could dream. Ibuka's view was totally different from what my father can ever see, so he was good for him, and he needed him."

Morita and Ibuka met late in the summer of 1944 when they were both assigned to a task force charged with developing a heat-seeking missile, code-named "Marque," in time to turn the tide of the failing war. Morita, the youngest member of the team at twenty-four, was a lieutenant in the Imperial Navy with a degree in physics from Osaka Imperial University. Though he was painfully thin and of medium height for a Japanese man, his features were striking; he had large, unblinking eyes, a high, aquiline nose, and a large, full mouth with pronounced lips. His hair, which would turn silver in his early forties, was still jet black, and he wore it parted down the middle. His hands were slender; there was about him an air of aristocratic delicacy that belied his stamina and intensity. Ibuka, thirteen years older than Morita, was a taller, heavier man with shovel hands and an ungainly manner that was a striking contrast to Morita's elegance. He wore thick eyeglasses and spoke with a heavy Tokyo workingman's accent. An electrical engineer, he was participating on the team as a civilian contractor to the military. At the time, he managed a measuring instruments company that had been supplying weapons and tactical systems to the navy since 1940.

For close to a year, the men spent intense periods of time together, brainstorming technology with other engineers and officers, and, later, by themselves, discussing Japan's future after the war. In his capacity as naval liaison officer, Morita visited Ibuka's company in the Tsukishima District of Tokyo, on the bay. Ibuka was producing radar devices under contract to the navy, and he showed Morita a room full of female students from the music academy in Ueno whom he had employed to adjust the oscillation of the transmitters to the pitch of a tuning fork. Years later, Morita would recall the moment as an example of the ingenuity that had first drawn him to the older man.

In September 1944, as the firebombing of Tokyo began, Ibuka

evacuated his company to the small town of Suzaka in Nagano Prefecture, one hundred miles northwest of Tokyo. After this, he rarely made the journey south all the way to Zushi for research committee meetings, but Morita traveled to see him several times, ostensibly in his capacity as naval technical officer overseeing production. During these visits, in the stillness of the Nagano countryside, the men seem to have deepened their mutual regard. Discussing the progress of the war, they shared their certainty that, despite the propaganda, defeat was imminent, and discovered that each other's views were based on information in shortwave broadcasts from the United States which both were monitoring illegally. Possibly, Ibuka shared with Morita his vision of the reconstructive value of technology in peacetime; whether the men discussed going into business together after the war is not known. Their last meeting before the defeat was on July 27, 1945, when they listened together to reports of the Potsdam Declaration calling for Japan's unconditional surrender. By that time, Ibuka had already resolved to return to Tokyo to begin again but seems to have decided not to tell Morita, possibly for no better reason than his customary reticence about making clear what was on his mind.

On August 6 and 9, 1945, the United States dropped atomic bombs on Hiroshima and Nagasaki. At noon on August 15, Emperor Hirohito went on the radio to inform his subjects that they must now "endure the unendurable" and lay down their arms. The emperor had never spoken publicly before, and across the islands of the Japanese archipelago virtually every citizen listened. Those who did not have radios of their own, particularly in rural areas, gathered at town halls or in the gardens of wealthy merchants to listen in silence to the emperor's reedy voice speaking unthinkable words through the static of the broadcast. Many wept openly.

Like his countrymen across the land, Ibuka, in his warehouse office in Suzaka, listened to the broadcast standing at attention in front of the radio. If he was feeling grief, humiliation, anger, or regret as he listened, these were not emotions he communicated. What impressed the handful of engineers standing with him was, on the contrary, his excitement that the war was at an end, an exhilaration many of them shared. Years later, Ibuka would recall with distaste the applications required for beginning any development project during the war years, and the bickering with the Ministry of Communications and War Office bureaucrats who knew nothing about engineering. More important, he and his colleagues were united by a passion for technology and invention: for them, the Americans, though they were the enemy, had never been the "hairy barbarians" portrayed in wartime propaganda. During the war years, Japanese engineers had based their work on American technology, which filled them with admiration and, for its superiority, chagrin. Their bible during this period was F. E. Turman's *Radio Engineering*, an American textbook that they ransacked for ideas and methods, struggling with the English, until it was dog-eared and underlined in multiple colors.

According to one of the men in the room with Ibuka that afternoon, Akira Higuchi, "Any engineer with heart was overjoyed by the war's end. We all felt that now at last we could take on some real work, not just weapons for the military but useful things, and we felt that developing real products would allow us to catch up and even move ahead of American technology." On his first trip to the United States in 1954, Higuchi went out of his way to visit Cambridge, Massachusetts, because he knew it was the home of General Radio, a transformer manufacturer well known in Japan during and after the war. When he finally located the factory on the Charles River, it looked deserted. Higuchi gazed at the old brick building from across the street until he felt satisfied that he had seen General Radio with his own eyes, and returned to New York.

Technology was Masaru Ibuka's lifelong passion. His father, a Christian from the island of Hokkaido, graduated from the Department of Electrical Engineering at Hokkaido University and went to work at a copper refinery north of Tokyo, in Nikko, where Ibuka was born in 1908. Following his father's death when he was two, he moved back and forth between his mother's house, first in Tokyo and later, when she remarried, in Kobe, and the home of his paternal grandfather, Motoi, who had grown up a samurai in the Aizu fief and had risen through the ranks of the civil service to the position of provincial deputy governor.

By his own account, Ibuka adored his mother, and was desperately lonely for her during the years when circumstances required them to live apart. In 1937, the year after his own marriage, she died at age fifty-three. His stepfather phoned him in Tokyo with news that his mother was seriously ill. Ibuka took the first train to Kobe, an overnight trip in those days, but arrived too late. For ten years, he would not set foot in Kobe. In a memoir, he wrote: "For years after I got out of college I was working so hard that I neglected her . . . if she were only alive now I'd bring her anything she wanted to make her happy, any treasure she could name."

As a second-grader in Tokyo, Ibuka was given an Erector Set, which taught him, in his own words, "the excitement of putting things together." In high school he became an avid ham radio operator, a hobby he enjoyed all his life. As a student in mechanical engineering at Waseda University, he built his own electric phonograph and designed a loudspeaker system for the athletic grounds near the Meiji Shrine in Tokyo. For his graduation project in 1930, he created something he called a "light telephone," which used high-frequency sound waves to control the intensity of a light. Ibuka succeeded in using the device to "send" a visible signal as far as a mile and a half. The success of this experiment earned him a reputation as "Ibuka, the student inventor." Subsequently, he adapted the same technology to create a product he called "dancing neon," which he patented in his own name. The product was submitted to the Paris World's Fair in 1933 and won the Gold Prize for inventions. The headline in a Tokyo newspaper read, "Basking in the light of international recognition: a genius inventor."

That same year, Ibuka took the employment examination at Shibaura Electric Company, now called Toshiba, and failed. Possibly, his focus was too internal to accommodate questions put to him by someone else. There would always be something obsessive about his intense interest in things, fixated to a point that would at times seem almost autistic. Fortunately, an entrepreneur named Taiji Uemura had heard about Ibuka's inventiveness and invited him to join his company, PCL, Photo-chemical Laboratories. Recognizing that he was not a person to be managed, Uemura gave his new employee the freedom to follow his own research interests, and Ibuka became engrossed in optical sound and its application to film. But as PCL developed a connection to the Toho film studios and began production of early "talkies," he lost interest. In 1936, he went to see his benefactor, and, explaining his distaste for film production, offered his resignation. Uemura persuaded him to transfer into a new company he had just formed to manufacture 16-millimeter sound projectors (the Japanese military was "expanding" into Asia and producing more propaganda films). Once again, Uemura gave Ibuka the run of the company, and even created a "wireless department" for him. Between 1936 and 1940, Ibuka continued his work on optical sound. As Japanese aggression in Asia expanded, he became interested in ways of combining electrical and mechanical engineering to create weapons applications, and went back to Uemura. As a result of this meeting, to provide Ibuka the opportunity to develop his new interests, Uemura founded Japan Measuring Instruments with capital he raised from the Lion Toothpaste Company.

Under Ibuka's leadership, the company prospered during the war by supplying the military with weapons systems and related products. Ibuka developed a vessel location system, which involved amplifying low-frequency disturbances in magnetic fields and which revealed, too late in the war to allow action, twenty-five American submarines submerged in the Gulf of Formosa. The company's engineers also developed a widely used telephone scrambler and the proprietary radar system that had impressed Morita. At war's end, Japan Measuring Instruments employed fifteen hundred people in a large facility located in a Nagano apple orchard. Choosing the site at a time of heavy food rationing, Ibuka had remarked, "At least we'll have apples to eat."

Ibuka knew more about the disastrous course of the war than he had learned from monitoring shortwave radio broadcasts: through his father-in-law, Tamon Maeda, a man who was to contribute significantly to Sony's early success, he was also privy to gloomy conversations taking place in diplomatic circles. Maeda was a well-known internationalist who had been attached to the Japanese embassy in Paris and had written editorials for the newspaper *Asahi Shimbun* before the war. When Japan attacked Pearl Harbor in 1941, he was in New York City working as the executive director of Japan House, and was among the first repatriates to be shipped home. During the war, under the Tojo cabinet, he was governor of the Prefecture of Niigata and, briefly, a member of the House of Peers. In the postwar cabinets of Prime Ministers Higashikuni and Shidehara, until he was purged in 1947, he was minister of education. Maeda had advocated mending the Japan–United States relationship since before the war; now, in its final moments, diplomats and politicians sympathetic to his view gathered at his summer home in Karuizawa, near the site of Ibuka's company, to discuss the future of Japan's relations with the United States following the inevitable defeat. Reports of these conversations had confirmed Ibuka's certainty that the war would soon be over.

In the silence that followed the emperor's broadcast, Ibuka announced his decision to return to Tokyo to start again, and, without inviting anyone explicitly, communicated his desire that others accompany him. This was a delicate issue. Ibuka was managing director, but not head of the company. The senior managing director, Kenzo Kobayashi, the Lion Toothpaste heir, was opposed to leaving. In his view, only a fool would abandon a viable business in the safety of the countryside, where there was food to eat, for the chaos of a devastated Tokyo.

Once Ibuka had made up his mind, he was unswervable. He pointed out that Japan Measuring Instruments was fundamentally a weapons manufacturer and would not be equipped to take advantage of the peacetime market. More important, the hub of new research and development in Japan would be Tokyo: any engineer who hoped to work on the cutting edge of technology would have to do it there. The debate ended with Kobayashi's decision to wait and see in Nagano. Ibuka asked his benefactor for permission to leave, and it was reluctantly granted. Seven of the company's engineers volunteered to go along, including Akira Higuchi, who, for the second time, left a steady job to follow Ibuka; in 1937, he had been lured away from a parts supplier called Nanao Wireless to join him at Japan Optical Sound. "I said 'Yes!' in two shakes and left without a second thought," Higuchi recalls. "It was as if we were communicating telepathically. I followed him then, and I never left him." This time, Higuchi left his wife and mother behind in Nagano for over a year until he was able to bring them to Tokyo.

On September 2, 1945, the Japanese surrender was signed aboard the USS *Missouri* in Tokyo Bay. Early in October, Ibuka and his seven engineers moved into the Shirokiya Department Store building in Nihonbashi, for 350 years the center of Tokyo's business district. Having no merchandise for sale, the store was housed on the street level of an otherwise abandoned seven-story wooden building that had been burned in the firebombing. Ibuka had prevailed on an acquaintance to sublet to him what had formerly served as the telephone switchboard room on the third floor: the engineers pushed the disconnected switchboard into a corner, moved a worktable and drill presses into the cramped space, and set up shop as the Tokyo Telecommunications Research Institute. (Akira Higuchi had suggested "Ibuka Telecommunications" and had been scolded by Ibuka for being "an idiot!") The name resounded grandly, but it was far from clear, and would remain unclear for some time, what the company could or should do, not merely to support itself, but to come anywhere near realizing Ibuka's vision for his new business.

History's great entrepreneurs have been endowed with the capacity to perceive sharply things that are invisible to ordinary eyes. Ibuka possessed this gift, and he could envision the company he wanted to build in detail even as he looked out across the nightmarish wasteland that was Tokyo in 1946. In January of that year, he set down his dream in a ten-page document he called "The Founding Prospectus." The original, in Ibuka's cramped, engineer's hand, has been preserved at Sony and may be viewed by special request, delivered to a reading room by a nervous freshman employee wearing white gloves. In the opening lines, Ibuka declares that his company will be designed by and for engineers: "My first and primary objective was establishing a stable workplace where engineers could work to their hearts' content in full consciousness of their joy in technology and their social obligation." Later he reiterates in words that have been preserved in the amber of Sony culture and can be quoted verbatim by employees even today: "Purpose of incorporation: creating an *ideal workplace, free, dynamic, and joyous*, where dedicated engineers will be able to realize their craft and skills at the highest possible level."

In a section he called "Management Policies," Ibuka revealed the idealism and naïveté that would later pose challenges to his more pragmatic partner. The following dictum about profit and growth, for example: "We shall eliminate any untoward profit-seeking, shall constantly emphasize activities of real substance, and shall not seek expansion of size for the sake of size." Elsewhere, in a section on service, he remarks on the absence of reliable dealers to repair radios damaged during the war and cautions: "We must place profit here as a secondary motive. Our service commitment should be pure and total, including even the preparation of a pamphlet that will explain to ordinary customers why their radios are in need of repair, in a manner they can understand." Ascending to even higher seriousness of purpose, he continues: "Some future tasks of our service department should include introducing the latest technology from abroad, maintaining a library of current texts on telecommunications, and establishing a lecture series to spread and extend general knowledge of electronics among consumers."

Given the national mood, which had never been bleaker, Ibuka's idealism was remarkable. The war had cost Japan roughly 25 percent of its total wealth. Most of the major cities had been destroyed, leaving millions homeless. Tokyo, like Hiroshima, Nagasaki, Nagoya, and Osaka, was in ruins: in 1946, an estimated 47 percent of its inhabitants had no roof over their heads. There was no public transportation; all of the city's trolley cars, and most of its buses, had been destroyed. Disease was endemic; in the winter of 1946, 20 percent of the urban population was afflicted with tuberculosis. That same winter, and the severe winter following, homeless people in the city's parks froze to death in their sleep.

Nearly seven million soldiers had been left stranded in China and Southeast Asia; more than half a million men came home to Tokyo and were met coldly, like the American veterans of a later war, by a nation that did not wish to be reminded of the dream from which it had been brutally awakened. There were no jobs. Soldiers lined the streets begging in their tattered uniforms or joined the outlaw gangs of *yakuza* with brass coins in their ears who ran the burgeoning black market in rice, fuel, and clothing. In 1946, annual per-capita income was seventeen dollars.

People felt more than simply deprivation; the humiliation of defeat was in the air. Japanese girls, known as "onlys," shacked up with GIs in return for nylons and canned goods from the military commissary. Adults scrambled with children for the sticks of chewing gum that Americans threw into the street from their trucks. According to a memoir by Kenzaburo Oe, sugar was so scarce that a whiff of an empty gum wrapper could make a man dizzy. It was a dark, lawless, despairing time. The novelist Osamu Dazai, a troubador of hopelessness and self-pity, became a national icon with the line, known to every Japanese, "I can offer no excuse for having been born!"

Paradoxically, it was also a jubilant time. The apocalypse in the form of the atomic bomb had come and gone, and there was life to be lived. The leftist critic Masahito Ara beautifully expressed the exuberant energy released by this realization in his February 1946 essay "A Second Prime of Life":

> When I first read that Dostoevsky had been condemned to death and then reprieved only a moment before his execution, I was filled with envy close to despair. It was not his genius that I envied so much as that abnormal experience of his that not one in a thousand or even a hundred thousand can have. . . . However, what we have just experienced with the Defeat has been in no degree less traumatic than that great Russian writer's experience. Until the emperor announced the unconditional surrender, we were prepared for collective suicide. During the air raids we had entrusted our precious lives to some corner of an air-raid shelter not fit to be called a garbage can. For years, we had been obliged to consider the severance of all bonds of love and affection for a draft notice the highest honor. . . . For those incapable of ending their lives with the slavish cry Banzai for his imperial majesty, it is no exaggeration to reflect that the thousand days now over were a living hell.
>
> Looking back, it is clear that this past year has been a year of miracles. We have seen hell, we have known heaven, we have heard the last judgment, witnessed the fall of the gods and witnessed before our eyes

> the creation of the heavens and the earth. We have accumulated incredible experiences rarer even than Dostoevsky's.

While the nationalist in Ibuka might have been affronted by Ara's Marxist view of the war, he would certainly have affirmed the critic's notion of "a year of miracles" and "a second youth," a time of optimism and renewal.

The first four years of Sony's history, from 1946 to 1949, are about a small company struggling to stay alive as it searched for a product of its own worthy of Ibuka's expectations. In the beginning, any means of making honest money had to be considered seriously: suggestions ranged from selling sweetened miso soup to building a miniature golf course on a burned-out tenement lot.

The company's first innovation was an electric rice cooker. In 1946, a crude toaster oven, which transformed unbleached flour into burned bread, was enjoying a vogue. As food was a national concern, and electricity was more accessible than fuel, Ibuka liked the idea of an electric cooker, but, as always, he was loath to imitate a product already on the market. Instead, he fitted the bottom of a wooden tub with an aluminum filament to create a rice cooker. If there was anything clever about the design, it was his idea that the rice itself would function as a timer switch; as the water steamed away and the rice dried, it would theoretically break the connection with the filament and turn off the appliance. Someone knew a place outside the city which sold wooden tubs and he arranged for a truckload, but the mechanism never functioned reliably and the cooker was unsellable. The abandoned tubs were stored on a wall of shelves in an early Sony warehouse. Standing before them during a television interview, Ibuka reminisced: "We bought more than a hundred of these tubs in Chiba and made them into rice cookers. The problem was, in those days you never knew what quality of rice you were getting. With good-quality rice, if you were careful, it came out fine. But if the rice was just a little off, if it was too moist or too dry, we'd end up with a batch that was soggy or falling apart. No matter how many times we tried, it just wouldn't come out right. This is such a simple device, there was nothing to fix! I remember sitting there on the third floor in Shirokiya day after day being fed rice that wasn't fit to eat. We simply couldn't make a product out of this, and finally we gave it up and were stuck with all these tubs."

Fortunately, there were radios to be repaired. After the war, radios were increasingly in demand as a source of music and world news. Many had been damaged during the bombing. And the military police had disconnected shortwave units to prevent access to American propaganda. Ibuka's team reconnected the shortwave coil and developed an adapter that converted medium-wave into all-wave radios. This innovation equipped sets to receive international news on shortwave and readied them to receive local broadcasting when it was reinstated.

On October 6, 1945, the journalist Ryuzo Kaji featured the company's work on radios in his column "Blue Pencil," in Japan's second largest newspaper, the *Asahi Shimbun:*

> We have welcome news that even the most ordinary radio sets can be modified to receive shortwave broadcast with a simple adjustment. Mr. Masaru Ibuka, formerly a lecturer in the Department of Science and Engineering at Waseda—and Minister of Education Tamon Maeda's son-in-law—has gone into business under the name of Tokyo Telecommunications Research Laboratory. With offices on the third floor of the Shirokiya Building in Nihonbashi, Mr. Ibuka has set out *quite apart from any commercial motive* [italics mine] to augment the use of shortwave receivers by modifying normal receivers or by connecting adapters. . . . When, in the near future, private broadcasting is licensed again and broadcasting begins on a number of different frequencies, it will be very difficult to tune through the "twist of wires" with a conventional radio set, but Mr. Ibuka assures us that the rebuilt sets or those fitted with his adapter will have no trouble picking up the new signals.

The positive bias of the article—"quite apart from any commercial motive"—is in part explained by the fact that the journalist was a friend of Tamon Maeda's, who had introduced him to his son-in-law. The *Asahi Shimbun*, a single page at this time, was widely read, and Kaji's column drew a line of customers to the Shirokiya with radios in their arms. Akira Higuchi recalls the repair business with glee: "People would bring in these deluxe American radios. Sometimes all the coils had been removed, but often we'd open them and find that a military policeman with a good head had simply clipped the shortwave coil in one spot. All we had to do was solder it together again and the radio was fixed and we could charge for it!"

But radio repair would not suffice to keep the company alive. Ibuka's mainstay during the first three years was a product that his engineers had developed during the war, a voltmeter driven by vacuum tubes. "Not only is this product highly admired in our own country," he wrote in the "Prospectus," but "the U.S. Occupation Army was so fascinated by it that they took one back to America to use as a benchmark, eloquent proof that we can hold our voltmeter up proudly to the entire world!" The inflated language conveys Ibuka's pride at having earned the admiration of American technicians: for many years to come, the postwar recovery would be fueled in large measure by a deeply felt need to catch up with American technology.

Ibuka calculated that he could meet payroll by selling ten voltmeters a month, but by the end of the first year, monthly production had reached thirty to forty. To achieve this volume, the engineers worked late into the night, letting themselves in and out of the department store on the first floor, which locked its doors at 6:00 P.M., with improvised passkeys. Akira Higuchi, who had been appointed factory manager, remembers being caught by security men working for the department store more than once.

Because the tools necessary to assemble the voltmeters were not available, the engineers had to make their own. The streets were littered with useful things that had been abandoned, including motorcycle springs, which they converted into screwdrivers. The vacuum tubes that powered the meters had to be hunted out of boxes of junk in black market stalls in Akihabara and Ueno, a district dubbed "Ame-yokocho"—Yankee Alley—because of the GIs who sold contraband from military bases. Only fifty in every lot of one hundred tubes performed to the voltmeter's specifications.

As business grew, the company moved from the switchboard room on the third floor to the entire seventh floor, the only other habitable floor in the ruined building. But in August 1946, after Morita had joined the company, Shirokiya decided to open a dance hall for Occupation troops and asked Ibuka to vacate. Higuchi remembers a line of auditioning dancers watching him and the others with amusement as they prepared to move. The question was where to go; commercial space in Tokyo in 1946 was hard to find. Through his connections at Yokogawa Electric, Ibuka managed to sublet two factory spaces, both on the outskirts of Tokyo. One, in Mitakadai, the Tachifuji plant, garaged fire engines. On sunny mornings Ibuka's men could drive the engines outdoors, but this was not permitted when it was raining, and the men had to work around them. Before long, the company was asked to vacate once again because they were consuming too much electricity at a time when it was still rationed. Late in the year, Ibuka went looking for a space large enough to accommodate the entire operation, and found a dilapidated wooden building that had escaped burning on Gotenyama Heights in the southern part of the city, near the harbor. Gotenyama had once been known for the beauty of its blossoming cherry trees, but in the winter of 1947, when the company moved in, it was a desolate ruins. Today, Sony's worldwide headquarters stands on the site of the wooden building known as "the factory on the hill."

Akio Morita listened to the emperor's broadcast on August 15, 1945, in his ancestral home in the seaside town of Kosugaya, south of the commercial city of Nagoya in central Japan. Following the bombing of Nagasaki on August 9, he had received orders to travel to Nagoya on navy business, and had been granted a day's leave to visit his family. As he left his unit in Zushi, he angered his superior officer by informing him that he would not be returning should the surrender be announced while he was away: Morita suspected, correctly, that officers would be ordered to commit ritual suicide by hara-kiri when the war was lost, and he was not prepared to die a martyr's death for the imperial cause.

Nonetheless, informed by his mother on the morning of August 15 that the emperor would speak at noon, he dressed in his officer's uniform and strapped on his sword in preparation for the broadcast. His younger brothers, Kazuaki, twenty-two, and Masaaki, eighteen, stood at attention at his side as the emperor announced the end of an era. The younger boys were also in the Imperial Navy. Kazuaki had been drafted while a student in economics at Waseda University, and Masaaki and his entire middle school class had enlisted. As the war ended, both Morita brothers were being trained to fly suicidal kamikaze missions at the navy's flight academy: if tears were shed in the Morita home that day, they were tears of joy.

The house of Morita was a prosperous merchant family who had been doing business in the Nagoya area of central Japan since the early seventeenth century. The family business was sake. In its brewery in Kosugaya, Morita Company distilled from rice a regional blend of sake called *nenohi*. The company also made miso, the fermented soybean paste from which the Japanese make their daily soup, and soy sauce. Morita's great-grandfather and grandfather had allowed the business to decline, devoting their time to building the family collection of Chinese and Japanese art and porcelains. Morita's father, recalled from his study of business administration at the newly founded center of Western learning, Keio University, had restored the business to prosperity: Morita and his two younger brothers and sister had grown up in affluence in a sprawling house in Nagoya's finest residential district, Shirakabe-cho, "white-birch park." Like the other families on the street, the Moritas had a large garden, a tennis court, a chauffeured car, and servants.

In Morita family tradition, the eldest son of each generation took the first name Kyuzaemon on becoming head of the family. Morita's father was Kyuzaemon Morita the Fourteenth. From the day he was born, January 26, 1921, Morita was raised to become the fifteenth-generation Kyuzaemon. At formal gatherings, the family hierarchy was reflected in the seating order, which, though never explicitly diagrammed, was inviolable: as the eldest son, Morita sat at the head of the long banquet table, at his father's side. Next came his uncles, his aunts, and finally, at the end of the table, his cousins and two younger brothers and sister. From the time he was six, Morita accompanied his father to the annual New Year's Day meeting at the company and sat with him on the dais as he delivered his New Year's speech to Morita employees. At age ten, he was required to attend board meetings and to sit in on private conferences at home between Kyuzaemon and company managers. As a middle school student, he participated in inventory checks, counting bottles stored in Morita warehouses. He was even taught to taste sake from the midwinter barrels to check the maturation process.* Summers, when the family returned to the house in Kosugaya to spend a month at the beach, Morita's father would halt the car along the way and, pointing to a family warehouse full of sake or soy, remind his oldest son, "All this belongs to you. You will be the boss!"

The middle Morita brother, Kazuaki, remembers vividly and with seeming equanimity the preferential treatment his elder brother received. As the second son in the family, he did not expect to be treated equally, and he moreover enjoyed the freedom to do what he liked with his spare time. There were occasions when he was also required to accompany his father on business, but it was somehow made clear, in the implicit way of Japanese families, that he was along for the ride. As it happened, Kazuaki would end up running the family business. The youngest brother, Masaaki, who was to have a distinguished career at Sony but would never head the company, was left alone to play with his model airplanes and dream of becoming a pilot.

While still in elementary school, Morita developed a fascination with mechanics. He began disassembling appliances at home, and his reading shifted from schoolbooks to technical manuals: before long, he was building his own vacuum tube radios and devising clock radio switches. Kazuaki observed this development with uneasiness. "At some point," he recalled, "I began to wonder if my brother really intended to take over the business as Father expected."

By Morita's own account, his fascination with electronics in particular dated from the day his father brought home an electric phonograph. His mother was a music lover; the children had grown up listening to recordings of Caruso, Efrem Zimbalist, Bach, and Mo-

* Interestingly, Morita would grow up a teetotaler, rare in his day for a Japanese man, and particularly unusual, and even problematic, for someone in the business world, where drinking was a ritual of prime importance. Perhaps Morita was expressing resentment at having been pushed so hard as a child in the direction of family tradition.

zart on a hand-cranked Victrola. The imported electric phonograph Kyuzaemon brought home one day was a luxury item, the first of its kind in Japan, for which he had paid six hundred yen, one-third the cost of a Japanese automobile. When Morita tried playing the records he knew from the family Victrola on the new machine, he was astonished by the superior quality of their sound. And it was the sound itself, he would later insist, not the music, which cast its spell on him. There is no knowing whether it was sound or music that inspired Morita at that moment of listening. Certainly he was obsessed with high-fidelity technology all his life. In later years, he would design for the house he built in Tokyo a "listening room" said to provide the finest acoustic experience imaginable in its day. It was here that musical friends of the Sony family—Herbert von Karajan, Lorin Maazel, Leonard Bernstein, Seiji Ozawa—would come, sometimes straight from the airport, to listen to Sony CDs of their own performances played on Sony CD players over Sony speakers.

When it was time to enter "higher school," as it was known in the prewar system (the approximate U.S. equivalent of the sophomore year of high school through the sophomore year at college), Morita asked permission to apply to the Science Department of the Eighth Higher School, and Kyuzaemon assented. At this and each subsequent juncture in the road that carried Morita away from his family business, Kyuzaemon showed a remarkable degree of understanding in view of how disappointing his eldest son's plans must have been to him. Perhaps, even this early, having observed the intensity and focus of Akio's interest, Kyuzaemon was on the way to resigning himself to the inevitability of a break in family tradition.

Having secured permission from home, Morita had to pass an entrance examination in subjects he had neglected while at middle school, English, classical Chinese and Japanese, and advanced mathematics. To prepare, with the help of tutors provided by his father, he took a year off. In 1937, he passed the examination and became the lowest-ranking student at his middle school ever to be admitted to Eighth Higher in science. In his third year, he chose physics as a major and felt that he had found his subject; Morita loved the lab especially, because it allowed him to feel that he was exploring how things worked. His teacher, Gakujun Hattori, urged him to continue his studies at university, and provided him with an introduction to an applied physicist named Tsunesaburo Asada, who was teaching at Osaka Imperial University. Morita visited Asada's lab and decided on the spot that it was the place for him. Kyuzaemon expressed "disappointment" that Akio had chosen science rather than economics but once again gave his son permission to follow his own interests.

During Morita's four years at Osaka, Japan was at war in the Pacific; by 1943, most students had been mobilized by the military, but those majoring in science were still being deferred. Morita's performance earned him a position as assistant to his professor, who by that time was at work on telecommunications research for the navy. By 1944, the tide of the war had turned and all deferments were canceled: it was only a matter of time until Morita received his draft notice, the "red slip," which by that late date was tantamount to a death sentence. Then he learned that graduates in physics were being allowed to continue their research if they qualified for a permanent career in the navy. Though he had no interest in a military career, the alternative was an assignment as a radar operator on a ship in the battle zone. Morita passed the examination and was assigned to the Naval Office of Aviation Technology at Yokosuka, on Tokyo Bay. It was from there that he joined the task force where he met Ibuka late in 1944.

As he stood at attention, listening to the emperor, Morita had already resolved to become a physicist, or possibly a teacher of physics. Within weeks of the surrender, an opportunity presented itself in the form of an invitation from Professor Hattori to join his new faculty at the Tokyo Institute of Technology. Morita asked once again for his father's permission and received it. In his 1985 memoir, *Made in Japan*, he wrote, "Since my father was still in good health and running the business, there really wasn't any need for me at the Morita Company *at the time* [italics mine]." The implication is that Kyuzaemon even now had reserved the right to recall his eldest son.

Morita was preparing to accept Hattori's invitation when he happened to see Ryuzo Kaji's "Blue Pencil" column in the Nagoya edition of the *Asahi Shimbun*, and learned that Ibuka had gone into business in Tokyo. Would Sony have happened, the counterfactualist might ask, if

Morita had failed to notice the column? Why, in any event, had Ibuka neglected to inform Morita of his plans? Was he unable to locate Morita or simply preoccupied? Or was he not yet as taken with the younger man as Morita was with him? There is no answer, only the historical fact that Morita learned about the new business indirectly and contacted Ibuka to inform him that he was coming to Tokyo straightaway and wished to do anything he could to be of service.

For six months, Morita honored his commitment to Hattori, teaching physics at the Tokyo Institute and working with Ibuka and his engineers in his spare time.* In January 1946, Douglas MacArthur's General Headquarters announced a purge: former officers in the Imperial Army or Navy would be removed from political office, government positions, and faculties at institutions of higher learning. Morita was quietly jubilant. Although actual notice of his dismissal from the Ministry of Education did not arrive until November 1946, he persuaded the dean to release him in March, suggesting it was in the school's interest that he not begin a new term in April.

Morita was now free to join Ibuka, but to make this move he would need his father's blessing. In April 1946, the would-be partners took the night train from Tokyo to Kosugaya. There can be no doubt that by this time Ibuka was determined to work with Morita: in hopes of enhancing the effect of his petition, he had prevailed on his father-in-law to go along. It was an uncomfortable journey, in an old railroad coach, with cold wind and soot blowing through the broken windows. At the Morita house, the young partners and their distinguished companion were served freshly baked bread (the Moritas also owned a bakery) and imported jam. Years later, Ibuka still talked about this exotic treat at a time when most Japanese were having difficulty finding sufficient third-grade rice to eat.

There is no record of what was said that April morning at the

---

* Morita was living with his youngest brother, Masaaki, in a house that belonged to the Iwama family, his neighbors in Nagoya. (The house had been abandoned since the time of the Tokyo firebombings.) Masaaki, who was a student in his brother's class, remembers fighting the temptation to steal a look at the exams that Morita wrote up at night by the light of a flashlight.

Moritas' ancestral home. What is clear is that Kyuzaemon released Morita from his obligation to head the family sake business. Kazuaki Morita remembers being in the house, but he did not attend the meeting that would dramatically alter the course of his own life as well as his brother's. Looking back, he says that he had the feeling that he was being prepared to take over in his brother's stead when, in 1941, his father had instructed him to major in economics. "It was a huge event, the eldest son had been inheriting the business for over three hundred years and fourteen generations, and now my brother was leaving. Still, I don't think my father was furiously angry about it because he had seen it coming gradually ever since Akio was a boy." In his brother's absence, Kazuaki Morita assumed responsibility for running the business, and served as CEO of the Morita Company, but only until Morita's own eldest son, Hideo, Kazuaki's nephew, was ready to take over. Not surprisingly in the context of the traditional Japanese family, Morita remained chairman of the company in absentia all his life. Even after the Sony empire had extended its reach around the world, Morita would take the time to travel to Nagoya to chair annual family meetings, and his word was always law. Though he never formally assumed the name Kyuzaemon, the vanity plate on the Lincoln Continental he kept in New York read "AKM-15," Akio Kyuzaemon Morita the Fifteenth.

In addition to giving his blessing to the new venture, Kyuzaemon Morita was moved, or persuaded, or some combination of the two, to invest. The initial figure inscribed in Sony history is ¥190,000, an amount equivalent to roughly $60,000 today. As cash was the scarce commodity in 1946, and as Kyuzaemon was taxed with rebuilding the business, it was likely not an insignificant amount of money to him, and his willingness to part with it reflects a traditionalist father's sense of obligation to the firstborn son. Subsequently, the business would experience financial difficulties more than once, and Morita would return to his father for additional funding. On each occasion, he issued stock to the family; before long, Kyuzaemon had become the principal shareholder. At one time, the Morita family owned 17 percent of Sony. Today, through its holding company, RayKay, and the Morita Foundation, the family controls a 10 percent share worth roughly $5 billion in the current market.*

On May 7, 1946, Ibuka's company was formally incorporated at a ceremony attended by twenty or so employees, friends, and advisers. The business was named Tokyo Tsushin Kogyo, Tokyo Telecommunications Engineering Company, Ltd., "Totsuko" for short. Tamon Maeda accepted the title of president, *pro forma*. Ibuka was the number-two man, managing director, and Morita followed as general manager. Although Ibuka had drafted his "Founding Prospectus" to read at the ceremony, he had entrusted it to his cousin Shozaburo Tachikawa, who, in the excitement, misplaced and forgot it.

That same year, year one of the postwar era, hundreds of new businesses were launched in metropolitan Tokyo, most of them to vanish as quickly as they appeared for want of capital. Tokyo Telecommunications Ltd. began life with a significant advantage in that its advisers were prominent men with powerful connections to the financial establishment. Maeda, who had nothing to do with actual business operations, was the key figure. In 1946, he introduced his son-in-law and his charismatic partner to a financier named Michiharu Tajima, a former classmate in the Law Department at Tokyo Imperial University. Tajima had been director of the Monetary Control Association during the war and, previously, president of the Showa Bank and a member of the House of Peers. He had also had a career with the Aichi Bank, headquartered in the Nagoya area, where he had become familiar with the house of Morita and Morita's father. Tajima in turn brought the fledgling company to the attention of Junshiro Mandai,

---

* Kyuzaemon raised funds for the new company by selling off ancestral land in the Nagoya area. Since 1993, when Morita's resignation following his stroke enabled the family to begin liquidating their holdings without risk of being charged with insider trading, Hideo and his mother have been repaying Sony's debt to the house of Morita. They have restored the 350-year-old ancestral home in Kosugaya, where the Moritas were village chieftains for generations, and they have rebuilt the neighboring Buddhist temple, the Hoshu-in, which was founded by Kyuzaemon the Tenth. "Cost has not been an issue," Mrs. Morita told me. "Morita knows what we are doing and he is thrilled to be able to honor his ancestors by repaying them for having given Sony its start."

president and soon to be chairman of the newly formed Imperial Bank, and prewar chairman of the Mitsui Bank and the Japan Bankers' Guild. Throughout his long association with Sony, including as chairman of the board from 1953 to 1959, Mandai used his influence with the banking community to secure funding for the company. According to Morita's memoir, Mandai virtually ordered Mitsui Bank executives to invest in new issues of Totsuko stock. By October 1947, as a direct result of his aggressive introductions and endorsements, Tokyo Telecommunications had increased its initial capitalization to the equivalent of $180,000. By the end of 1950, its capital had doubled again. At the time of his death at age seventy-five in 1959, Mandai willed his Sony shares to his alma mater, Aoyama Gakuin, a bequest that soon amounted to the largest endowment of any private school in Japan.

The fourth Sony elder, Rin Masutani, was a wealthy entrepreneur who had championed Ibuka since his early days at Photo-chemical Laboratories. Finally, there was Kyuzaemon Morita, the largest shareholder and a man with very broad contacts of his own. With these eminences in harness, and with Ibuka's driven, restless vision now grounded by Morita's pragmatic intuition, Tokyo Telecommunications Engineering Company, Ltd., was poised to achieve great things in the spring of 1946.

2

IBUKA THE MUSE: FROM "TALKING PAPER" TO TRINITRON

Masaru Ibuka's great gift was foreseeing product applications for new technologies and inspiring his engineers to overreach themselves in achieving the goals he set for them. Without exception, the Sony engineers who knew him worship his memory and characterize their years of working with him as the most creative and gratifying period of their careers.

The paradigmatic relationship was between Ibuka and Nobutoshi Kihara, a mechanical engineer whose product innovations in response to Ibuka's direction resulted in more than seven hundred patents in his name. It was Kihara's impression of Ibuka as a "practical visionary" which prompted him to apply for a job in April 1947. In February, about to graduate from Waseda University with a degree in mechanical engineering, he noticed a card on the department bulletin board: "Students wanted. Masaru Ibuka. Tokyo Telecommunications Engineering Company, Ltd." Having worked part-time assembling radios and phonographs, he recognized the company as the manufacturer of superior turntables, needles, and radio dials. He had also heard Ibuka lecture at Waseda, on the future of electronics, and had been impressed by his emphasis on innovation and usefulness.* In his résumé

* Waseda was Ibuka's alma mater, and in 1946, as a means of making extra money, he had accepted an appointment as adjunct lecturer in the Department of Mechanical Engineering.

Kihara wrote: "I can make shortwave receivers, five-tube superheterodyne radios, and hi-fi amplifiers." Akira Higuchi interviewed him for the job, and remarked that his interest in electronics was unusual for a mechanical engineer. Kihara replied, uneasily, "I like working with electronic devices," and later worried that he had ruined his chances. The following day he received notice of employment.

By all accounts, the Ibuka-Kihara connection appeared to be telepathic: Ibuka would convey excitement about a possibility he envisioned vaguely, and Kihara would translate his musing into an actual prototype. "Ibuka-san truly understood the mentality of an engineer," Kihara explained. "He never instructed me to do anything. Instead of ordering, he'd discuss things with you and make suggestions—he'd tell you what he hoped for. To all of us, that was a very special gift. When you receive an order you have to do what you've been told. But I was always like Ibuka, I hated imitating others. Even if he had given me an order, there's a stubborn place in me that wouldn't have listened!"

Like many of Sony's engineers, Kihara seems to have been motivated to a remarkable degree by the desire to please the man he considered his mentor and muse: "Ibuka would say that a cassette version would be handier than a reel-to-reel, or that a smaller and lighter machine would be easier to use and easier to sell. He was always thinking aloud about something. And I was always listening. And when we had finished work on a new product, I'd start to think about making that lighter version he had mentioned. Whenever I had time to spare I'd experiment on my own and rig something up. I'd have five or six of these prototypes stuffed away under my desk—experimental models. Then one day Ibuka would say: 'Kihara, a cassette-style portable model would be very handy to have, don't you think?'—and the next morning I'd show him something close to what he was imagining from my secret pile. And he'd be overjoyed. 'Kihara, this is just what I wanted!' he'd say. And off we'd go again. He was like a happy kid at those moments, and I loved seeing him that way. I loved making him happy. As an engineer, that was my greatest pleasure."

Over the years, the continuing reprise of this duet between the man with the dreams and the ingenious technician led directly to a dazzling array of Sony products: the company's first magnetic tape and magnetic tape recorder; transistor radio and television; the world's first videotape recorder (VTR) for commercial broadcast use; the first home videotape recorder, Betamax; 8-millimeter video movies; the video still camera known as "Mavica"; a color video printer; and an entire catalog of smaller and lighter variations on these basic products. Cumulatively, the innovations attributable to the Ibuka-Kihara team were the principal driver of Sony's growth from 1949 into the mid-seventies.

In the beginning, there was sound. In June 1949, on one of his frequent customer calls to NHK, the national broadcasting service, Japan's equivalent to the BBC, Ibuka stopped next-door at the U.S. Office of Civil Information and Education—the Occupation had taken over NHK's operations and facilities—and chanced to hear a tape recorder, an American model made by Webster Chicago, the first he had ever seen. The quality of the playback took his breath away—the Americans in the office were amused by his excitement—and after examining the machine quickly, he seems to have decided on the spot that this was the product on which he could found his company's reputation. Kihara remembers being summoned to Ibuka's office that afternoon and listening to a brief description of "light brown tape, maybe plastic, probably magnetic, that reproduced sound of unimaginable quality as it wound onto an open reel at approximately"—Ibuka's forefinger traced revolutions in the air—"about nineteen centimeters a second."

Tape recorders had been developed in Germany in the thirties by Grundig and Telefunken, and had been used to record and disseminate Nazi propaganda during the war. In the United States, after 1945, Ampex had led the market in developing the machines; Minnesota Mining and Manufacturing (later the 3M Company) was the principal supplier of magnetic tape. The technology had been reported in Japan but not yet applied.

Ibuka asked to borrow the machine so that his engineers could study it, and although the Americans declined to let it out of their hands, an officer agreed to bring a recorder to Gotenyama for a demonstration. Everyone at Totsuko gathered to record messages and to appreciate the quality of the playback. Unfortunately, the person Ibuka most needed to support the R and D effort, the company controller, was skeptical. Junichi Hasegawa was an accountant who had been sent to Tokyo by Morita's father with instructions to oversee and protect the Morita family's investment. Until his arrival, accounting and tax affairs had been handled by Ibuka's second cousin, Shozaburo Tachikawa, nominally the "manager of business affairs." At some point, as a result of Tachikawa's failure to depreciate properly the numerous prototypes generated by Ibuka's search for a viable product, the Tokyo tax office had paid the company an unpleasant visit. The news had reached Morita's father in Nagoya and had disturbed him sufficiently to dispatch Hasegawa. It is a curious fact of Sony history that the founders' authority increased as the company grew; in 1949, when only forty-five people were employed, it was still necessary to persuade others above them in the hierarchy.

Ibuka and Morita calculated that the project would require a development allocation of three hundred thousand yen (roughly $100,000 at current value), but Ibuka had implemented and abandoned so many schemes that his credibility was low; and Hasegawa took the dour position that the tape recorder would not repay what it would cost to bring to market. To persuade him, Ibuka and Morita resorted to an extreme measure: they took him to a black-market restaurant and laid on a feast and even a quantity of beer, still a rare and expensive luxury at the time. At some point during the long evening, Hasegawa approved the development budget; resisting the Sony founders under these, or indeed any, circumstances, would never be an easy thing to do.

The new company's eventual success at developing magnetic tape and a tape recorder with little else to guide them than an instinctive cleverness about circuitry and mechanism, and with scarcely any material resources, is a vintage example of the ingenuity and determination that drove Japan's postwar recovery. Ibuka and Morita agreed from the outset that they should produce both tape recorders and tape, to benefit from the demand for tape that the machines would create. From their reading, they had learned that the recording tape developed in Germany was plastic coated with magnetic powder. In a prewar book on magnetic materials, Kihara found mention of producing ferric oxide for use as magnetic powder by heating oxalic ferrite. At the end of a long day in the black-market district of Kanda, searching on foot, with Morita as his guide, for a pharmaceutical wholesaler with a supply of the chemical, he returned to Gotenyama with two large jars full of oxalic ferrite. As there was no electric furnace available, Kihara roasted the yellow powder in a borrowed frying pan. At a certain temperature, the powder began turning black and brown; the black substance was ferrous tetroxide, the brown was the ferric oxide he wanted. Kihara became expert at removing the pan from the heat just before the two compounds oxidized into a useless metal polish called colcothar.

By mixing the ferric oxide with shellac, Kihara was able to concoct a magnetic coating, but he discovered that he needed a finer powder than he was able to produce. At the time, a company called Papilio Cosmetics was running newspaper ads which claimed that its face powder was the finest on the market. Morita went to see the president, whom he had never met, and learned that the powder Kihara desired was finer than any cosmetic.

Finding a suitable base for the tape was also a problem. Plastic was unavailable in Japan. For months, the team experimented with cellophane, but after two passes through the recording head, the cellophane would stretch and distort the sound. They tried paper next, but no available grade was strong enough. A cousin of Morita's worked at the Honshu Paper Company in Osaka, and Morita went to see him and persuaded the company to produce for Totsuko one hundred pounds of extra-smooth craft paper strengthened with hemp. In Tokyo, the paper rolls had to be cut into quarter-inch strips by hand; Kihara fixed two razor blades a quarter inch apart, and the paper was painstakingly drawn between the blades.

There remained the problem of how to coat the paper. An airbrush was too powerful. The effective tool turned out to be fine brushes made from badger hair which cost eight hundred yen each. A strip of quarter-inch paper, thirty yards long for a single small reel, was laid out on the factory floor and all hands went down on their knees and carefully brushed on the magnetic paste. The paper tape was finally made to work, albeit the quality of sound was far inferior to 3M's plastic tape.

If you ask him today, Kihara, now the president of the Kihara-Sony Research Institute, is pleased to oblige with a demonstration. In front of your eyes, like a chef preparing a sauce, he will cook up a batch of ferric oxide, removing the original frying pan from the heat at the right moment with a flourish, and separating the brown powder into a dish. Adding clear shellac, he coats a paper strip using a brush of badger hair. He has prepared another strip beforehand, which has dried and is ready to use. Passing the paper through a simple recording head, he repeats into a microphone the standard Japanese phrase for a sound test, "*Honjitsu wa, seiten nari,*" Today, we have fine weather! He runs the paper strip back through the machine. At the scratchy sound of his own voice a smile lights up his face, and you feel certain that you are seeing his smile in 1949 on the day the "talking paper" worked.

In January 1950, the Totsuko team led by Kihara completed the Model G (for "Government") Tapecorder, the first to be produced in Japan. The machine weighed in at one hundred pounds and was priced at ¥160,000, a substantial sum of money at a time when a college graduate working in government was earning ¥70,000 a year. Confident that the market was ready, Totsuko produced a first run of fifty, exceeding substantially in the process of development and production the original allocation of ¥300,000. Hasegawa the controller was concerned; and before long he was beside himself as, month after month, the machines failed to sell. Morita demonstrated the Model G to everyone he knew, driving around in his Datsun truck to businesses and universities to record people's voices speaking and singing. Everyone found the machine amusing, but no one offered to buy such an expensive toy. After six months and mounting desperation inside the organization, Tamon Maeda, realizing that Japan's postwar judicial system was suffering from a shortage of courtroom stenographers, arranged a demonstration at the Supreme Court, which resulted in an order for twenty units.

Meanwhile, Ibuka learned from Maeda that the Ministry of Education had budgeted for the purchase of audiovisual equipment for all of Japan's schools, and he began musing to Kihara about a lighter, more practical machine. Kihara produced sketches for a smaller tape recorder in one night, and a prototype in ten days of work at an off-site retreat. The Model H (for "Home") weighed only thirty pounds, and sold for ¥80,000. Morita took the machine across the country, showing teachers gathered in classrooms and auditoriums how it could be used in a variety of ways to improve the quality of teaching. Thanks largely to his persuasiveness, it began to sell.

Ibuka was not satisfied. He wanted an even smaller and lighter machine, less expensive and handier to use. Kihara went to work again. The Model P (for "Portable") was the company's first cash cow, selling more than three thousand units in seven months. The Model M (for "Movie"), originally designed to record location sound for movies and modeled after a machine that had been manufactured by the Stancil and Hoffeman Company for use by paratroopers in the Korean War, became an even better seller. At twenty pounds, the reel-to-reel machine was light enough to be suspended from the shoulder by a strap and was therefore useful for man-in-the-street interviews, which were just coming into vogue when it appeared on the market late in 1951. In a popular magazine strip in the *Mainichi* newspaper, a reporter named Densuke wore the tape recorder on his shoulder to sample public opinion; and the Totsuko machine became widely known as the "Densuke," a name the company trademarked. For years afterward, broadcasters referred to all portable tape recorders as Densuke.

In 1952, the focus of Ibuka's excitement and expectations shifted abruptly from the magnetic tape recorder to transistors. In the United States for the first time, to observe how American consumers used their tape recorders, he learned that Western Electric, the parent company of Bell Laboratories, where semiconductors had been discovered in 1948, was offering a technical license for manufacturing transistors in return for a royalty. The agreement stipulated a $25,000 advance payment against royalties, a sum of money which could not be remitted from Japan without approval from the Ministry of Trade and Industry, MITI. On his return to Tokyo, Ibuka requested approval and in his own words, was "laughed out of the room." It wasn't that the MITI bureaucrats failed to grasp the potential importance of the transistor; by this time, the largest Japanese electronics manufacturers, Matsushita, Hitachi, and Toshiba, were developing semiconductors under technical license to RCA. Instead MITI asked Ibuka how a tiny company that had never made vacuum tubes expected to manage a brand-new and complex technology!

Meanwhile, in New York, a friend of Ibuka's had been lobbying Western Electric, and eventually, in a letter to Ibuka, the company indicated its willingness to discuss a technical license. In August 1953, Ibuka dispatched Morita to New York to arrange for a license contingent on approval by the Japanese government to release $25,000. Despite the fact that he spoke almost no English at the time and had to rely on an interpreter, he persuaded the vice president of licensing at Western Electric, Frank Mascarich, to grant him the license in spite of misgivings. Mascarich at eighty-nine vividly recalls his impression of Morita in 1953: "I wasn't terribly pleased with the arrangement. After all, he signed the license with a proviso that he would obtain government approval for the advance later. But he was so persuasive, and so anxious to proceed with his plans, and after all, it took considerable time and expense for him to travel to the United States from Japan, that I decided to give him some of the technical information so that he could take it back with him and immediately embark on his project to manufacture transistors." The information that Mascarich gave Morita that day was a two-volume compilation by Bell Laboratories scientists titled *Transistor Technology*, which was not available to the general public. Over the next two years, as Totsuko engineers struggled to develop high-frequency transistors, this chronicle would be their only guide.

At the time the agreement was signed, Mascarich had taken care to inform Morita that, in the opinion of Bell Laboratories, the germanium contact point transistor would be suitable only as a power source for hearing aids: at the time, though many attempts had been made, no one had succeeded in commercially producing transistors capable of the higher frequencies that would be required to power, for example, radios.

To the Japanese engineering community of the 1950s, Bell Laboratories spoke "in the voice of God," but Ibuka was undeterred. He knew that hearing aids could never be a big business in Japan: deafness was considered shameful, and people whose hearing was impaired were at pains to conceal it. Besides, he had already resolved that the next challenge the company should address was developing a transistor with the capacity to power a miniature radio. In part, he was drawn to the project by what promised to be its difficulty: the company's 150 employees included some 50 engineers and scientists with college degrees, and Ibuka felt certain that something substantial if not overwhelming was needed to keep them motivated.

Ibuka assembled a transistor team led by Kazuo Iwama, a young geophysicist from the University of Tokyo and Morita's brother-in-law. Iwama's family and Morita's had been neighbors in Nagoya and had grown up together in the same privileged environment. In May, just as Totsuko was incorporated, Iwama had married Morita's younger sister, Kikuko. Morita had recruited him the following month. Iwama would have a distinguished career at Sony, succeeding Morita as president in 1976. Less visible than Morita outside the organization, he is venerated inside Sony culture as a cautious, levelheaded rationalist who introduced the rigor of scientific method to Sony's R and D and process.

While Iwama began studying *Transistor Technology*, Ibuka went to work on MITI. Dr. Makoto Kikuchi, who was working at MITI at the time, remembers meetings at the ramshackle wooden building in Gotenyama at which Ibuka would lecture the bureaucrats about Sony's progress with transistors and about his intention to transform the marketplace with miniature radios. "We intend to proceed with or without you," he would conclude, "but if you approve our deal with Western Electric and give us some development money, you will look smart!" Ibuka's campaign was reinforced by a political shift inside MITI's Department of Electronics late in 1953 which disposed the agency more favorably, and permission was granted.

In January 1954, Frank Mascarich hosted Ibuka and Iwama on a tour of Western's Allentown, Pennsylvania, assembly plant. Many years later, in an interview on Japanese television, Ibuka recalled the moment with evident relish: "Late in 1953, the ministry finally granted us a permit, and I went flying off to America with the late Kazuo Iwama, to Western Electric, and that's where we saw transistors being manufactured for the first time. It was some complicated business, and I got pretty worried and asked Iwama if we maybe weren't over our heads here? But he'd been studying like crazy for months and was full of confidence. He asked question after question, and each one of them was loaded with technical terms in English that he'd managed to learn somehow—I couldn't believe my ears, and the people at Western Electric were amazed too. They wanted to know if we'd already manufactured transistors!"*

Iwama remained in the United States for three months to study the technology. All day he observed factories and laboratories affiliated with Western Electric, Bell itself, and Westinghouse. At night, in his motel room, he wrote up his notes from memory and mailed them back to headquarters in Tokyo, nine large pages in mid-January, eight pages more on February 19, nine pages on February 21, five pages on April 7, another on April 9, eight pages on April 13, and four more on April 15. Each page is covered top to bottom with diagrams, technical terms in English, and minutely detailed instructions on every aspect of the production process that Iwama observed, all inked in his meticulous hand. Some of his notes are on hotel stationery, most are sky blue aerograms. The forty-four original pages, known as *The Iwama Report*, have been preserved at Sony and may be viewed on request. They convey the intensity of focus and the determination to succeed that drove not only Iwama but the entire organization in those early days. Indeed, these closely covered pages radiate an energy that helps explain Japan's postwar resurgence.

By June 1955, relying on the data he had gathered in America, Iwama and his team succeeded in producing its first alloy-type transistor. The next difficult step was increasing the frequency. After six months of experimenting, Iwama achieved a transistor capable of generating several hundred megacycles by using a process called "phosphor doping," which Bell had tried and rejected. Now the problem was mass production. Test runs produced only one to five functional transistors in a hundred, a yield that was far below the minimum for commercial success. A jubilant Ibuka nonetheless ordered production to

* NHK's "Birth of the Transistor Radio" aired on September 18, 1980.

begin. In his unorthodox view, so long as the process had been demonstrated to work even once, research on improving yields could be conducted simultaneously with production. During the first year, depending on the effectiveness of separate processes such as slicing the grown crystals and etching them, yields varied significantly from day to day. Ibuka had targeted ten thousand transistors—for ten thousand radios—in the first year, and Totsuko produced eight thousand.

Meanwhile, Kihara was designing circuitry for a miniature radio the size of a cigarette case; in August 1955, the company announced its model TR-55, the first transistor radio in Japan. Ibuka was hoping to be first in the world, but in the United States a company called Regency had beaten him to market by a month with a radio powered by transistors developed at Texas Instruments.

Ibuka wanted something smaller than the TR-55. In March 1957, Sony released the world's first "pocketable"* transistor radio, which sold 1.5 million units and established Sony as the market leader. In truth, as the device was slightly too large to fit inside a normal shirt pocket, Morita ordered custom-made shirts with outsized pockets for his domestic and U.S. sales forces. By 1957, driven by the radio business, Sony's revenues had grown to $2.5 million, and the company employed 1,200 people.

The focus of Ibuka's interest was as volatile as it was intense. When his staff heard him announce, "It's time to graduate!" they understood that he was onto something new and would not look back. In the summer of 1957, Sony's current chairman, Norio Ohga, a student in Germany at the time, was asked by Morita to serve as Ibuka's European guide and arranged a tour of German tape recorder factories. As the tour began, Ohga was nonplussed by the discovery that Ibuka's consuming interest in tape recorders had vanished, displaced by his new preoccupation with transistor radios. At visits to Telefunken factories, Ibuka stood around vacantly, his mind elsewhere, while Ohga covered the awkward silence with questions of his own. Eventually, he canceled a

* The company claimed to have coined the English word, which actually dates to the eighteenth century.

number of their appointments and took Ibuka to see Philips in Holland and some transistor operations in Belgium.

Ibuka's failure to mask his lack of interest was typical. When he was happy, he was spontaneous and unabashed about conveying his pleasure abundantly in a way that gratified people and inspired them. But he was similarly uninhibited about letting his displeasure loom. He was easily bored—his secretary used to caution young engineers on their way into his office to limit themselves to thirty minutes—and when he lost interest he would cut off the offender in midsentence or, worse, as on the European factory tour, turn away and lose himself in reverie.

Capable of impulsive generosity, Ibuka was known to many in and outside the company for his consideration and kindness in troubled times. As a consequence of his single-minded focus on his own thoughts and feelings, he could also be rude and tactless, and more than once created awkward situations in the wide world outside Sony. A young assistant at the time remembers his own embarrassment when he accompanied Ibuka to a dinner in his honor at a private club for Italian industrialists in Rome. A teetotaler with no patience for the cocktail hour, Ibuka called for a plate of spaghetti before dinner and tucked in after sprinkling it liberally with sugar (whether he was at home or at the Italian embassy, Ibuka enjoyed his spaghetti sweetened). Having removed the edge from his own hunger, he sat through the formal dinner impatiently and excused himself as coffee and brandy were being served. His aide whispered that the guest of honor might offend his hosts by departing before the meal was over, but Ibuka brushed him aside and returned to his hotel.

In Sony hagiography, Ibuka's tactlessness is interpreted as evidence of his innocence and purity of purpose, and it does appear that he was a genuine naif. Dr. Makoto Kikuchi, who moved from MITI to manage Sony's Central Research Laboratory from 1974 to 1989, recalls a day when his employer showed up at the laboratory unannounced, as he was wont to do, and asked for a tour. Kikuchi showed him a cylinder being developed for use in a videotape transport, and explained that it had been honed and polished to the world's smoothest metal surface. Ibuka abruptly reached for the cylinder and licked it: "He had to try it with his own tongue, to see how smooth it was, and his eyes were flashing and he was happy, like a child. 'Let's have some fun, let's test this ourselves'—that, in a word, was Ibuka."

The memory of a night spent at Ibuka's house reinforces the impression of a willful child at play. An audio engineer named Katsuaki Tsurushima was summoned to the Ibuka's summer home in Hakone to demonstrate four-channel sound to a guest, the great Russian violinist Leonid Kogan. When the demo was finished and Kogan and his wife had departed, Ibuka led the young engineer downstairs to a room filled with mechanical toys. There was a doll whose hands and legs moved at the sound of clapping, a puppet called "Crazy Legs" whose magnetized shoes caused it to describe a random dance, and, Ibuka's favorite, a robot soldier who reversed directions when struck by a beam of light from a toy rifle fired from thirty feet away. Ibuka, in his sixties, played with extravagant delight until it was time for dinner. As Tsurushima was preparing to make the drive home, Ibuka pointed to the ribbon of taillights stretching to Hakone Pass, and urged him to spend the night. His motive turned out to be mahjong, a game that requires four players; Tsurushima completed a foursome, which included Ibuka's wife and his driver, whom he had taught the game to ensure that an extra player would always be at hand. Hours later, when it was time to go to bed, Ibuka showed his young employee to a room warmed by water from a hot spring piped beneath the floor and invited him to sleep as late as he wished the following morning, a Sunday. Tsurushima was awakened at 7:00 A.M. by someone moving on the other side of the rice-paper door to his room. As he stirred in bed, Ibuka's voice informed him that breakfast was ready any time. Tsurushima understood that he was to get up at once and did so; when breakfast was over, Ibuka suggested that they return to their game of mahjong, which continued through lunch and into the early evening.

If the episode evokes Ibuka's childlike capacity for exuberant play, it also illustrates the tenaciousness that was so often manifest in his business life. Ibuka was not endowed with a capacity for compromise. There was about him rather an obsessive single-mindedness, a dominant exclusivity of focus. A typical moment occurred when a young manager, Kunitake Ando, currently the president of Sony's Personal

IT Network Company, presented Ibuka with a road map of sales and marketing strategy for Trinitron color television. Having just returned from two years in the United States studying management at Bucknell College, Ando was working hard to introduce the company to mid-range strategic planning. Removing his blueprint from his briefcase, he handed it to Ibuka across his desk. Ibuka glanced at the document, crumpled it into a ball, and threw it into the wastebasket. "This is not what I wanted," he exclaimed. "I wanted better-looking color bars." Flabbergasted, Ando nonetheless found the presence of mind to object that color bars were not his job. Once he had grasped that Ando was a strategic planner, Ibuka retrieved the plan from the garbage and listened to an explanation.

At work, despite the certainty of his vision and his distinctly eccentric individuality, Ibuka relied heavily on those around him. On Morita he depended increasingly to find the means of financing Sony's growth and to carry the company out of the lab and the workshop and into the world. His secretaries were charged with responsibility for managing his business affairs and, to a degree unimaginable in the United States, his personal life. His first secretary, who was roughly Ibuka's age and who came to him by way of the entrepreneur who was his first champion, Taiji Uemura, was a devoted, ruthlessly efficient busybody named Wakiko Hara. She was known (and feared) throughout Sony as Hara-mama-san, or Mother Hara. When Tajima died and Ibuka replaced him as chairman in 1972, Mother Hara took the opportunity to retire, announcing that she was not comfortable referring to Ibuka as "Mr. Chairman" after so many years of calling him "Mr. President." Her successor, Hiroko Kurata, whom she trained for a year and who had begun her career at Sony as Akira Higuchi's secretary, served Ibuka from 1972 until illness forced him to retire in 1994, and continues to preside over his abandoned office.*

---

* When Ibuka fell unconscious from an attack of arrhythmia on an April morning in 1992, his wife's first phone call was to Kurata at her apartment. Kurata summoned an ambulance, then rushed to Ibuka's apartment in Mita to look after him. Currently, Kurata is the doyenne of the secretarial office at Sony headquarters. Together with Chairman Norio Ohga's secretary, Sato Ashizawa, she chooses personal secretaries for top management and determines their

Arriving at the office in the morning, Ibuka would begin his business day by changing into the gray jacket with red piping which Morita had asked his friend Issey Miyake to design for Sony employees. According to Kurata, he would drop his suit coat where he stood and it would have fallen to the floor had she not been there to catch it. The Issey Miyake jacket, with removable sleeves for summer wear, was a badge of membership in the Sony family and was as good as a credit card at restaurants and stores in the Gotenyama vicinity. While other executives used it occasionally, Ibuka put it on every day: it signified Sony and it was Morita's idea, and, for those reasons, it mattered.

Ready for work, Ibuka would ask Kurata what he was to do. He had noted his schedule on a card that he kept in his shirt pocket, but he preferred to hear the details from her directly. When meetings took him outside the building, he would phone in to ask about his next appointment. If it was 11:30 and he was free until 1:00, Kurata would instruct him to return to the office so that he could have a quick lunch with someone who had been waiting to speak with him. If there were only an hour between meetings, she would suggest that he invite so-and-so for lunch and recommend a restaurant on his way to the next appointment. When Ibuka was out at night entertaining customers or friends at Japanese restaurants or at Maxim's, which Morita had imported from Paris and installed in the Sony building in the Ginza, Kurata would remain at her desk until he phoned in to say he was going home and no longer needed her.

Life at home was troubled. In 1936, twenty-eight years old, Ibuka had married Tamon Maeda's second daughter, Sekiko. The marriage was arranged by a popular novelist and music critic, Kodo Nomura, whose wife and Ibuka's mother had run a kindergarten in Tokyo together, and who had known the Maeda family for years as neighbors in the summer resort Karuizawa. When Ibuka was living in Tokyo with his widowed mother, Nomura had taken care of him like a surrogate

---

annual bonuses. She is one of a handful of women in the organization, including Ms. Ashizawa and Morita's longtime secretary in New York, Hiroko Onoyama, with the rank of *buchō*, or general manager.

father. Later, he would become one of Sony's earliest investors, responding more than once to appeals from Ibuka for emergency funding.

Sekiko was a talented and sensitive woman who had attended art school and was an accomplished painter in the Japanese style. She loved embroidery, ceramics, and choral music. She was also well known to be temperamental and very high-strung. From the beginning, she was overwhelmed by the pressure of running a household and by the social demands on her as Ibuka's wife. On New Year's Day, for example, people of Ibuka's social station were expected to hold an open house for employees and friends. Ibuka looked forward to the ritual, but hosting a daylong entertainment was more than Sekiko's nerves could handle. During these difficult years, her principal confidante was the redoubtable Mother Hara.

There were three children: a daughter, Shizuko, born in 1937; a second daughter, Taeko, born in 1940; and a son, Makoto, born in May 1945 in the country town to which Ibuka had relocated his business during the war. From about age six, Taeko, who had seemed a robust, happy child, began to exhibit signs of a learning disability, which prevented her from attending a normal school. Ibuka, who adored his middle daughter, was devastated. Later he would establish a facility in northern Japan to care for the mentally disadvantaged, House of Hope, where Taeko has lived since 1973, and would say about her, "She is the cross I bear and the light of my life." Sekiko lacked the emotional stability to care for her daughter, and turned her over to her elder sister. The atmosphere in the house darkened and became difficult to endure. Relatives remember Sekiko as chronically hysterical, "an unmanageable handful."

There was another woman, a distant relative by marriage who had been a childhood friend of Ibuka's in Hokkaido, Yoshiko Kurosawa, the daughter of a railroad developer and one of three sisters known as the "beautiful Kurosawa girls." When Ibuka was growing up, his mother, Sawa, a great beauty herself despite the sadness she carried in her face, had often said that Yoshiko would make the perfect wife for her son, and Ibuka had been smitten with her. He chose, nevertheless, to marry Maeda's daughter, possibly out of interest in her father and his potential value to his career. During the war years, Ibuka lost contact with Yoshiko; after the war, in her own unhappy marriage to a doctor, she had contacted him in Tokyo and they had begun a relationship. The family retains the memory of a letter from Yoshiko to Ibuka, detailed and passionate, discovered by Sekiko—or left for her to discover—which led to Ibuka moving out of his home in Meguro in the summer of 1956 and into an apartment with Yoshiko. Sony's elders used their connections with the press to prevent a scandal and largely succeeded. Mother Hara and Morita's wife were actively involved in the cover-up. Ibuka and Yoshiko lived together secretly until they were married in August 1966. Thereafter, they lived together as man and wife, by all accounts happily, until Yoshiko's death in 1994. Ibuka died three years later, in December 1997.

For many years after the separation, Sekiko remained in the large Meguro house, which Ibuka had built in 1952. She was now a wealthy woman: when Ibuka moved out, he had left behind in her name a generous portion of his Sony stock. In 1987, she purchased a plot of land in Tokyo's most affluent residential suburb, Den'en Chofu, and built a small masterpiece of classical Japanese architecture with a thatched roof and perfect garden. She named it Pomegranate House, after one of her own still-life paintings, which she installed at the entrance.

Ibuka's son, Makoto, who was twelve when his father left the house, loved and pitied his mother, and for many years, until after his own marriage, refused to speak to Yoshiko, to whom he still refers as "Mrs. Ibuka." Concerning his father, he is more forgiving: "If my father had been more flexible, he might have stayed in his marriage and kept Mrs. Ibuka as his mistress, but he wasn't clever in that way: he had to have everything straightforward and out in the open."

Conversations with Makoto suggest that Ibuka was a distracted and an uncomfortable father who rarely counseled his son even after he came to work at Sony. After graduating from Rikkyo University in physics, and taking a master's in physics at Brown, Makoto went to work as a computer systems engineer at IBM Japan for eight years. In 1981, he hinted to his father that he was feeling stymied in his career and would like to come to Sony, but Ibuka failed to respond with an invitation. Makoto sought counsel from his father's second cousin Tachikawa, one of the original seven engineers who had accompanied Ibuka to Tokyo in 1947 and a man he considered an uncle. Tachikawa suggested that Morita should become his sponsor, and in the end it was Morita who went to IBM Japan to say that Sony would like to hire Makoto. Thereafter, Morita became Makoto's principal adviser and coach; Ibuka remained uninvolved in his son's career. Currently, Makoto Ibuka is securely installed away from the heat of competition at headquarters as senior managing director at the company where his father began his career, Sony-PCL, now a fully owned subsidiary.

On his eighty-fourth birthday, asked to name his proudest moment in Sony history, Ibuka responded, "Trinitron."* Sony's color television picture tube is one of its most valuable assets: at the end of 1998, 180 million Trinitron picture tubes had been sold worldwide. The first twelve-inch Trinitron was released in 1968, but the development process, perhaps the most painful experience in Sony history, began in 1961.

Color television had been available in America since the mid-fifties. The default technology had been developed by RCA, the market leader; three separate electron guns, one for each primary color, fired their beams at a focusing mechanism called a shadow mask, which converged them on their way to the picture screen. For complex technical reasons, radiance was compromised as the beams passed through the shadow mask and resulted in a darkened screen image with muddied resolution. The lackluster picture and premium pricing had discouraged sales: In 1960, only one color television set was purchased in the United States for every fifty sets sold. That same year, in Japan, color television accounted for only three hundred of the nine million sets purchased. Nevertheless, there was no question in anyone's mind that the potential market was huge.

By 1961, though its black-and-white television business was thriving, Sony had fallen seriously behind the competition and still the

* The name was created by the leader of the development team, Susumu Yoshida, and refers to the three cathode ray tubes used to generate the color picture.

company had no plans for entering the color television market. Yet dealers were beginning to ask when they could expect Sony's color entry; and there was pressure from the sales division to purchase a ready-made system from another manufacturer which could be packaged and marketed as a Sony product. Ibuka refused, and would continue adamantly to refuse, to copy the shadow mask system. His insistence on originality no matter what the cost seems to have been genuine, at the heart of his vision of himself and his company. While there is ample room to question how innovative Sony products actually managed to be, it is clear that Ibuka's legacy of innovation as an ideal has been preserved as a principal incentive in Sony culture.

In March 1961, Ibuka and Morita attended a trade show sponsored by the Institute of Electronics and Electrical Engineers at the New York Coliseum. They were accompanied by Kihara, who had come along to demonstrate his prototype transistor video recorder. It was Kihara's first trip abroad, and as he wandered the floor enjoying the unfamiliar sights and sounds, he noticed at an inconspicuous booth a television screen that was very much the sharpest and the brightest he had ever seen. Excitedly, he found his employers and hurried them across the hall to see for themselves. Morita went into action on the spot; he learned that the color tube, called a Chromatron, had been developed for the military as an IFF (identification of friend or foe) display by the Nobel Prize–winning physicist E. O. Lawrence, the father of the cyclotron, and was owned by a small subsidiary of Paramount Pictures called Autometric Laboratory. Before he left the booth, Morita had phoned and secured an appointment with the president of the company for the following morning. Kihara accompanied him to a run-down building on West 45th Street in Manhattan, where Morita used his labored English to initiate a negotiation that would lead to a technical license from Paramount to produce "a Chromatron tube and color television receiver utilizing it."

The company had purchased a color television technology it would never succeed in mass-producing. Early in 1963, a young physicist named Senri Miyaoka was sent to New York for two weeks to acquaint himself with Chromatron at the Autometric Lab, which was to be closed as soon as the technical transfer had been accomplished. Miyaoka remembers the armed military guard at the entrance, and the "squalor" of the windowless basement room where ten American engineers were still somewhat glumly at work on the system. Though he spoke little English, he enjoyed the Americans, who communicated their frustration with the technology. One of them went so far as to express his gratitude to Sony for releasing him from a hopeless task. "We had no professionals in picture tubes," Miyaoka recalled, "and though we sensed that bringing Chromatron to the assembly line would be a formidable challenge, its uniqueness made it irresistible."

For two years, the Chromatron team labored to develop a commercial prototype and a process technology for producing it. Lawrence's Chromatron tube utilized only one electron gun. Color occurred when the beam reached stripes of phosphor on the surface of the screen. On the way, the beam passed through a grid of high-voltage wires, which aimed and accelerated it. Controlling the focus and acceleration of the beam at high voltage inside the vacuum of the picture tube was only part of what turned out to be a production nightmare. By September 1964, a Chromatron prototype with a seventeen-inch screen was ready, but mass-production problems had not been eliminated and test runs were not promising. Ibuka alone was confident; he insisted, however unrealistically, that the product would succeed. He had the product announced and displayed it in the Sony showroom, creating market excitement and pressure. Sony invested in an entirely new facility at Osaki Station in Tokyo to house the Chromatron assembly line, and the prototype was rushed into production. The yield was dismal: out of one thousand picture tubes, only two or three were usable; the others suffered from "color variance," faces going green at the outer edges, greens merging sickeningly into reds. Ibuka once again asserted his dictum that improvement must be achievable if even one working picture tube could be produced, and refused to shut down. Now 150 people were at work on Chromatron, which Ibuka had made the company's top priority. The machines sold in Japan at the retail price of ¥198,000 ($550), but each set cost over ¥400,000 to produce. Eventually, thirteen thousand sets were sold at a huge loss to the company.

Morale was at a record low. Miyaoka remembers sitting in the company cafeteria, wondering aloud to his colleagues whether it was time to move on to a company such as Toshiba or Panasonic which understood picture tubes. In fact, no one left: to quit when the company was in distress would have been tantamount to abandoning one's family in a crisis.

It was a dark time, the more unsettling to everyone because tension between Ibuka and Morita was manifestly in the air. No one ever witnessed an argument, but people were aware that Morita urgently wanted to cut losses while Ibuka would not budge. Miyaoka's boss, the physicist Susumu Yoshida, remembers a meeting when Morita angrily accused the gathered engineers of taking advantage of Ibuka's commitment to the technology to indulge their curiosity in the problem-solving process, costing the company money it couldn't afford. In November 1966, Iwama called Yoshida into his office and grimly informed him that the company was "close to ruin" and that Chromatron would have to be abandoned by the end of the year if production yields could not be improved. In Miyaoka's memory, the worst moments occurred when people from "Morita's camp," planners and accountants, began showing up at the brainstorming sessions at which he and the other engineers were grappling with Chromatron. These outsiders seemed like "spies," and it was assumed that they were reporting back to Morita what they had heard with a negative bias.

In the autumn of 1966, Ibuka finally announced that he personally would lead a team to search for an alternative to Chromatron. Susumu Yoshida traveled to the United States to inspect a General Electric product called "Porta-color" and was impressed. The thirteen-inch set used a shadow mask system with three electron guns aligned (in the RCA system, the guns were offset in a delta configuration). He was also surprised by advances RCA had made in screen brightness, and chagrined to learn that RCA was producing twenty thousand color sets a month compared to Sony's one thousand. In Tokyo, Yoshida's report was a source of consternation: Sony was already years behind the competition in a robustly growing market; perhaps there was no choice but to adopt the shadow mask system.

The decision was to continue looking for a distinctive Sony solution. The engineers involved were unwilling to allow five years of work to end with imitation: Sony had been founded by Ibuka precisely in order to be a company for engineers. As Morita struggled to calm Sony dealers, he was governed by his respect for Ibuka. To finance the additional cost of R and D, working through banking contacts that had been established by the directors, he secured a loan from Japan Development Bank, Sony's first development loan, of ¥650 million (roughly $2 million at the time) at 6.5 percent interest. The eventual success of Trinitron, still invisibly distant, would enable the company to repay the loan in three years.

The thirty engineers on Ibuka's team began exploring multiple approaches to color simultaneously. Ibuka moved from group to group with questions and encouragement, but nothing developed until Yoshida wondered aloud to Miyaoka what might happen if three cathodes were used to generate three electron beams from a single gun. Miyaoka was skeptical, but relieved to have something new to try; by this time he was uncomfortable about receiving his paycheck, mortified by the thought that his salary was being earned by Sony's audio products. He rigged a single gun with three aligned cathodes and was, by his own account, astounded when he produced a screen image in color, blurred but bright, unaccountably bright. He reported his results to Yoshida, who passed them on to Ibuka.

Saturday of that week (the Japanese workweek was six days), at a few minutes past four in the afternoon, Miyaoka was summoned to Ibuka's office. The timing proved to be critical: Miyaoka was an avid cellist who looked forward all week to the Saturday rehearsal of his community orchestra. By dashing for the train when the bell rang at 4:30, he could just manage to be on time for the 6:00 P.M. rehearsal in the Tokyo suburb of Fujisawa, where he lived.

Ibuka had seen the seeds of success in Miyaoka's report and was now intent on hearing from the twenty-nine-year-old physicist, in the presence of his immediate boss, Susumu Yoshida, that the single gun and three cathodes was a viable approach. Miyaoka had been experimenting for less than a week and was far from certain that a focused picture could be produced with the single gun. But on this afternoon, it was music that was on his mind, and he could tell that Ibuka was not going to dismiss him until he had heard what he hoped to hear. At the last possible moment, in desperation, Miyaoka declared that the system could be made to work commercially, and, hastily excusing himself, left for his rehearsal. Monday morning, Ibuka announced that the company would henceforth concentrate all its resources on developing the single-gun picture tube.

In the months that followed, Ibuka galvanized his engineers into a tireless effort. Miyaoka would arrive at the lab at 7:30 in the morning and find Ibuka waiting for him at his desk to quiz him about his progress the day before. Sony now employed seven thousand people worldwide, and here was the company's founder and president waiting for a freshman employee when he arrived at the lab. "Finding him there should have made me uncomfortable," says Miyaoka, now the president of Sony's polytechnic college, "but he was so involved it was just exciting. What we achieved in those days was because of him and for him."

Ibuka was obsessed with Trinitron. When someone from the team appeared at his office, he would wave them in no matter what he was doing, and then was likely to wander off with them to the laboratory for hours. Sometimes he would reappear as abruptly as he had left, but only long enough to fetch Morita so that he could also have a look. Hiroko Kurata remembers dreading the appearance of Miyaoka or anyone else from Trinitron, particularly late in the day, which was likely to mean that she would have to wait at her desk for hours into the evening, until Ibuka finally returned or telephoned to say she was free to leave.

By February 1967, the picture had been focused. Late that summer, an engineer named Akio Ohgoshi proposed replacing the shadow mask with an "aperture grille." The grille permitted more electrons to pass through it as the beam headed to the screen and resulted in a brightened picture. Finally, in April 1968, Ibuka called a press conference to introduce Sony's twelve-inch Trinitron color TV. The team had assembled ten units for the introduction, but these engineering models were still a long way from being ready for the assembly line. Yoshida and his team anticipated that they would need another year of work before the product would be ready for mass production. They were flabbergasted to hear Ibuka conclude the press conference with the promise that ten thousand units of Trinitron would be ready for market by October. The following month, Morita made the same announcement in New York.

This was Ibuka's arbitrary but effective approach to accelerating Sony's entry into the market. When the team protested, Ibuka recommended that everyone should study the history of the Manhattan Project for useful hints. To Yoshida, he said only, "I'll never do this to you again so please indulge me this once," and asked him to become head of production responsible for meeting the October deadline. Yoshida recalls that he was furious and tried to decline the assignment. In any event, he was a physicist and not a process designer. "But Ibuka was going to have his way no matter what," he now reflects with a smile. "He took me out to play golf, he phoned over and over again. One day he even showed up at my home! What could I do?"

The team working frantically under Yoshida referred to him as the General—*Shogun* Yoshida—and lived in fear of him. According to Miyaoka, anyone expressing doubt about success in meeting the deadline was transferred from the team. In September, Miyaoka himself discovered a defect in the cathode wiring which would cause the tube to darken after several months of use and asked Yoshida to extend the deadline. The next day, he was transferred back to development, with instructions not to mention the defect to anyone, including Ibuka. After Trinitron reached the market, Yoshida recalled the defective sets and replaced the faulty cathode; he chose to spend the extra money rather than compromise the marketing deadline.

As the first run of twelve-inch Trinitrons came off the assembly line on deadline, Ibuka bowed to Yoshida and his group, saying, simply, "Thank you." Tears filled his eyes.

Ibuka "graduated" from Trinitron and became the driving force behind the development of Betamax technology; but in the mid-seventies he began to disengage from Sony's day-to-day business and became a passionate advocate of educating children while they were still very young, a subject to which he may have been led by his heartbreaking experience with Taeko. As early as 1971, he wrote *Kindergarten Is Too Late*, a book that sold more than a million copies in Japan and became a minor bestseller around the world. Subsequently, he published fully a dozen more books, including *Zero Years Old* and *The Brilliant Fetus*, in which he emphasized the importance of in utero stimulation. He befriended Shinichi Suzuki, creator of the Suzuki method for teaching very small children to play musical instruments, and Dr. Glen Doman, founder of the Institutes for the Achievement of Human Potential and author of well-known books on teaching infants reading and mathematics. And he availed himself of every opportunity to lecture people on his favorite subject, including, in later years, meetings of the Sony board.

If Ibuka needed help in letting go, it may have come from the difficulty he now began to experience in keeping up with new technologies. He was an avid student, but there were ideas he was unable to fathom, or perhaps declined to understand, with characteristic stubbornness. According to Makoto Kikuchi, then director of Sony's central laboratory, "He'd call me up and ask me to come over and explain, for example, the notion of entropy in thermal physics. And the trick was, you had to make him *feel* that he understood. It was very subjective, and you had to keep him interested until he felt that he had grasped the idea." More than once, failing or refusing to grasp a technology, Ibuka would reject it out of hand, and his unrelenting rejection could create problems for his engineers and for the company. An example was optical laser technology, in particular the general junction laser, which was at the heart of the compact disc that Sony developed jointly with Philips beginning in 1978. Another of Ibuka's blind spots was parallel signal processing, a technology essential to computers, and, equally critical, integrated circuits. Kazuo Iwama, the geophysicist who was in charge of all Sony technology until his death in 1981, was distressed about the implications of Ibuka's increasing ignorance of technological advance as the seventies progressed. At one point, he summoned Kikuchi to his office, closed the door, and, explaining the source of his consternation, asked him to try his best to convince Ibuka of the importance of the Computer Age. According to Kikuchi, who spoke in English, "Ibuka knew many things about amplifiers, television, and video. But, very frankly, he had no understanding of the essential problems of engineering and the future of engineering as it related to information technology."

3

AKIO MORITA: DISCOVERING AMERICA

On his first trip to the United States in 1953, Akio Morita was overwhelmed. He was not prepared for the scale of things, the giant cars and wide roads, the distances, the soaring buildings. The power of the booming economy felt crushing. To a friend from his high school days in Nagoya who was working in New York, Shigeru Inagaki, Morita confided his doubt that a small Japanese company had any chance of surviving in this giant country. The journey had its moments: signing the provisional agreement at Western Electric and a visit to Coney Island.* Nonetheless, as he departed Idlewild Airport for Europe, despite his determination to bring Sony to the West, his confidence was shaken.

Germany, much further along in the postwar recovery than Japan, was no more reassuring. He visited Volkswagen, Mercedes, and Siemens, among others, and was disheartened by what he experienced as postwar Germany's robust health. At a restaurant in Düsseldorf, a waiter served him a dish of ice cream garnished with a tiny parasol and

* Coney Island had resided in Morita's imagination as a magical place since he first heard about it as a child from his uncle Keizo, an international traveler. When he moved his family to New York in 1963, he promised his sons to take them there but apparently never managed to find the time. Masao Morita, lamenting the stroke that felled his father in 1993, told me ruefully that among the many things they had always planned to do together and now never would was ride the giant roller coaster at Coney Island.

informed him obligingly that the paper bauble was made in his country. Morita never forgot his chagrin at that moment as he reflected that the world's consumers associated "made in Japan" with trinkets and cheap imitations.

From Düsseldorf, he traveled by train to Holland. As he crossed the border and began seeing windmills that reminded him of Dutch paintings and people riding bicycles and working in the fields, he felt more at ease. His destination was Philips Electronics, and he was surprised to discover that this great organization known around the world was headquartered in the old world town of Eindhoven. Here the scale and pace of life felt manageable. Even the statue of Dr. Philips in front of the train station reminded him of the bronze statue of Kyuzaemon Morita the Tenth which stood in his ancestral village of Kosugaya. In a letter to Ibuka, he wrote, "If Philips can do it, perhaps we can also manage." Akira Higuchi and others recall that on his return from this first expedition Morita spoke incessantly about the importance of building an international brand, and they attribute his sharpened focus to inspiration from Philips.

Shortly after his return, Morita proposed finding a name for the company which could be pronounced and recognized outside Japan. Tokyo Tsushin Kogyo signified nothing and was a tongue twister to boot. The English equivalent, Tokyo Telecommunications Engineering Company, was hardly better. There were a number of international organizations using three-letter logotypes, IBM, RCA, AT&T, and so on, but TTK had already been taken by the national telephone company and also seemed to be an infelicitous combination of English letters. As it turned out, the answer, or most of it, was already imprinted on the boxes of Soni-tape that the company had been marketing since 1950. The name had come from the Latin *sonus*, for sound. Morita now combined *sonus* with the English *sonny-boy*, an expression that conveyed to him the youthful energy and irreverence he wanted at the heart of the company, to create Sony.* This combination had the

---

* Because the vowel in *sonny* bears no resemblance to the long "o" in "Sony," as pronounced in English, native speakers of English cannot be expected to make the association Morita intended.

virtue of simplicity—the company logo and name would be the same—and the additional merit of meaning only Sony in every language.

The company began using the Sony trademark on its products beginning with the TR-55 transistor radio in 1955, but did not officially change its name to the Sony Corporation until January 1958. The delay was due in part to opposition from the company elders, Michiharu Tajima in particular. Tajima was an archconservative from the old school who demanded formality and strict adherence to social convention (in 1948, he was appointed general director of the Imperial Household Agency and assumed responsibility for managing the lives of the imperial family). Once he had accepted the position as adviser to the company in deference to his former classmate, he insisted on being consulted about every decision. Finally, Ibuka had to ask his father-in-law to suggest to Tajima that he refrain from attending weekly technical meetings, where his presence intimidated the engineers from speaking their minds. Now Tajima demanded to know how a Japanese company could possibly choose a foreign name that could not be represented in Chinese characters and had to be spelled phonetically.* Morita pointed out, no doubt politely, for age and seniority would always count with him, that Canon Camera was thriving around the world, and he stood his ground. Eventually, the elders gave in.

In 1955, Morita returned to the United States with Sony products in hand, tape recorders, stereo microphones, and the transistor radios that he was counting on to penetrate the American market. He knew he was going to need a distributor. The previous year, Akira Higuchi had been sent to New York to sell a Sony version of a dynamic microphone that performed on a par with American models but was less expensive. The company had high expectations for the product, but when Higuchi approached individual retailers, they refused to talk to him. Finally, someone took the trouble to explain that dealers in the United States did not buy products off the street. Forewarned, Morita made the rounds of agents and distributors, but found little enthusi-

---

* Japanese is written with a combination of characters—ideograms—borrowed from Chinese and a phonetic syllabary.

asm: it was a time when the national trend was toward larger cars, larger homes, and much larger TV and radio consoles designed for larger living rooms. Then a purchasing agent at the Bulova watch company saw the radio and informed Morita that he would take one hundred thousand units, provided that he could sell the radios under the Bulova name. "I'm sure you understand," he told Morita. "Nobody has heard of Sony."

One hundred thousand units was a huge order, worth more than the total capitalization of the company at the time. Practically speaking, it was too large: the limit of Sony's production capacity at the time was ten thousand transistor radios a month; increasing capacity to the required extent would have increased the cost of goods and cut deeply into profits. But profit wasn't the critical issue; what mattered to Morita was establishing the Sony name around the world. To compromise now before the company had a foothold in the American market was to invite defeat. Morita sent a cable to Tokyo explaining his intention to decline. Ibuka wired back at once: the board agreed that the company could not afford to reject the order; Morita was to accept Bulova's terms. What followed is not certain. According to Sony's official history, *Genryu*, not always reliable, after a week of cabling back and forth, Morita telephoned Tokyo, and "as the phone bill ate into his precious supply of dollars, [he] reasoned with his colleagues. Once he had their consent, he went to inform the prospective client." What seems more likely is what Morita implies in his memoir: that he was unable to persuade the Sony board long distance and made the decision to decline on his own authority, by no means the last time that Morita would defy his own board of directors. In fact, as his power grew in tandem with Sony's growth, he became less restrained about employing it autocratically. In this case, he seems to have been acting not as executive vice president accountable to Ibuka and the board, but as a founding partner whose father was, and would remain, the company's largest individual shareholder. If Tajima had been upset by the choice of Sony for the company name, he was beside himself when he learned that Morita had rejected the Bulova offer and threatened to resign his seat on the Sony board. At Ibuka's request, Maeda and the other elders mollified him, and in 1959, he became chairman. Years later, Morita would say that turning Bulova down was the best business decision of his career.

Through Ibuka's friend Shido Yamada, Morita was introduced to a "manufacturer's representative" named Adolph Gross, who liked the Sony radios and offered to help Morita sell them; this meeting led to a five-year contract with Gross's company, Agrod, to import and distribute Sony radios in the United States. Although the Sony history styles Agrod as "an audio giant," it was in fact a shell company that Gross had created to legitimize himself as an importer of British hi-fi speakers and German turntables. Gross was the only employee, and there was certainly no distribution network per se. What he did have to offer was a connection to a bona fide electronics distributor with its own transcontinental sales network. Delmonico International was managed by three business colleagues who had left Olympic Radio, Adolph Juvilier, Herb Kabat, and Albert Friedman. In due course, Gross helped Morita contract with Delmonico to serve as principal distributor in the United States for a period of five years, beginning late in 1957. There was of course an angle—Adolph Gross loved angles—which involved, according to Gross's accountant at the time, Irving Sagor, the payment of excise taxes. Until it was repealed in the sixties, the federal government collected an excise tax of 10 percent on electronics and other luxury items, such as perfume, which were manufactured outside the United States. The loophole that Gross "optimized" to the benefit of all parties was that the tax was payable by either the importer or the distributor. Agrod, acting as importer, would pay the 10 percent on Sony's FOB price of $21 per unit, add on a dollar per unit for itself, and pass the product to Delmonico, which would retail it for $39.95. In this way, Sony got its manufacturer's price, Agrod earned a bonus dollar per unit, and Delmonico saved a dollar by not having to pay 10 percent of the $40 retail price. Several years later, when Agrod had gone out of business, Sony America received a citation from the IRS for payment of what it claimed was the balance of excise tax due on a product retailing for $40. By this time, several hundred thousand Sony transistor radios had been sold, and the claim was for a considerable sum of money. Irving Sagor, who by this time was working for Morita, proved that Agrod had been acting as an official importer and won a dismissal in federal claims court.

Between 1955 and 1960, Morita returned to New York several times a year. He stayed at seedy hotels near Times Square, did his own laundry in the sink, and ate most of his meals at the Horn & Hardart Automat, which saved him not only money but the necessity of conversations in English. Eventually, he moved upmarket to a Stouffer's on the top floor of 666 Fifth Avenue. Later he would speak nostalgically and with some pride about his days of privation in New York. Not that he was ever poor: the problem was that MITI had imposed a limit of $500 per trip on the foreign currency that could be taken out of Japan.

Over time, Morita developed what appears to have been a close friendship with Adolph Gross. In his memoir he recalled an evening when Gross took him to see *My Fair Lady.* Morita loved Broadway musicals, and he was jubilant about seeing the smash hit of the 1957 season, the more grateful to his friend because he knew he must have paid dearly for the tickets. He was therefore astonished when Gross fell asleep as the lights went down and slept soundly in his $100 seat throughout the show.

They were an unlikely pair. Morita, in his mid-thirties, was driven, tireless, eagerly inquisitive, very sharp, impeccably well mannered, and, due to his pampered upbringing, somewhat naïve about the world. Gross, close to sixty when they met, was, according to Sagor, a Runyonesque character who spoke lower Broadway and chewed constantly on the stump of a dead cigar. An incorrigible *hondler* who loved making deals and tired of them as soon as they were advantageously concluded, Gross became Morita's first mentor and guide to the mysteries of American business. Morita repaid him with gratitude and affection, calling him his American father. Perhaps he was drawn to the older man's experience and savvy in much the same way as he had been drawn initially to Ibuka. Gross's death of a heart attack in London in 1958 was a blow to Morita. Thereafter, he provided for Gross's widow, Dorothy, for the rest of her life.

Sony released its smallest transistor radio to date, the TR-63, in

March 1957, and the product was an immediate success in Japan. The radio was priced at ¥13,800, an amount equivalent to the monthly paycheck of the average white-collar worker, and sold briskly nonetheless: since 1955, Japan's GNP had been growing at 10 percent annually, and with inflation low and savings high, the consumer economy was booming. Sales of the TR-63 in particular received a boost from the release by Victor Records of a hit tune satirizing the life of the white-collar worker, "Thirteen Thousand Eight Hundred Yen." Victor promoted the record in leaflets dropped from airplanes which Sony dealers collected and displayed in their windows.

In the United States, sales of the TR-63 through Delmonico were also building. In late November 1957, Morita had to charter a JAL cargo plane to carry in additional product for the Christmas rush. By the end of that first year, twenty-five thousand radios had been sold in the American market.

But with Adolph Gross out of the picture, the relationship with Delmonico deteriorated. The company began pressuring Morita to deliver a radio it could sell in larger volumes at a lower price. The competition had models on the market priced at $15 to $20 compared to $40 for the TR-63. Morita would not consider lowering his premium pricing, nor was he willing to produce a cheaper model. For one thing, the radios as priced were highly profitable to Sony and they were selling. According to Irving Sagor, the Sony model was known as "the transistor radio that worked." As important, Morita was determined to undo the stigma associated with "made in Japan" by establishing Sony as a manufacturer of high-quality products that commanded a premium price. Given his commitment to this goal, it was only a matter of time before Delmonico's lobbying for lower prices and steep discounts had alienated him.

At the end of 1959, Delmonico stepped over the line by announcing in the professional press, without a word to Morita, that it would distribute Sony's first transistor television, a product that had been announced in Japan only late in December and was not due to go on sale for another six months. In New York in January 1960, Morita learned that Delmonico had established a price arbitrarily and was even taking orders without authorization from Sony. Accompanied by Edward Rosiny, Adolph Gross's lawyer, he visited the Delmonico offices in Long Island City and suggested to managing partner Albert Friedman that it was time to terminate the distribution agreement. Friedman reminded Morita that Delmonico was owned by Thompson-Sterret, a public company; because the Sony contract represented money to the shareholders, a settlement would be required. Rosiny responded by pointing out that Delmonico was in breach of contract and liable for appropriate redress for having damaged Sony's reputation in the United States by unlawfully soliciting dealers.

The negotiation, conducted by Rosiny with Morita looking on, lasted until mid-February, and was complicated by the fact that Delmonico was in possession of close to $1 million worth of Sony radios. Delmonico agreed to terminate for a settlement fee of $300,000. Rosiny lowered this figure in stages, first to $200,000 and eventually to $100,000. Morita was relieved and prepared to make payment, but Rosiny insisted on returning to the table a final time and won an additional $25,000 concession, in order, as he explained to Morita, "to earn [his] $25,000 fee out of Delmonico's money." From this moment on, for the next twenty years until his death in 1978, Eddy Rosiny would be Morita's teacher of jurisprudence and Sony's principal negotiator and litigator in the United States. Rosiny initiated Morita into the complexities of the American contract, and, later, beginning in the early seventies, was his guide through the even more alien territory of American litigation.

The return of Delmonico's inventory of radios had been included in the settlement, and Morita wanted the goods safely back in his hands as quickly as possible. Irving Sagor, who had been managing Agrod's failing business since Gross's death in 1958, volunteered the use of the Agrod warehouse at 514 Broadway in Manhattan, on the corner of Spring Street in what is now SoHo, and arrangements were hastily made. Morita hired a local trucking service to move the goods, and they arrived from Long Island on a freezing morning in February. With the help of Charlie Farr, Agrod's warehouse manager, and a Japanese student attending Columbia on Sony's first scholarship, Morita and Sagor unloaded thirty thousand transistor radios in individual boxes, working all day and night until 3:00 the following morning. As the men sat drinking coffee wearily at a desk in the Agrod office, someone tripped the burglar alarm and a security guard rushed in and held them at gunpoint, thinking he had surprised an international gang of thieves. He left only after Sagor had convinced him that he belonged there by opening the safe and showing him documents.

Morita had already decided what he was going to do with the radios; he had resolved to establish a Sony subsidiary in the United States to sell Sony products directly into the American market. Ibuka and the others in Tokyo were skeptical. A few of Japan's largest trading companies, such as Mitsui and Itochu, were operating in America, but no one in the electronics industry had attempted to deal directly into the U.S. market. The standard wisdom held that American middlemen were essential; it was this assumption that had led Morita to Agrod and ultimately to Delmonico in the first place. But Morita insisted that Sony would never master the sophisticated American market—and he had never had a moment's doubt that a leading position in the United States was what he must achieve to realize his vision for the company—without an opportunity to learn American marketing from direct experience. As always, his enthusiasm and confidence prevailed. Sony had already submitted an application to the Finance Ministry for permission to bring $500,000 to the United States for future use. Early in February, the application was approved and the funds remitted to New York through the Mitsui Bank.

On February 20, 1960, Morita established the Sony Corporation of America—inside Sony it would be known as Son-Am—a wholly owned subsidiary capitalized at $500,000, and opened for business in the rat-infested Agrod office-warehouse at 514 Broadway. To oversee the business in his absence, Morita imported a former high school classmate named Masayoshi Suzuki from headquarters in Tokyo and installed him in New York as executive vice president. "Big" Suzuki, as he was known in New York, had been at Sony only a year, dragooned by Morita from the Mitsui Trading Company, where he had been dealing in oil in Hong Kong and the Near East. Irving Sagor was persuaded to give up his CPA practice to manage Sony's books and would remain in the company for thirteen years. A former partner of Adolph Gross's named Milton Thalberg came in to organize advertising and marketing. Kazuya Miyatake, Sony's Columbia student, was pulled out of school and made warehouse manager. A salesman named Hiroshi Okochi on his way home from Europe to Sony Japan was intercepted in New York and appointed national sales manager; and a young, fast-talking Sony pistol with good English named Hajime "Jimmy" Unoki was rerouted from a tour of Africa and appointed branch manager responsible for sales in the New York metropolitan area. Sagor's secretary and several engineers brought from Tokyo to service the radios Morita intended to sell completed the team of thirteen people.

Morita was president, and from this moment on was commuting to work from Tokyo, spending ten days each month in New York. Nominally the number-two man in the main company, but with actual responsibility and discretionary power equal to Ibuka's, he more than had his work cut out for him in Japan. But the focus of his vast energy was Sony America, and during the early years he kept both hands on every aspect of the business. The space he had inherited from Agrod was a storefront in what had been a textile warehouse building, narrow and deep, with two rows of desks in the office on the street and the warehouse behind: Sagor remembers Morita rushing from desk to desk to keep current and in control, poring over the books, shouting into the telephone in his rudimentary English, or composing progress reports to Tokyo which had to be keypunched and sent out by telex at night. The following January, on his fortieth birthday, the staff presented him with a fireman's hat: in the atmosphere of excitement and urgency he created, the office generated a lot of heat. By summertime, Okochi and Thalberg had established one-room sales offices in Los Angeles, San Francisco, Chicago, and Dallas, and were meeting their sales goal of six thousand radios per month. Japan was still on the six-day workweek, and the Japanese were at the office on Saturdays; everyone worked long hours, often until midnight or later. On their way home, they would stop to eat at one of the all-night delis on Broadway. Morita developed a taste for matzo-ball soup.

In fact, the environment at Sony's New York office was distinctly Jewish. According to Irving Sagor, "With all the Yiddish that was flying back and forth between us, it was only a matter of time before young accountants sent from the Tokyo head office began asking for the *tsetle* when they wanted to see bank receipts." At about this time, when ads for Levy's rye bread began appearing all over the city—"You don't have to be Jewish to love Levy's rye"—Morita would point them out and nod in agreement.

Over time, as he extended Sony's operations in America and Europe, Morita's inclination to employ Jews to grow and manage the business is evident and arresting. Adolph Gross, Irving Sagor, and Edward Rosiny, his first consiglieri, were only the beginning. There was Ernest Schwartzenbach, the first American president of Sony America; Harvey Schein, Schwartzenbach's successor; Paul Burak, a partner at Rosenman and Stern who replaced Rosiny following his death in 1978 and has been the company's principal legal adviser ever since; Walter Yetnikoff, who managed Sony's music business worldwide and was a catalyst in the Hollywood acquisitions; and Dr. Ron Sommer, now chairman of Deutsche Telekom, who was brought to America to run Sony Electronics and served as president of Sony Europe in the early 1990s. Finally, and most dramatically, there was Michael "Mickey" Schulhof, Morita and Ohga's protégé. A physicist born of cultured European parents and raised in New York, Schulhof enjoyed, between 1987 and 1995, a position of privilege and authority as a foreigner which seems likely to be inscribed in Japanese business history as unique.

Though no one inside Sony has anything illuminating to offer about this unmistakable pattern, it cannot have been coincidental. Granted, Morita did not choose his earliest contacts in New York. They just happened to be Jewish businessmen. But he must have felt surprisingly at home in the company of men like Gross and Rosiny, for in later years it is clear that he consciously sought out Jews to bring to Sony, or at least entertained a Jewish bias when presented with options. Irving Sagor has a "vague memory" that in an orientation he required of Japanese executives bound for the United States, Morita counseled the Sony men to recruit Jews whenever possible. According to Sagor, "He seems to have felt that Jews were smart, imaginative, and very compatible with the Japanese in temperament and ways of looking at the world."

Asked to account for Morita's and his own predilection, Sony chairman Norio Ohga, while he declines to acknowledge explicitly that the bias exists, points out matter-of-factly, as though he were invoking an axiom of natural law, that Jews "tend to be very sharp at business and especially good in money matters." The essentialism of this view, which I have heard him express at various times, might be construed as anti-Semitic, but I have never sensed anything malevolent behind it. It has always seemed to me rather that he was exhibiting a Japanese tendency to essay an insular and naïve stereotyping as an approach to dealing with people other than themselves (Ohga's sentiments recall the fascination that Shakespeare anticipated his insular Elizabethan audience would project on Othello the Moor). Certainly, the Japanese have a history, particularly since the war's end, of anti-Semitism as vicious as any in the world. But to label Morita or Ohga anti-Semitic would be to misunderstand and malign them. At Edward Rosiny's funeral in February 1978, Morita stood at his coffin in the funeral home with his head bowed for a long time, and to others in the room, his grief was palpable. For years after, he traveled to the cemetery in Long Island to lay flowers at Rosiny's grave on the anniversary of his death. Clearly, as evidenced by the deep and lasting friendships that Morita and Ohga maintained with Rosiny, Schein, Yetnikoff, Schulhof, and others, there was something that drew them powerfully to Jewishness as a cultural quality they recognized and appreciated. The importance of the family and the respect for learning and tradition come to mind as likely ingredients of this attraction. Perhaps it also had to do with the sense of being foreign which is integral to Jewish consciousness.*

In August 1960, Morita sent Big Suzuki back to Tokyo six months after he had arrived (at the time he lured him away from his other job he had promised his wife that her husband would not have to be out of the country for more than two weeks at a time) and replaced him with another boyhood schoolmate, Shigeru Inagaki. Like Suzuki and

* Morita and Ibuka both loved the Broadway musical *Fiddler on the Roof.* Ibuka saw it for the first time with Irving Sagor and exclaimed in amazement as they left the theater, "They're just like us. They seemed Japanese!" Morita saw the show repeatedly and eventually arranged for a Japanese production to be mounted in Tokyo, where it became a perennial hit.

countless others, Inagaki had been quickly persuaded to leave his job to join Morita at Sony. Thirty-nine at the time, Morita's age, he was an aggressive salesman who had been working in the United States for seven years as a sales manager at a camera company called Kaloflex, and had a solid grasp of wholesale-retail dynamics in the American marketplace. As a childhood friend, he could also be counted on for unquestioning loyalty and, given the boldness of the undertaking and the risk involved, his dedication was important to Morita. Inagaki served the company faithfully for fourteen years, but Morita never saw fit to promote him to president. "He was a tireless worker," Sagor told me, "and he would follow orders until he bled to death." Possibly Inagaki failed to demonstrate the creativity and charisma that Morita expected from the leader of his American outpost. Given the difficulty he would have in finding the right person once he was ready to replace himself as president, this must have been cause for considerable regret.

Extending Sony into America was a process that would involve Morita in numberless confrontations on every level of his business and personal life. From the beginning, in spite of his awkward English, he handled himself in these situations with a mastery that was even more remarkable in view of the native Japanese distaste for confrontation. Perhaps in times of crisis in New York, Morita was able to tap into the sense of superiority which he had been raised with as eldest son and heir. Possibly, as his son Hideo insists, he was simply a great performer, able to hide and override the discomfort he may well have been feeling when engaged in, or simulating, American combat. Morita never said explicitly that the American approach to resolving conflict affronted his Japanese sensibility. He did, however, more than once characterize American business as "brutal," and in several essays over the years compared American and Japanese behavior, highlighting what he perceived as the gulf between them. In 1966, Sony's newly formed division of international trade placed an unusual classified ad in a Tokyo newspaper. It appears to have derived from Morita's experience in New York: "Wanted," it began, "*Japanese* men up to thirty years of age who can PICK A FIGHT in English!"

There can be no question that Morita was an amazingly quick study, and, from his earliest dealings with Americans, effective. For example, in 1961, negotiating for rights to produce the disastrous Chromatron, he found himself across the table from Charles Bluhdorn, chairman of Gulf and Western, which, as parent company to Paramount, controlled the patents. When the deal was signed, Bluhdorn told Morita, "I have been in business many years and have never been outmaneuvered. But today you have outsmarted me." Morita prized this sort of compliment above all others. "You see," he gloated to a young assistant, "I am a kind of Japanese Jew!"

Morita's first encounter with Union Carbide, the battery maker, was another early example. Morita received a letter in New York via Tokyo which threatened to bring suit against Sony for including a Union Carbide nine-volt battery in each radio it was importing. Morita visited the company with Sagor, and listened politely as the executive in charge explained condescendingly that a license to buy his batteries in Asia did not entitle a small Asian company to turn around and import the batteries into the United States, and that he expected Sony to cease and desist at once. Morita looked the man in the eye and replied tersely that he was missing the point: the issue was not the single battery in each radio, but the demand for additional batteries which Sony was creating with its radios. With that, he gestured to Sagor, rose, and walked out of the room. Sony never heard another word from Union Carbide, but years later Morita negotiated a joint venture with the company to produce batteries in Japan.

Within a year of founding Sony America, Morita committed Sony to becoming the first Japanese company to sell stock in the United States, and spent a nightmarish six months laboring to fathom and then assimilate American business practice. In 1960, as trade between Japan and the United States was being liberalized, the Finance Ministry began accepting applications from Japanese corporations wishing to sell American depositary receipts. ADRs represented actual shares of stock held in Japan (much as paper currency was once backed by government gold reserve) and traded in dollars through brokerage houses in the United States. Morita quickly persuaded Ibuka and the board of directors that they must seize this opportunity. The obvious reason was purely financial. Sony had spent heavily in the 1950s, developing transistor production capacity—by 1960, it was producing one million units a month—and building two new plant facilities and a research laboratory. Its principal bank, Mitsui, under pressure from Junshiro Mandai, the first chairman, had provided short-term loans but lacked the funds to become a major shareholder: in 1960, while 23 percent of Japanese industry was held by banks, Mitsui owned only 8 percent of Sony. The opportunity to sell shares in the world's richest financial market was tantalizing. For Morita, an even bigger prize would result from being first. If Sony could pioneer the U.S. market, the event was certain to draw national attention to Sony America: building Sony into a national presence was more than a long-range goal, it was an obsession.

In the fall of 1960, Sony applied to the Finance Ministry along with one hundred other companies, and was by far the smallest of sixteen businesses to receive approval, including Toshiba, Hitachi, Yawata Steel, and the giant trading companies Mitsui and Mitsubishi. Sony may have made the cut because it had already begun to establish a reputation in America. The Bank of Tokyo agreed to serve as trustee bank in Japan; Nomura Securities and Smith Barney became the underwriters in New York. Yoshio "Terry" Terasawa from the New York office was Nomura's man on the project. He had just returned from his honeymoon when he joined the team in February 1961 and had to work so hard that he scarcely saw his wife for the next four months. Smith Barney was represented by one of its partners, Ernest Schwartzenbach, a Swiss-German Jew with an evolved interest in Japanese culture who developed an abiding respect for Morita during the months they worked together day and night.

Morita personally led the Sony team responsible for meeting Securities and Exchange Commission requirements. SEC lawyers and accountants wanted to see Sony documents, including contracts, which had to be translated into English. There were on-site inspections of Sony plants. And Price Waterhouse conducted an audit, which revealed a major problem: Sony's books were not consolidated. In Japanese accounting practice, subsidiaries were treated as discrete entities and commonly used as caches for hiding losses or storing profits; consolidating everything in one balance sheet was a mammoth task that Sony was at first reluctant to undertake. Nor was bookkeeping the only incongruity that came to light. American lawyers were unsettled to discover a clause in many of Sony's contracts specifying that both parties would discuss any changes that affected one or the other party's ability to fulfill the terms of agreement. How could there be a contract, they wanted to know, which did not provide sanctions or penalties in case of default? Morita later recalled struggling through interpreters to explain the Japanese social convention whereby a man's word counted for more than a binding instrument. The Americans were unhappy to see that Ibuka and Morita had signed personal guarantees for a number of Sony's short-term loans, offering their residences as collateral. They were even more disturbed that the banks had nowhere guaranteed in writing that the loans would be renewed. At each uphill turn in the discussions, the burden of explanation fell to Morita; frequently, the sessions trailed off into frustrated or angry silence. One day, in Schwartzenbach's suite in the Imperial Hotel in Tokyo, where the team was working day after day until one or two in the morning, Morita declared the project "terminated" and stormed out of the room. Terasawa chased him down the corridor in desperation, fearing for his job, and implored him to return. Clearly, the effort to engineer a cultural bridge from one business world to the other was taking an emotional toll. Later, Morita would learn to mask his feelings more successfully, but efforts to reconcile Japanese and Western conventions of business and legal practice would always tax him.

On June 7, 1961, thanks in large measure to Morita's vision and endurance, Sony placed for sale on the American market twenty thousand American depositary receipts representing two million shares of common stock priced at $1.75 per share, a trifle higher than the closing price of Sony stock on the Tokyo exchange. The ADRs were bought up in two hours. Terasawa and the other salesmen at Nomura instructed their Japanese secretaries to answer the ceaseless phone calls "Sony—sold out" and left the office for the day. The closing price was $2.40. On June 15, Morita was handed a check for $3 million. As he had hoped, the story of Sony's pioneering adventure was widely reported in the American press. According to his wife, Yoshiko, the success of Sony's first public offering on the New York Stock Exchange was for Morita the happiest moment in his business career.

The publicity generated by the stock offering led Morita to thoughts of a Sony showroom in New York. He had watched shoppers strolling up and down Fifth Avenue and had sensed their affluence: given the premium-quality image he wanted Sony to project, Fifth Avenue was the place to be. When Irving Sagor dutifully pointed out that Sony America, with only Sony's line of radios to sell, could hardly justify, or, for that matter, afford the expense of building and maintaining a Fifth Avenue showroom, Morita told him not to worry, that he would see to it that Sony headquarters in Japan would pay for building and maintaining the property. Morita wasn't expecting profits from Sony America in the short term. For some time to come he would insist, despite protests from the board, that Tokyo subsidize the operation. According to Sagor, "In those early years, there was never any pressure on us about earnings. The pressure was to build a marketplace for Sony, and that is what we tried to do."

In October 1962, Sony's showroom at Fifth Avenue and 47th Street opened to the public. From a flagpole above the entrance, side by side with the Stars and Stripes, Morita had hung a Japanese flag, the first to unfurl in New York City since before World War II. This symbolism in itself had created controversy that the media had covered. More than four hundred guests attended the opening ceremony, and hundreds more would stop in the next day and every day thereafter. Morita was there with Yoshiko and would later write that it was at this moment, as he watched the American crowd milling around the centerpiece in the display, Sony's five-inch micro-TV, about to go on sale, that he decided the time had come to begin living with his family in New York. It was a tactical decision: Morita reasoned that to sell to Americans effectively, he would have to know more about them and how they lived. It was time to settle the frontier.

4

MORITA THE DAZZLER: THE MAN BEHIND THE MASK

In June 1963, Morita moved his wife, three children, and a maid from Tokyo into a twelve-room apartment at 1010 Fifth Avenue at 82nd Street, opposite the Metropolitan Museum of Art. The company had sublet the furnished apartment from the violinist Nathan Milstein, who was moving to Paris for two years, the period of time that Morita planned to reside in New York. The New York office had searched hard for an appropriate location: image would always be a high priority for Morita, and he had been advised by his new friends at Smith Barney that the Upper East Side was the place to be. The rent was $1200 a month; all in all, including living and entertainment expenses and private-school tuition for the three children, the family sojourn in New York would be expensive. The budget, Morita insisted, would have to come from Tokyo headquarters.

Ibuka was reluctant to let him go. Sony was now a $77 million business employing more than six thousand people. As president of Sony Sales, at the time a separate subsidiary, Morita was building a domestic sales network to compete at home with Matsushita, Hitachi, and Toshiba, and was overseeing Sony's expansion into Asia and Africa through the international division. The Chromatron color TV experiment required additional financing, and funding Ibuka's brainchildren would always be Morita's responsibility. His prolonged absence would also deprive the home office of assets not reflected in a balance sheet:

his galvanizing charisma and energy. No Japanese organization had ever sent its number-two man, certainly not its prime mover of people, to live abroad at such a critical time or, indeed, at any time. There was also the expense to consider. In the end, as Morita knew he would, Ibuka agreed. Morita made and kept a promise to spend a week or so in Tokyo every two months. In 1985, he would write, "I stopped counting my trans-Pacific trips at 135, a long time ago."

If Morita's move to New York was unsettling to Sony, it was also a profound disruption in the lives of his wife and their three young children. When Morita announced his decision to Yoshiko, it is unlikely there was much discussion: notwithstanding the Western-worldly aura he projected, he was a thoroughly traditional Japanese husband, stricter, if anything, than his own father. "We were not allowed to discuss things with my father," Morita's eldest son, Hideo, told me, speaking in English: "My father is the person who gives us orders, not a person to discuss. It is always one-way. We have to get orders from Father what we have to do. Right or left. There is no way I can answer back to him."

Morita returned to Tokyo to tell the children about the move in person, promising them a trip to Disneyland on their way to New York, then moved into the Milstein apartment by himself, leaving Yoshiko behind with the children to close up the Tokyo house. Alone in the city as he awaited the family's arrival in June, he spent his time preparing a campaign for Sony's new micro-TVs, commuting to the office on lower Fifth Avenue by bus, the better to observe the American consumer close at hand. He also began shopping for private schools for his children. Many were reluctant to accept three Japanese children who spoke no English—Akio Morita was not yet a name that would inspire special treatment. There were two Japanese-language schools operating in New York at the time, but Morita was determined that the entire family should become immersed in American culture. During this stay, and afterward, he chose to socialize with Americans, resisting the inclination among resident Japanese to band together.

With help from a partner at Smith Barney, Morita eventually found his way to St. Bernard's, a private boys school on 98th Street between Fifth and Madison avenues which had been founded in 1904 by two

Englishmen from Cambridge. R.I.W. Westgate was the perfect figure of an English schoolmaster—"St. Bernard's is one-third bright boys, one-third wealthy fathers, and one-third very pretty mothers" he was fond of saying—and, having previous experience of the largesse of Japanese parents, admitted Hideo, who was ten and would be entering fifth grade, and Masao, eight, who would be going into third. Coats and ties were required. Tuition was expensive, $2,000 per student. For his six-year-old daughter, Naoko, Morita chose the Nightingale-Bamford School.

In June, Morita met the family in California and took them to Disneyland. Masao, who had assumed from watching American TV programs dubbed in Japanese that Americans spoke his language, was overwhelmed when he discovered his mistake. The week after they arrived in New York, Morita threw the boys into the water to sink or swim at Camp Winona in Maine, hoping they would be forced to acquire adequate English for school. According to Masao, he and his brother spent most of the year at St. Bernard's in the dark, except in math class, which they experienced as rudimentary. A file surviving at St. Bernard's notes that the Morita boys also distinguished themselves in shop and geography class, and were at the top of everyone's invitation list because they could be counted on to arrive at birthday parties bearing the latest "electronic gadgets."

While 1010 Fifth Avenue was not exactly the heartland, neither was Morita principally interested in the man in the street, though he would master the art of selling to him. It was always his position, at least partly as a function of his personal values, that Sony's advantage was to be created by marketing quality products at a premium price: the focus of his attention was the urban, affluent consumer. At the same time, intending to create markets and control them, he was actively interested in the business establishment itself.

Morita's move to New York was prompted in part by his desire to win peerage for Sony and himself. He had discovered that New York society, in striking contrast to Tokyo, was couples-oriented. Invitations came addressed to Mr. and Mrs. Akio Morita, and a single man created a seating problem at dinner parties. Considerations of this kind were foreign: in Tokyo, then as now, wives were not included in business socializing, which took place at expensive bars or in private rooms at exclusive Japanese restaurants, but rarely at home. In New York, Morita had learned that entertaining was important, and that a man was expected to host people at home with his wife at his side.

Once they were installed in their Fifth Avenue apartment, the Moritas worked diligently and with remarkable success at building a social life. They joined Edward Rosiny's country club in Long Island so that Morita could meet people while indulging his passion for golf; they invited guests to dinner at the apartment at least once a week; they went so far as to host cocktail parties, the American social ritual that is perhaps most alien to the Japanese sensibility.* Yoshiko Morita, who had no entertaining experience at home and arrived in New York speaking almost no English, quickly earned a reputation as a polished hostess. In time, "Yoshi" Morita, as she was known to her American friends, the fashion editor Diana Vreeland, Mrs. David Rockefeller, Children's Television Workshop founder Joan Gantz Cooney, and others, would make a place for herself in the inner circle of New York high society, and would become an intimate and confidante to Bill Blass, "Jimmy" Levine, and Leonard Bernstein. In September 1993, David Rockefeller arranged for her appointment to the board of trustees of the Museum of Modern Art; and when the museum's new annex opens in 2002, it will house an Akio and Yoshiko Morita room (over the years, Yoshika has acquired an important collection of modern American paintings).

Sony's new market entry in the early sixties was micro-TV. Living in New York enabled Morita to wade into the launch with his full power as a salesman. Nights at home he spent in front of the TV studying commercials, and was amused and impressed by Volkswagen's humorous campaign organized around the notion of a lemon. As it happened, another St. Bernard's parent, a man named Arthur Stanton, was VW's

* The novelist Yukio Mishima, who was aping Western behavior during this same period for reasons of his own, threw cocktail parties at his Tokyo home for the perverse pleasure of observing Japanese guests attempting to conceal their dismay as they struggled to balance drinks and plates of canapés while standing in a crowded living room smoking cigarettes and making conversation all at once.

first distributor in New York, and arranged a lunch meeting for Morita with the illustrious creative director William Bernbach of Doyle Dane Bernbach, VW's advertising agency in America. Morita and Bernbach hit it off at once. The Sony showroom was near Bernbach's office, and he had admired the Sony products on display. But when Morita told him that Sony's budget would be $500,000 a year to start, Bernbach replied that the agency minimum was $1 million. Morita assured him that the Sony name would shortly be bigger than Volkswagen and persuaded him to accept the account. Doyle Dane Bernbach went on to create a series of print ads illustrating the benefits of "Tummy Television," a portly gentleman in bed at night enjoying a lightweight five-inch Sony propped on his belly; "Telefishin', the nine-inch anyplace Sony TV" in a rowboat on the lake; and the ambulatory four-inch "Walkie-Watchie." The light humor of the ads drew attention, and consumer awareness grew. When reporters phoned to request a visit from someone to explain the product, they were treated to a sales pitch from Morita himself and, inspired by his showmanship even in English, stories appeared in *Time*, *Newsweek*, and *Fortune*. The micro-TVs began to sell. Despite a price of $250 at a time when a twenty-seven-inch set was available for $150, sales climbed steeply: by 1969, Sony had sold one million units. Interestingly, despite the Japanese flag flying on Fifth Avenue, most consumers, including actual customers, remained unaware that Sony was a Japanese company. Morita was uneasy about the possibility of a negative reaction, and did what he could to sustain the misapprehension. The required "Made in Japan" label, for example, was positioned on the product as inconspicuously as possible, in the smallest permissible size; and more than once, Sony edged below the minimum, causing U.S. Customs inspectors to turn back shipments.

Morita had arranged to live in New York for two years. But in the summer of 1964, just over a year after the move, his father died following a long illness, and he pulled his sons out of camp in Maine and moved the family back to Tokyo. His brother, Kazuaki, was running the family business in Nagoya, but that did not mean he was head of the family. That position, and the authority to make all important decisions, would always belong to Morita himself as eldest son.

Though he was now once again commuting from Tokyo, living in New York in an apartment Sony rented for him on 58th Street between Fifth and Sixth avenues—later he would move into the Museum Tower building on 53d Street—Morita continued the networking that would establish him before long as the most-recognized and best-connected Japanese businessman in the Western world. By the late sixties, alone among Japanese businessmen, he was on the international advisory boards of Pan American, IBM, and Morgan Guaranty Trust, and had developed lasting relationships with a broad spectrum of American corporate chieftains: James Watson, Jr., of IBM and its subsequent chairmen, Frank Carey and John Opel; Pat Haggerty of Texas Instruments; Michael Blumenthal, secretary of the treasury under Jimmy Carter and chairman of the Burroughs Corporation; Donald Kendall, chairman of Pepsi-Cola; David Rockefeller of Chase Manhattan; Dr. Edwin Land of Polaroid; and many others. Peter Peterson, currently the chairman of the Blackstone Group and former secretary of commerce under Nixon, recalls taking Akio and Yoshiko Morita in the early seventies to play golf at Augusta National, the club of choice for business leaders. Peterson was amazed to discover that Morita was known to everyone: "Rube Mettler was there, the chairman of TRW, and there must have been eight or ten other CEOs of major companies, and Akio had met them all and had dinner with them. The guy must have had about ten meals a day while he was staying here!"

Throughout the seventies and eighties, Morita augmented his presence and influence inside the global business establishment. He was a member of the Wisemen's Council, officially the U.S.-Japan Economic Relations Group, a panel of eight businessmen representing the interests of Japanese and American business, which was funded jointly by the Carter and Ohira administrations and was increasingly the center of attention at U.S.-Japan business meetings like the Shimoda conference. Beginning in 1973, he was also a member of the Trilateral Commission founded by David Rockefeller, a group of powerful businessmen who came together twice a year to work on issues affecting

Japan, the United States, and Europe. In April 1992, he became chairman of the commission's Japan delegation. His cochairmen were Rockefeller, a close friend, and Otto Graf Lambsdorff, a member of the German Parliament.

Morita's fame and social skill were unquestionably an invaluable asset to Sony. For one thing, it wasn't very long before he was known as the man to see, more than anyone in an official government capacity, for counsel and help in doing business in Japan. As early as 1965, when Patrick Haggerty came to Tokyo seeking approval to manufacture integrated circuits in Japan, it was Morita he approached about a union with Sony; the result was Texas Instruments, Japan, a fifty-fifty venture. In 1968, after failing repeatedly to find a new partner for its giant record business inside the industry, Goddard Lieberson of CBS turned to Morita, who quickly saw the merit in a record company and rushed Sony into a joint venture, the first step in the direction that would lead to Hollywood. In 1971, James Roche, chairman of General Motors, came to Japan to negotiate a 35 percent purchase of Isuzu Motors, a highly sensitive deal, which was viewed in Japan with national uneasiness as a threat to the domestic automobile industry. Morita knew Roche from the Morgan Guaranty Trust board, and helped him orchestrate what turned out to be a successful acquisition, arranging meetings with the then head of MITI and eventual prime minister Kiichi Miyazawa. Another of Morita's connections through the Morgan Guaranty Advisory Board on which they both sat was Donald MacNaughton, chairman of Prudential Life. In 1976, MacNaughton was in Japan looking for joint-venture partners and getting nowhere. On the last day of his trip he called on Morita and described the difficulty he was experiencing in talking business with leaders of Japan's life insurance industry. Morita inquired casually if Sony might be of help. In fact, since his first trip to Chicago in 1957, when he had gazed upward in awe at the Prudential tower, Morita had planned someday to include a financial institution in Sony's portfolio of businesses. Only recently, he had abandoned thoughts of acquiring a bank after satisfying himself that it would not be possible under Japanese law.

Going down in the elevator after the meeting, MacNaughton wondered aloud if Morita's offer had been in earnest. When this remark was reported back to Morita, he called MacNaughton at his hotel to assure him that Sony was actively interested in a partnership. The eventual result of their meeting in New York the following week was another highly profitable departure from electronics, the Sony-Prudential Life Insurance Company.

Morita's association was not limited to businessmen. Every U.S. ambassador to Japan during the seventies and eighties—including former Senate majority leader Mike Mansfield of Montana—left the country with fond memories of evenings at the Moritas' home in Tokyo and an indefinable but present sense of loyalty to Sony. Otto Graf Lambsdorff recalled conversations with Japanese politicians arranged by Morita for his edification: "We would be sitting in a Japanese teahouse, with the Japanese prime minister, or the former prime minister, or the future prime minister, and the German ambassador, and here we are sitting in this elegant private Japanese atmosphere and discussing politics and international business. These were memorable evenings."

Henry Kissinger shared similar memories. He had been introduced to Morita at a Sony party in Washington in the fall of 1971 when he was head of the National Security Council under Nixon, and over the years, the men had developed a high regard for each other. Whenever Kissinger was in Japan, Morita invited him to attend a breakfast group that met once a month at the Okura Hotel and included top executives and the familiar array of past, present, and soon-to-be prime ministers. Kissinger recalled that these breakfasts were "beautifully organized, beginning at 8:00 sharp, not 8:01," and was grateful for the rare opportunity to observe the inner workings of the Japanese establishment. Morita also organized dinners at his home to which he invited younger people, emerging figures in business and government, providing Kissinger an opportunity to sample the tenor of the times. Kissinger reciprocated, inviting Morita to evenings in New York or weekends at his country home in Connecticut and introducing him to influential people in American business and government. There were other, more tangible favors. Kissinger mentions an "unfortunate misunderstanding" by a Japanese television network which he asked Morita to resolve, and which he did resolve quickly without ever explaining how. In 1979, when Morita wanted access to the Chinese government to prepare the Chinese market for Sony, Kissinger arranged a meeting with Deng Xiaopeng. "If I had thought, tomorrow he's going to ask me to promote Sony, I might have thought twice about it," Kissinger told me. "But he was very wise because he was always promoting Sony without promoting Sony."

Peter Peterson is perhaps the best example of an influential businessman with rich experience in government whose friendship with Morita over the years has been invaluable, directly and indirectly, to Sony. Peterson first met Morita in the mid-sixties when, as CEO of Bell and Howell, he visited Ibuka and Morita in Tokyo at the insistence of Edwin Land, who had described them as a "great team you must meet if you are going to Japan." The men got to know each other better while serving together on the Pan American International Advisory Committee. In 1970, an American television manufacturer, Emerson, filed a collective antitrust suit against Sony, Panasonic, Sanyo, and Sharp for dumping TVs into the American market. The following year, Peterson joined the Nixon White House as assistant to the president for international economic affairs. Morita immediately paid him a call to persuade him of the injustice of accusing Sony of dumping when its TV sets were priced 30 percent higher than those of other Japanese manufacturers. Kunitake Ando, who was Morita's assistant at the time, accompanied him to his meeting with Peterson, and recalls that Peterson acknowledged the apparent injustice of including Sony in the suit, but was careful to make clear that he was not in a position to interfere. He did recommend, and facilitated soon after, a meeting at the White House with Morita and other Japanese CEOs to discuss trade issues.* Later, when Peterson had left the government to head Lehman Brothers, Morita asked him for guidance in designing an executive compensation plan for Sony, still unheard of in Japan. In 1988 and 1989, the private investment bank Peterson founded in 1985, the Blackstone Group, played a central role in Sony's acquisition of CBS Records and

* Morita fought the Emerson suit aggressively for ten years until the federal district court ruled in Sony's favor. The decision was appealed, and Sony won again, two and a half years later, in the federal court of appeals.

Columbia Pictures. In 1992, he accepted Morita's invitation to become the first American not employed by Sony to sit on the Sony board (Michael Schulhof had joined the year before). Since 1993, Morita has been incapacitated by a stroke, but Peterson continues to be involved in Sony's affairs: currently a senior adviser to the corporation, he heads the Compensation Committee for the American subsidiaries and sits on three of the four American boards.

What accounted for Akio Morita's unique ability as a Japanese businessman to establish and sustain beneficial relationships with people like Peterson? There is striking agreement among those who knew him over time that he was special because he was someone who seemed to understand them, and, as important, whom they could understand. Peterson puts it as well as anyone: "When it came time for Akio to do business in the United States, whether it was joint ventures or licensing or whatever, he could pick up the phone and talk to almost any businessman in America. And instead of it being 'Who is this again?' and interpreters and all that sort of thing, Akio knew these people at the human level, at the personal level. And let's be honest: to many American businessmen, the Japanese business culture is foreign, they don't feel comfortable with Japanese businessmen, and they don't know them to a large extent as human beings. But they did know Akio in that way, and therefore when he called, people listened."

Henry Kissinger was more theoretical: "First of all, the Japanese in my experience are not great communicators. They tend to operate within their consensus, and when they get dropped out of the consensus and get into a dialogue with other cultures it's tough because they don't feel they have the authority to make independent decisions. So, even for many of us who have Japanese friends whom we value, the problem of communication is very difficult. Morita could conduct a dialogue, and while he was a very patriotic Japanese and a firm defender of the Japanese point of view, he could communicate it in a way that was meaningful to non-Japanese. . . . He was probably the single most effective Japanese spokesman I ever met."

Peterson also emphasized Morita's humor and his warmth: "The Japanese, as a group, tend to be relatively reactive, I would say, relatively passive, socially. Akio was a passionate, gregarious personality.

He loved to laugh and talk and sing. I'll never forget when they introduced that karaoke system, he had me and a colleague singing songs with him . . . something else about him that I think characterizes most immensely successful people was his energy level. I remember even playing golf with him at Augusta. He'd get up at 6:30 in the morning. 'Let's play eighteen holes. Why don't we play a second eighteen holes.' I'd be dragging around after thirty-six holes. He had this tremendous energy, so when he walked in, people were delighted to see him because they knew it would be a good time."

By all accounts, Morita had the gift of incandescence. People observing him in action at various moments in his life were left with a similar impression that he "lit up" a room with his presence, literally "glowed." I have observed this effect myself. In February 1992, his penultimate year of health, Morita put in an appearance at the World Economic Forum in Davos, Switzerland, which I happened to be attending as an observer. He was by no means the only luminary present: Mandela and de Klerk were there, and so were Chinese premier Li Peng with a retinue of lethal-looking teenage bodyguards, and the prince of Wales, Henry Kissinger, former Federal Reserve head Paul Volcker, Budesbank president Karl Otto Poehl, and many others. But it was Morita who turned heads as he strode across the outer hall greeting his friends, and Morita who stole the show in plenary sessions. Speaking from the stage in his earnest, imperfect, yet vivid English, he assured the audience that they would find the Japanese market open to them, but only if and when they had something appealing to offer the Japanese consumer. Then he smiled his outrageous, radiant smile and suggested that the competitiveness of American industry might be improved if its leaders remunerated themselves less richly. Clearly this proposal was a slap in the face, yet it was delivered with such warmth and collegiality, such familiar ease and good humor, that the American CEOs who had paid $14,000 to hear him rose as though ensorcelled to applaud him as he left.

The memory of Morita's brightness survives as part of his legacy to Sony. When I asked Norio Ohga, the current chairman of Sony, to explain why he had overlooked the front-runners in line to succeed him as president, he wrote down on a pad in his elegant hand an expression that means "to shine dazzlingly, as of the sun"—*san-san to kagayaku.* "The leader of Sony must have radiance," he said, and though he didn't bother to add, "like Akio Morita," there was no mistaking that Morita's presence had been invoked.

What people experienced and succumbed to and remembered about Morita was of course his luminous charisma. He was a striking man, with silky hair that turned silver in his early forties and that he parted down the middle in the manner of a Meiji dandy, and his bluish eyes, rare for a Japanese, that gave rise to rumors that among his ancestors lurked a White Russian. His curiosity about people seemed boundless, and all those who met him, however fleetingly, were left with the impression that they were his friend. Michel Galiana-Mingot, the former head of Sony France, remembers the day he escorted Morita on an inspection tour of Sony's new plant in Bayonne. As he was driving away from the plant, Morita noticed a white house across the surrounding fields and asked to stop. The occupants, he explained, were Sony's neighbors and he wanted to meet them and pay his respects. The master of the house happened to be in the ice cream business, and Morita spent an hour with his family sampling ice cream from all over Europe while a cavalcade of official cars and policemen on their motorcycles waited outside.

Morita's social energy was unflagging. For twenty-five years, beginning in the mid-sixties, until he was felled by a stroke in 1993, he presided over Tokyo's most international and luminary salon, bringing together for parties and dinners celebrities from the worlds of business, government, and the arts. His house in Aobadai, an exclusive enclave of industrialist residences near the Shibuya District, had been designed to make foreigners feel at home. There was only one room with a floor of tatami mats, across the hall from Morita's study upstairs, which was sometimes used to serve an intimate meal to Japanese guests or to offer them a ceramic bowl of the frothy, bitter green brew of the tea ceremony. The living room was "contemporary": Italian leather couches and chairs, glass and chromium tables, recessed lighting in the matte black ceiling. In one corner stood Morita's prized 1921 Steinway baby grand, which had been converted into a player piano, one of only three that have been built. A floor-to-ceiling alcove housed Morita's collection of one thousand piano rolls in individual wooden boxes, coded performances that brought Paderewski, Rachmaninoff, and Lipatti to life when they were threaded into the DuArte mechanism. The room was equipped with floor and ceiling speakers.

The adjacent dining room looked out on the lawn and Japanese garden at the rear of the house. Morita had commissioned a thick lacquer table, which seated sixteen and could accommodate as many as twenty if necessary. His initials, AKM—Akio Kyuzaemon Morita—were emblazoned on the cutlery, the dinner napkins, and the gilt-edged plates. On the sideboards and table, figurines from Yoshiko's Steuben glass collection gleamed in the candlelight.

The legendary hospitality at the Morita house was founded on Yoshiko's genius as a hostess. She cooked American, French, Italian, Chinese, and Japanese food, and gave meticulous instructions to her terrorized maids on how to serve each kind of meal. Her style in general was elaborate abundance. For wine selection, she relied on the sommelier at the Tokyo branch of Maxim's. She kept a file card on every guest who came to the house; a glance at the entries opens a small window on the social scene chez Morita:

> Andre Watts (black American pianist):
>
> April 16, 1970: comes to dinner with his mother. He may be a mama's boy; Mother doesn't like fish.
>
> Leonard Bernstein:
>
> The truly unusual foreigner, he prefers sushi and sashimi. We prepared sushi for him and a steak for his assistant. I also had a steak for him just in case, but he never touched it; all evening he wolfed down sushi and sashimi with one hand; with the other, using a chopstick in place of a baton, he conducted recordings of his own concerts.
>
> Mr. and Mrs. Robert Ingersoll:
>
> Joined for a weekend at the Hakone house. When I mentioned we would be there without maids, he brought pancake mix and prepared a lovely breakfast for us all. Remember to have milk and eggs on hand when he comes again, for the pancakes.

Andy Williams:

He loves corn chowder and roast beef. The night after his final concert in Japan he came to the house looking exhausted. But when he saw the table his face lit up; by the end of the meal he was feeling so much better he sang "Moon River" for us at the table.

Mr. and Mrs. Fischer-Diskau:

April 16, 1970: he takes a simple fillet steak only; no pepper or spices. I overheard them whispering about whether they would be able to use chopsticks. So I joked: I'm sure you know how to enjoy a simple steak. Then I slid back the doors to the dining room a little early so they could see the Western table setting; their anxiety about being invited to a Japanese home seemed to disappear.

To the Americans and Europeans who knew him over time, in striking contrast to the "typical" Japanese businessmen they encountered in their professional lives, Morita seemed neither reticent, uncomfortable, aloof, enigmatic, nor inscrutable.* On the contrary, he was a dynamo of dazzling energy, gregarious, passionate, spontaneous, charismatic, and fun-loving. Most appealingly, he seemed very like themselves.

But was he really as effortlessly at ease in the company of his foreign friends as he appeared, as familiar and at home with their way of perceiving the world and behaving in it? Did Morita truly understand his foreign friends and associates, and did they truly know him, in Peter Peterson's words, "at the personal level"? There is no unambiguous answer. But there is evidence to suggest that Morita had to labor hard to achieve what may have been the illusion of familiarity. There is

* The poignant truth is that "typical" Japanese are almost certainly not the people they appear to be when dealing with foreigners outside Japan in languages other than their own. As a nation, Japan is second only to the United States in its linguistic insularity: abroad, the Japanese tend, even in the global nineties, to recede into a formal, tight-lipped impassivity that is actually a kind of resignation about ever making themselves understood. This unfortunate withdrawal is often misinterpreted as lifelessness. "Japanese humor," someone informed me knowingly, "is an oxymoron." In fact, Japan has one of the world's great comic traditions, both in literature and in performance. When they are at home among themselves, all Japanese love to laugh.

even room for speculation that Morita's lifelong, tireless campaign to install Sony in the West required a painful personal struggle to reconcile a foreign sensibility with his own and that, notwithstanding his success at building Sony outside Japan, he was never able to resolve that tension satisfactorily. From this vantage, it is possible to perceive Morita in his fraught relationship with the West as a paradigm of Japanese modernization in the twentieth century.

According to the current head of the Morita family, Morita's eldest son, Hideo, his father was a consummate performer all his life and no one, foreign or Japanese, not even Ibuka or his immediate family, ever saw him unmasked: "I am very much like him; I am the closest, even my mother tells me so, and we are very good at acting. This is the only way we have to encourage ourselves. If you want to play a king, you have to be like a king all the time. My father was very good at that, and so am I. He had an image of himself as the head of the Morita family, the fifteenth generation, and also as the president of one of the fastest-growing companies in Japan. And also he had to 'act'—I'm sorry to use that word but I can't help it—*he had to act as the most international-understanding businessman in Japan*. Actually, I don't think that was true, but he had to act that way until he had a stroke. He tried very hard, and he worked and he studied hard to play that role. I admire that. *But it was never real. He could never be good at any of those roles, including as a husband, including as a father!*"

To be sure, Hideo Morita's assessment is subjective, however deeply felt, yet what he has to say seems less surprising than predictable. Morita was raised in a traditional Japanese family in which all things were understood implicitly, where communication, when it occurred, was oblique and equivocal. With this background, how can the aggressive outspokenness and confrontation at the heart of American business behavior have come naturally to him? How, indeed, can he have been naturally at ease with American social style in general?

Despite the gregariousness, which Hideo implies was a performance for his Western friends, other Japanese who knew him at home perceived Morita to be preternaturally shy. The journalist Hiroshi Yamaguchi, for many years a confidante, remembers a first meeting he arranged in the mid-1960s between Morita and Noboru Goto, the scion of the Tokyo railway and department store fortune. Goto, who had taken over his father's empire and was building his network of personal connections, had requested a meeting with the man whose prominence in Japanese business circles far exceeded the size of his upstart company. Morita accepted Goto's invitation, conveyed by Yamaguchi, to meet for dinner at one of the exclusive Japanese restaurants called *ryōtei*, in the geisha quarter of Akasaka. Yamaguchi accompanied Goto, and Morita brought the industrialist Yoshitami Arai. For the entire evening, choosing to converse with their intermediaries, neither Morita nor Goto spoke a word to the other across the table. Yamaguchi recalled with amusement that Morita would not even look Goto in the eye. "As I sat there watching him, I realized how shy he was."

However he may have appeared to others on his international rounds, at home Morita was a strict traditionalist. The polite Japanese word for "wife"—*oku-sama*—means, literally, "the personage at the rear of the house." Morita expected Yoshiko to come forward when he needed her at his side to manage their social environment and to remain in the background, where he was certain she belonged, at other times. It was an open secret that Morita was an amorous man who indulged his taste for romance on the same global scale as he conducted his business affairs. Hiroshi Yamaguchi was for years his principal fixer. "When Yamaguchi-san called home with some excuse," Masao Morita explained with a smile, "my mother always knew he was lying even though she pretended to take him seriously." Yoshiko was expected not to see what did not concern her and to keep her own counsel about what she may have known. A naturally forceful person, she declined to play the role of the acquiescent wife on more than one occasion, creating havoc in the household and requiring Morita momentarily to make amends. At a Ginza bar one night, Yamaguchi cautioned Morita about Yoshiko's outspokenness, suggesting that eventually it would harm him. Morita agreed, and asked the journalist to speak to Yoshiko on his behalf. Later that night, at home in Aobadai, Morita poured a snifter of brandy for his friend and fled the room while Yamaguchi explained to Yoshiko that a Japanese woman can be a liability to her husband if she is too aggressive in her behavior. Supposing him drunk, Yoshiko listened politely, then drove him home in her own car to his apartment nearby. When Yamaguchi got out of the car, she noticed that he walked straight to the building on steady legs. The next day, she phoned a mutual friend and told her angrily that she would not have permitted "the villain Yamaguchi" to lecture her in her own living room had she realized he was sober.

As a father, Morita recreated the family hierarchy that he had experienced in his own childhood. Hideo grew up with the implicit understanding that he was to be head of the family. At family gatherings, he sat at the head of the table with his father. His brother Masao's place was at the far end of the table with the lesser cousins. From the age of six, Hideo was required to attend family meetings. On excursions to Kosugaya, Morita would point to the family brewery and explain, echoing his own past, that it would all belong to Hideo one day. Morita went so far as to show his firstborn son his grave site alongside his own in the Buddhist temple next to the Morita ancestral home. Shortly after the family returned from New York, when Hideo was in junior high school, he helped his father prepare the "mock-up" of a grave in the garden of their first house in Tokyo, in the fashionable Roppongi District: "We went in it together. It was my grave and my dad's but not Masao's. I know Masao hated it, but he had no right to say anything because it was our family tradition. I felt guilty personally, and I still do, but I can't help it."

Masao, three years younger than Hideo, suffered and continues to suffer. He says he complained to his father only once, on a skiing trip to Vermont in 1963 when the family was living in New York. Masao remembers walking through the snow with his father from the ski lodge to their cabin after dinner and tearfully demanding to know why Hideo was always given special treatment. Morita told him this was not something a child could be expected to understand, adding that all children were equal in the eyes of their parents. Masao was not reassured: "I grew up with an inferiority complex. I had to prove to myself that I was not that bad even though I was always the second kid and never the first. At school I worked really hard and always got better grades than my brother. My father noticed. He'd say, 'You got a good grade!' but it didn't help."

From conversations with his sons, it appears that Morita was not only domineering but capable of sending mixed signals that confused them cruelly. Hideo recalled an episode from his boyhood which still troubles him: "The first term of my second year in junior high, I got very bad grades just before winter vacation. The day before I had planned to go skiing with my friends, my dad came back very late at night, and he was mad about my grades. So he called me in the middle of the night [Hideo was at a dormitory in a private school in Tokyo] and told me I was not allowed to go skiing. I said 'OK, I'll study during the holiday and stay at school.' After he called me, he called Masao and he said the same thing. Masao said, 'I'm sorry, my dad, I will study very hard this winter, after I come back from skiing. And I will do my best next term, so may I go to ski?' My dad said yes. Then he called me again: 'Masao says he's going to study hard after he comes back from skiing. He's even going to bring his schoolbooks on the trip. He cried and begged. If you wanted to go skiing, and I know that you love to ski, why didn't you try hard to convince me?' But I was told ever since I was a kid that I was not allowed to talk back to my father. What he said was his decision. That is the way he brought up the eldest son. And I've been confused ever since. I don't know what to do. You see, the head of the family in my family . . . he can do anything he wants."

When they were ready for the ninth grade, Morita sent his sons to boarding school, Masao to World Atlantic College in Wales and Hideo to Copford College in Colchester, England. There Hideo was tagged with his English nickname, "Joe," an abbreviation of the only Japanese name his English schoolmates knew, the infamous Imperial Army general and wartime prime minister, Hideki Tojo.

Masao remained abroad. Following graduation from Georgetown University in 1978, he worked at J. P. Morgan for three years, first on Wall Street and later in the City of London. In 1981, having been away for ten years, he returned to Japan to apply for work at Sony as he had always planned. His father, chairman since 1976, spoke to him only once, asking whether he fully understood the implications of being at Sony as a Morita. Doing as well as others would not be nearly enough, but neither would doing far better than others—90 percent of the people, Morita told his second son, would be waiting for him to falter. With these words of encouragement, Masao took the job, and was placed in Audio Planning, a department managed by Nobuyuki Idei, Sony's current president.

Meanwhile, Hideo had been marking time until his father decided he was ready to take over the Morita sake business in Nagoya, his destiny as heir apparent. After England and two years at Asia University in Japan, he had transferred to the University of California, Davis, which he hated, to study agriculture. Back in Japan in 1975, he completed a final year at school and, still waiting for his father to order him to Nagoya, went to work at CBS/Sony Records, a joint venture run by Morita's chosen successor, Norio Ohga, a man whom Hideo and Masao had known since childhood as their "very strict uncle Ohga." Hideo quickly became the playboy of CBS/Sony, traveling around the world to audition the best-looking acts he could find and actually signing the Nolan Sisters, cute British siblings whose records sold well in Japan.

In the spring of 1978, Hideo made front-page news with his sudden marriage to Yuki Okazaki, a popular singer, dancer, and actress with whom he had been secretly involved for over a year. "I am the eldest son of one of the richest families in the world," Hideo explained matter-of-factly, "and she was a top star in Japan. It was as if Rockefeller Junior was going out with Mariah Carey. People loved it." Morita's response was to disinherit Hideo on the spot, forbidding him to enter the house. As if that were not enough, in another example of his quirkiness as a father, he telephoned Masao, who was at Georgetown, and informed him that he was to think of himself from that moment on as future head of the family. In Hideo's view, the resolution of this episode was the cruelest blow that his father ever dealt his younger brother: "First he calls Masao and tells him, 'OK, from now on we think your brother died, you'll be the eldest son so you take over the family and all the family business.' Two years later I got divorced; I had nowhere to go, so I go home and tell my dad and he takes me back. He didn't say anything, he just opened his hand. Next day, I am the eldest son again, and Masao is back where he started. I don't think my dad said a word about it to Masao."

To prepare Hideo for taking over the sake business, Morita had him transferred from the record company to the accounting department at

Tokyo headquarters. Masao, speaking in English less fluent than his brother's, managed nonetheless to convey his feelings at the time with poignant clarity: "Sony was the only thing I had. My brother always has a place to go. I had to work so hard, and I was always feeling pressure—why does he have to come to Sony to make the pressure more? I told my father: 'I don't think it's good if both of us are here—so if he wants to stay, I'll leave.' I didn't want to compete with my brother at my father's company. I didn't want to be in a position to choose or be chosen. Why should I compete? I have always had some inferiority complex with my brother. That's why I didn't want him to be at the same ballpark."

To Masao's relief, Hideo left Sony after one year in accounting and a second year in finance to enter the family business. He began as vice president of RayKay, the family holding company based in Tokyo. Since 1996, Hideo "Joe" Morita has been president and CEO of both RayKay, Inc., and the Morita Company, Ltd. His uncle Kazuaki had been running the sake business since his father had joined Ibuka in Tokyo, for more than forty years; but when Akio indicated that his firstborn son was ready, Kazuaki stepped aside. Hideo implies, without saying so, that this step was not easy for his uncle. Sensitive to complex feelings that remained, in the manner of the traditional Japanese family, unexpressed, he invited Kazuaki's son, his cousin, to rejoin the company from his job at a food seasoning company, Aji-no-Moto, which produced monosodium glutamate, and promoted him to *semmu*, or senior vice president. "I felt an obligation to repay Kazuaki for managing our family business for all those years. So I'm prepared to work with my cousin and to look after him for the rest of his life. Or he looks after me, or whatever. But that doesn't change the fact that it's going to be my company. In my family, my father's decision counts. He is the head, and it happens that he also made his brothers rich. Their rule is: 'What would your father say?'"

When Morita dies, Hideo intends to take the name Kyuzaemon the Sixteenth. He is also the father of triplets, two boys and a girl. The firstborn child, the eldest by several minutes, will be raised to become the next Kyuzaemon.

Masao remains at Sony and continues to climb the corporate ladder.

After three years at Sony of Canada, Ltd., a subsidiary then co-owned by Sony and the Canadian entrepreneur Albert Cohen, an intimate of his father's for many years, he joined the office of corporate strategy and planning, which reported directly to "Uncle Ohga." In 1996, under the new Idei administration, he was appointed president of the Personal and Mobile Communication Company and corporate vice president. In 1998, he was moved to an executive position at Sony Music Entertainment (Japan), Inc. Masao acknowledges that he has received privileged treatment from Ohga and, more recently, Idei, who was stationed in France when he was at school in Wales and provided him a home away from home. But he continues to feel pressure: "For me, working for Sony is a very, very serious matter. I know my father built this company. I know how much he loves this company. And I know I am his son. I don't want to disappoint my father . . . but there will always be a big difference between us: For my father, this is Morita's Sony. But I will always be Sony's Morita."

The question remains: Was Akio Morita, the strict Japanese traditionalist, as effortlessly at home in the West as he appeared to be, or was he simulating Western behavior? At the very least, there is evidence that Hideo Morita was accurately describing the effort it cost his father to achieve the effective communication for which he was so admired. The following excerpt is from a televised dialogue with a well-known Japanese commentator, Saburo Shiroyama, about the differences in Japanese and American management style:

> *Morita:* Grammar and pronunciation aren't as important as expressing yourself in a way that matches the way Westerners think, which is very different from our thought process. You have to switch off your Japanese way of seeing things, or they will never understand what you are saying. First of all, they want to hear the conclusion right away; in English sentence structure, the conclusion comes first.
>
> *Shiroyama:* I've heard that one of our prime ministers was on his way to the United States for the first time and asked you for advice, and you suggested the critical thing was to start right out with a "yes" or "no" followed by a brief explanation—

Morita: When they ask questions or express an opinion, they want to know right away whether the other party agrees or opposes them. So in English, "yes" or "no" comes first. We Japanese prefer to save the "yes" or "no" for last. Particularly when the answer is "no," we put off saying that as long as possible, and they find that exasperating.

Shiroyama: But in Japan, as you explained, we don't come out with "yes" or "no" but prefer expressions like "I'll take this under consideration." Our feeling is that vagueness in these cases is less offensive. So when you're in America you must be clear, and when you return to Japan you must be vague—is it hard to switch back and forth?

Morita: It's more difficult than you can imagine.

Despite their impression that he was one of them, this small critique reveals Morita appraising Westerners from the opposite side of a cultural divide, working hard to decipher the puzzles that stood between him and successful communication, but aware of a separation.*

There are other indications that Morita was conscious of the tension created by two incongruent worldviews, one that came naturally, and another that had to be acquired, or at least imitated, if he was to achieve international success. In 1971, he decided to found a men's club for established leaders and rising stars in Japan's business and financial worlds. He wanted a neutral ground where businessmen from across the spectrum of industry could gather informally to share their views on the economy. This was a radical notion, for it was, and remains, a commonplace that Japanese business is organized in strictly vertical families known as *keiretsu,* but Morita made it happen. According to Hiroshi Yamaguchi, who knew him as well as any man, he had another motive: becoming chairman of the Keidanren, the Federation of Economic Organizations, Japan's most powerful corporate entity. To achieve that high office, normally reserved for leaders of estab-

---

* There is evidence that Morita was at pains to conceal from his foreign friends any indication of distance between them. This partially explains his consternation at the appearance on Capitol Hill of an unauthorized English translation of a collection of his critical speeches, *The Japan That Can Say No.* See page 217.

lishment businesses, a category in which the upstart Sony was not included, he knew he would have to extend his connections to the power elite of business and high finance.

Historically, private clubs based on the British model have not thrived in Japan for several reasons. One is the strictly hierarchical nature of the society: men of different ages and, accordingly, social standing, cannot spend time comfortably together. The question of affiliation, supremely important in the Japanese order of things, also poses a problem. Normally, one's affiliation is to a company or group of companies, a government ministry, a university, or a branch of the military.* Until Morita tried and succeeded, no one had established a club in which membership was based on success and reputation in the business community rather than specific affiliation.

The first step was finding someone with the appropriate cachet to serve as chairman of the board of trustees. Morita felt that he needed someone older than himself—he was forty-nine at the time—and prevailed on Jiro Yanase, five years his senior and head of his family business, Yanase Motors, still Japan's leading importer of foreign luxury automobiles. With Yanase in place, Morita proceeded to recruit a core group of the old guard, men of Meiji (born before 1912) who had become leading figures in business and finance: Hiroki Imadato, the financier; Shun Ishihara, president of Nissan Motors; Eishiro Saito, chairman of Japan Steel; and Tadakichi Aso, son-in-law to Japan's most famous prime minister, Shigeru Yoshida. For a location with the required status, Morita chose a space on the basement floor of a fashionable Ginza District building, which also served as headquarters of the publishing company Bungei Shunju. To manage the club, he persuaded a former geisha from Tokyo's Yanagibashi District to give up her own bar near the Nikko Hotel in the Ginza. This "Mama-san" was

---

* In the early sixties, Morita was a member of a group of businessmen whose shared affiliation was the Imperial Navy. The group included Kanichiro Ishibashi, the founder of Bridgestone Tire; Keizo Saji, chairman of the Suntory Group; the journalist Yamaguchi; and the chairman of Nomura Securities, Tsunao Okumura. Yamaguchi recalled a typical outing, when the group traveled to Nagoya to play golf and then adjourned to that city's toniest bar, Natsume, where they surprised and possibly offended other patrons by singing Imperial Navy songs from World War Two.

not only charming and a great beauty; by virtue of her training as a geisha, she could be relied upon to maintain discrete silence about anything she might overhear in the course of an evening. Eventually, the club extended membership to forty businessmen for an initiation fee of ¥1,000,000, at the time about $3,000. A number of the senior members were avid *go* players and would show up to drink and play most nights of the week; the financier Imadato snuck out of the hospital, where he was recovering from a heart attack, to play a game. Ibuka showed up from time to time, often in company with his close friend for years, Soichiro Honda, the founder of Honda Motors. The younger regulars were Morita's personal cronies: the CEO-owner of Wacoal, a man named Koichi Tsukamoto, who was known as the Japanese brassiere king; the proprietor of Sushi-ko, Tokyo's most exclusive sushi bar, an avid *go* player who sometimes arrived bearing lacquer trays heaped with sashimi and sushi for everyone; and Hisami Matsuzono, who had made his fortune selling yogurt drinks and was the owner of a national baseball team and a closet full of French impressionist paintings. The club survived until 1990, when it was closed because a much younger membership had taken to playing blackjack for high stakes and Morita worried that the nightly gambling would attract unfavorable attention. In later years he was, in any event, rarely there, unable to find time in a schedule that took him increasingly out of Japan and around the world. It was also a fact that Yoshiko had always disapproved; she was not fond of the former geisha who ran the establishment and was furthermore convinced, not without cause, that Morita was never up to any good when in the company of the brassiere king and his other pals.

To the business sociologist, Morita's private club was a treasure house of discoveries, not least among them its name, Club Amphi, after the English *amphibian*. On the way to his summer resort in Karuizawa with Yamaguchi in his chauffeured car, Morita had happened on the word in an English dictionary and proclaimed it exactly right. Inscribed on a brass plate set in the wall above the bar were the following words: *We Japanese businessmen must be amphibians. We must survive in water and on land.* It is unlikely that Morita's reference was simply to the best and the worst of times in the business cycle. The irreconcilable environments he had in mind when he pictured Japan's business leaders as amphibians were surely Japan on the one hand, with a set of cultural values and practices rooted in native tradition, and the Western world on the other, an entire universe of values and conventions which could be mimicked, but, as Morita would eventually conclude on the basis of his own experience, never entirely assimilated.

5

SONY'S FIRST AMERICAN: LESSONS IN LOGIC FROM HARVEY SCHEIN

If establishing social relationships in America required Morita to adjust his mental and emotional circuitry, building the business in a foreign land proved to be an even tougher cultural challenge. It is here, inside the business, that the immense difficulty and frustration in attempting to align two disparate cultures is most evident.

Morita was far ahead of his time in committing in the early sixties to what is now the hackneyed notion of "global localization," maintaining control of global strategy at world headquarters while entrusting national subsidiaries to local managers who are optimally sensitive to the needs of their own markets. Where Sony America was concerned, the problem was finding American managers who were at once superior businessmen in their own environment and capable of understanding Morita's goals and strategies for Sony. This delicate balance has not been easy for any global company to achieve, and continues to elude most Japanese organizations in particular. The electronics giant Nippon Electric Company is a case in point. For years, NEC's former chairman, Tadahiro Sekimoto, one of Japan's most theoretical business thinkers, articulated a strategy that called for foreign subsidiaries to be managed by local nationals. The reality, however, even today, is that most of NEC's international businesses are run by Japanese executives dispatched from Tokyo.

Under Morita, Sony had conspicuously more success than the

competition: the top executives of many of Sony's subsidiaries are "foreigners." Nevertheless, particularly in the American companies, where the stakes have always been highest, achieving "mutual understanding" with Tokyo has been a vexing challenge fraught with disappointments.

From 1960, when he founded the American subsidiary, until 1966, Morita personally oversaw the U.S. business as president of Sony America, and, driven by his energy, it slowly but steadily grew. While he was away in Tokyo, which was often, his childhood friend Shigeru Inagaki had nominal charge of the office and steered the business with advice from Morita's friend and Sony's American lawyer, Edward Rosiny. In 1966, when Ernest Schwartzenbach retired from Smith Barney, Morita persuaded him to take over as the first American president of Son-Am and promoted himself to chairman. Morita trusted and felt easy with the refined Schwartzenbach; unhappily, his ability to manage the business satisfactorily had scarcely been put to the test when he drowned in an accident in 1968.

Once again, Inagaki and Rosiny were managing the business as a steering committee of two. Morita took back the title of president but was largely absent. The timing was particularly unfortunate because Sony was preparing to launch Trinitron in the United States. Significantly, despite the pressing need for resident leadership, it took Morita nearly three years to replace himself. He was certainly ready by this time and indeed committed to relinquishing the tiller of the American business to an exceptional American manager. More accurately, in light of what followed, he was prepared to give up a measure of control: Morita was essentially an autocrat who knew how to delegate but was never content to abdicate his power. In any event, he seems to have been unable to find a qualified American manager willing to take the job. The word on the street was that Americans who accepted employment in Japanese companies, even at executive positions, could not hope to achieve parity with Japanese executives. It was even assumed that those who did go to work for the Japanese had accepted the compromise because they lacked the competence necessary to secure American jobs. In 1969, Rosiny went looking for a manager to build a national sales network for Trinitron, and found Ray Steiner at Sylvania. According to Steiner, when he heard that Rosiny was representing "a Japanese company," he was tempted to walk away. Fortunately, his admiration for Sony products outweighed his reservations and he joined the company and later, briefly, served as president.

In June 1971, Morita dispatched his brother-in-law Kazuo Iwama to New York as the new president of Sony America. Whenever a significant change in responsibility occurs, Sony tends to reconfigure management broadly, a reflection of the critical importance the organization assigns to each individual's place in the web of personal relationships. In this context, June 1971 bought a sea change: Masaru Ibuka moved from president to chairman, but would remain actively involved in corporate decision-making for another five years; Morita became president, effectively the unopposed leader of the company; Iwama stepped into the position just vacated by Morita as executive vice president. Iwama was now the third most powerful individual in the organization. He was plant manager at Shibaura, where Sony's semiconductors were produced, and general manager of the second manufacturing division, responsible for all of Sony's television production. He was in addition the company's chief scientist and overseer of new product R and D. He was, in other words, the last person Sony could afford to lose to America for an extended period of time. Morita's choice reveals how committed he was to finding the right person to be president of Sony America. In fact, he resisted sending Iwama for over a year, discussing a number of other Japanese executive candidates with Inagaki, who vetoed them all. Naturally, Morita would have overridden Inagaki's reservations had he seen fit. When Inagaki himself finally requested Iwama, whom he greatly admired, the request was initially denied. Finally, Morita agreed reluctantly that Iwama should go.

All things being equal, it isn't entirely clear that Iwama would have been a wise choice for the job. His passion was science and engineering, not sales and marketing. And though he was a charming man with a demonstrated ability to win respect and affection from foreigners, he was not inherently a social creature. He was, however, brilliant, full of energy, and committed totally to Sony's success. Perhaps most important of all, he was family.

With his characteristic attention to image, Morita moved the

Iwamas into an apartment at 985 Fifth Avenue, one block south of where he had lived with Yoshiko and the children. To announce his arrival, he arranged a reception and buffet dinner at the Hotel Pierre just opposite Central Park, inviting three hundred of his personal friends, including many prominent business leaders, and in his welcome speech asked them to help Iwama feel at home.

During their two years in New York, Iwama and his wife, Morita's younger sister, Kikuko, worked diligently to maintain Sony's social presence in New York. Once a week, following Morita's example, they had guests to dinner: the Conrads of RCA, CBS's president Frank Stanton, Western Electric and IBM executives. Normally they invited four couples and served a menu recommended by Yoshiko. Following an appetizer, which Yoshiko had often served to foreigners with success, a steamed custard called *chawan-mushi*, there was roast beef and vegetables, salad, cake, and coffee. "Small portions of roast," Yoshiko counseled, "so everyone can enjoy it all." Beginning in their second year of residence, the Iwamas also began hosting Sony employees and their families on assignment from Tokyo, four couples at a time, as well as American employees, such as Irving Sagor and his wife. By all accounts, Iwama communicated considerable warmth in spite of his famous reticence, and was generally well liked by the Americans.

It isn't clear that he contributed very actively to the running of the business: though he was only in Tokyo for one week every three months, he spent much of his time in New York poring over technical faxes from headquarters and seemed preoccupied by product development issues in Japan. As Sony began readying its first video recorders for the American market, there was building pressure on him to return.

Late in the summer of 1971, an unexpected opportunity presented itself in the person of Harvey Schein, a senior executive at CBS whom Morita knew and respected and who confided abruptly that he was leaving his company and was available. Morita felt certain that in Schein he had found the exceptional American manager he needed to implement his vision, and, in many respects, Schein was the right man, aggressive, shrewd, acutely cost-conscious: during slightly more than five years as president and CEO, from 1972 to 1978, he grew Sony

America from a $300 million to a $750 million business. Managing him, however, would prove to be more difficult than Morita imagined. The relationship began comfortably, but before long it became strained and ended abruptly with hard feelings on both sides. Some of the trouble was the inevitable result of two powerful egos colliding. At bottom, the conflict was rooted in incongruities that went deeper than individual differences. As such, the relationship was paradigmatic.

Harvey Schein had first entered Morita's field of vision when he came to Japan in 1966 to renegotiate Columbia Records' prewar merger with Nippon Columbia. MITI had recently announced that it would henceforth permit fifty-fifty joint ventures between Japanese and foreign companies. Schein hoped to arrange a 50-percent piece of Nippon Columbia for CBS, which owned the Columbia label worldwide. A protégé of William Paley, the CBS chairman, he was at the time the youthful president of the company's International Records Division, responsible for Columbia joint ventures around the world. When his conversations with Nippon Columbia stalled, Schein tried to interest other Japanese record companies in a joint venture but was unable to make headway. He was feeling thwarted and miserable when a former employee of Nippon Columbia whom he had hired away introduced him to Morita, then executive vice president of Sony. They met for lunch at the Takanawa Prince Hotel, just down the hill from Sony headquarters in Gotenyama. According to Schein, "Before we had finished our soup, we had the outline of a deal for a fifty-fifty record company. He seemed to speak English fluently, and he grasped ideas instantly. I couldn't believe what I was hearing from him about how easily this was going to happen; he said less than a year, and in less than a year we had the company and we were breaking open a barrel of Morita sake for a banzai toast."

Sony was still fifteen years away from developing the compact disc with Philips Electronics: at the time, 1967, there was no visible reason to enter the music business. But Morita's prescient eye saw beyond the horizon to precisely the convergence of hardware and software that would become the company's animating strategy. CBS/Sony Records was, in that sense, the first step leading to Sony's subsequent acquisition of CBS Records in 1988 and Columbia TriStar Pictures in 1989.

To conduct the negotiation for Sony, and then to manage the new business, Morita appointed Norio Ohga, then thirty-seven, the man he would presently choose as his successor. As Ohga proceeded to build CBS/Sony into the company's most profitable subsidiary, Schein was feeling unappreciated at CBS. A man he had hired to be his adjunct lawyer, Clive Davis, was placed in charge of the International Group, including the International Records Division, which Schein had built and wanted to continue managing. CBS had moreover vetoed his proposal to form a second joint venture with Sony to sell electric guitars and keyboards around the world. Morita had been cautiously receptive, suggesting that CBS should first acquire a musical instruments company with a more resonant international image than Fender Guitar, which it already owned. Schein promptly arranged a lunch meeting with Henry Steinway, whose company was on the block for $21 million, and acquired Steinway Piano for CBS for $19 million. Morita was now disposed to consider the new venture, and Ohga added his enthusiastic support and even urged the purchase of the Kawai Piano Company as well. CBS president Frank Stanton, however, was not happy about including Sony in the new venture. A series of awkward meetings followed, and it became clear to Schein that his proposal had been rejected. On his way to California with Morita on the CBS corporate jet, he announced that he intended to leave CBS: "Morita says, 'I don't want you going to the competition.' At the time, Schwartzenbach has drowned and Iwama is the president, but Ed Rosiny is actually running the company and it's time for new management. And Morita sits there for a while, and then he says, 'The only problem is, it's illegal in Japan to give stock options to executives, and I don't know how I'm going to make you happy.' So we sit there in silence on the plane for a while, with our own reading. And about an hour later, I say to Akio, 'How about if Sony lends me some money, and I'll buy Sony stock with it? And I could have the option of repaying the loan or handing back the stock?'"

On Labor Day weekend 1972, Morita phoned Schein at Candlewood Lake in Connecticut, where he was staying with his family, to say that Sony was ready to make a deal. The details were formalized over breakfast at Morita's apartment in New York the following week.

Schein asked for a loan of $1.1 million to be used for purchasing Sony stock, and an annual salary of $135,000, which would make him the highest-paid employee in the company, including Morita himself. He also wanted a twenty-year consulting agreement to be triggered by his departure from the company under any circumstances, services rendered to require no more than forty hours annually at times and places designated by himself. Morita agreed. Many years later, Norio Ohga would remark with a smile, "Harvey was the greatest negotiator of contracts for himself I ever met."

Before turning Schein loose in New York, Morita brought him to Japan for a month to "learn the Sony way." This was the first and the only time he would bring an American executive to headquarters for such an extended period: clearly, he considered his investment in Schein long-term. Schein remembers being driven straight from the airport to the Moritas' home on Lake Hakone for a weekend with the family. "I had a stomachache when I got there, I must have been nervous, and Mrs. Morita, Yoshi-san, massaged my stomach for me in the most amazingly tender way—it was like I was part of the family." During his stay, accompanied by a brilliant interpreter named Sen Nishiyama* and Morita's assistant, Kunitake Ando, Schein toured Sony facilities from Hokkaido in the north to Kyushu in the south. Ando recalls being asked by Schein after a visit to a Sony retailer whether Japanese dealers dipped into the till and had to be closely watched. He was also impressed by the explanation Schein offered for his own paranoia, that he had been raised in his parents' store in the Bronx "to the sounds of National Cash Register."

---

* Raised in the United States and educated in Japan, Nishiyama was a true bilingual with a rare gift for simultaneous interpreting in and out of both languages. Beginning in 1961, he had worked for the American embassy as interpreter to Ambassador Edwin Reischauer, who was fluent in Japanese himself but preferred to use Nishiyama in public. The two men developed a familiar, humorous style together, at moments rather like a stand-up comedy routine, which made them famous and very popular all over Japan. On Nishiyama's retirement from the embassy, Morita had persuaded him to come to Sony as Ibuka's interpreter and chargé d'affaires. Ibuka objected to Nishiyama's creative approach to interpreting. When he swore at someone in Japanese, he wanted the imprecation rendered in English and became angry if Nishiyama tactfully omitted it.

In October 1972, Harvey Schein went to work as president and CEO of Sony America. In 1966, the company had moved from lower Fifth Avenue to an old three-story warehouse building on Van Dam Street just across the 59th Street Bridge in Queens; one of Schein's first projects was to move Sony back to Manhattan, where, in his view and Morita's, it belonged. His choice of location, after considering the newly finished Squibb Building, was 9 West 57th Street—"9 West" as it became known inside Sony—another new building with a striking concave structure and sweeping views of Central Park just to the north. Early in 1973, Schein moved himself and the company's eight hundred employees into Sony America's new offices on the forty-second and forty-third floors.

Schein's business style was very American, outspoken and confrontational. By his own account a "hard-ass" with a cold eye on the bottom line and a temper that flared quickly when he sensed hedging or didn't see results, Schein bruised American egos and bewildered and terrified resident Japanese executives. He was, in other words, unsentimental in the extreme, and it is precisely his unsentimentality that Morita dryly recalls in his memoir: "His approach was not Japanese but based on pure, hard, straight, and clear logic. . . . The problem with the logic game, however, is that it leaves little room for the human factor."

Kenji Tamiya is one of many senior executives at Sony today who took their lumps from Harvey Schein but still speak of him admiringly as a man of integrity who taught them a lot about business. "He used to say his mandate was profit," Tamiya recalled, "and since Morita had picked him and not someone else, he was going to do it his way!" Tamiya was one of the young internationals Morita began to recruit and cultivate beginning in the mid-sixties as he extended Sony's operations around the world. He spoke Spanish and English reasonably well, had a good head for business, and was an affable, appealing young man who was normally at ease in the company of foreigners. After five years in South America, he had been sent by Morita to establish Sony Hawaii in 1968 and then on to Chicago for three and a half years as branch manager. In 1971, he had been moved again, to Los Angeles, where he first met Schein when he came to California on a tour of the U.S. offices. Early in 1974, Schein phoned Tamiya to say he wanted to move him to New York to oversee product planning and procurement for all of Sony America. Thirty-seven at the time, Tamiya had been on assignment abroad for more than ten years and was eager to return to Japan; besides, he disliked New York and found New Yorkers offensive. When he "politely declined," Schein offered, over the telephone, to make him a vice president and promised him an office with a view of Central Park, a company car, and a parking space in the building. Overwhelmed by Schein's spontaneity and directness, unthinkable in a Japanese organization, Tamiya accepted the promotion in spite of himself. "Come to think of it now," he told me, "Harvey didn't mention salary." There was a reason. When Schein reported to Iwama that he intended to promote Tamiya and wanted to give him a substantial raise, Iwama had approved the promotion but vetoed the raise, maintaining that Tamiya was still too young for the salary that Schein had in mind. Schein argued that the custom in America was to "pay the job, not the individual," but Iwama would not listen. This collision was one of a number which strained their relationship.

Tamiya relocated to New York in April 1974, reporting to Ray Steiner, and vividly recalls the tense atmosphere in the New York office during Schein's tenure. "He was a *no-nonsense guy*," he told me, speaking Japanese inlaid with English, "energetic, very sharp, and very strict about *cost-control.* Every executive in the office was terrified of him, Japanese and Americans alike, because if you didn't have the right answers when you presented to him, he'd beat you to a pulp. Ken Iwaki [in charge of U.S. corporate planning] used to memorize his presentations before he went in to see him. The only one he never bothered in public was my boss, Ray Steiner." According to Tamiya, Steiner had warned Schein privately that he would walk off the job and never look back if Schein ever got out of line with him, but promised in return not to oppose Schein in front of the Japanese executives even when he disagreed with him. As a result of this truce, Schein never yelled at Steiner, and Steiner was curiously neutral around Schein: Tamiya recalls that he and others who worked for Steiner wondered why their boss didn't help them more when Schein was punishing them. Between Schein and Steiner, notwithstanding their public behavior, there was real enmity; Schein maintains even today that Ray Steiner built his career at Sony by currying favor with Iwama. Tamiya claims that the stress of mediating between them brought his weight down forty pounds and gave him an ulcer. Ironically enough, Schein advised him when he got sick that he should protect his health by learning to vent his frustrations, as he himself did. This prescription was more than Tamiya could manage. Instead, he relied on Tagamet, newly on the market at the time.

Schein began stepping on toes at once. On examining Irving Sagor's books, he discovered accounts receivable that had been outstanding for two or three years, and no reserve for bad debts. He wanted what he called "a range of reasonableness," a cushion based on "the Joseph theory" (his rhetoric tended toward the Talmudic) that things would go badly. He insisted on creating a 4 percent reserve, ruffling feathers in the process. Next he turned his attention to account management. Learning that Macy's of San Francisco, Sony America's largest customer, was delinquent, he ordered Steiner, senior vice president of Sales and Marketing, to "cut them off." When the president of Macy's called to complain, Schein informed him he would get his audio equipment when he paid his bills, and Macy's stopped being a problem.

Schein scrutinized every aspect of the U.S. operation, was dismayed at much of what he found, and loudly communicated his dismay. Sony's new Trinitron plant in San Diego was an example. When Schein learned of the project, which was nearly completed when he took over, he was appalled that Sony had chosen to locate a manufacturing operation in California, the state with the strictest environmental regulations in the country. His irritation grew as he found himself dealing with some of the problems arising from the decision. California's Environmental Protection Agency, for example, required that the water draining from the plant be purer than the water piped in. In an agitated conversation with Morita, Schein pointed out that Sony was not in the water purification business. In his opinion, which he made certain he conveyed, the real reason for the choice must have been the comfortable weather and the availability of superior golf facilities in the San Diego area.

Schein was wrong in this case. Morita had been led to San Diego by his determination that Sony's first manufacturing factory in the United

States should be a nonunion shop. Before settling on San Diego, he had dispatched a young Sony man, Masayoshi "Mike" Morimoto, to study the union situation up and down the West Coast. In Seattle, Boeing was unionized. A longtime friend of Sony's, Techtronix, Inc., maintained nonunion operations in Beaverton, Oregon, but Portland was smaller and farther away from Mexico, where Morita planned to order inexpensive subassemblies. Hewlett-Packard and Burroughs maintained nonunion operations in the San Diego area.*

Morita believed that powerful unions were responsible for what he viewed as the inferior quality of American manufacturing; moreover, controlling costs would be critical. Sony's managers in Tokyo had opposed his decision to locate in America, where labor was expensive, and had proposed Korea as an offshore base of operations. But Morita insisted that Sony would have to establish an American presence in order to dominate the U.S. market. He was also planning a defense against charges of dumping, which were based on the difference between the U.S. and domestic price for the same product: producing in California was likely to invalidate comparisons.

Something else Schein may not have realized was that the original plan called for assembly only, a much cleaner process than manufacturing the Trinitron picture tubes. In December 1971, Sony leased a property from AVCO Development Company twenty miles north of San Diego and hired AVCO to build a facility. The plant opened in August 1972 and was profitable within three months.

In November, the plant manager, Junichi "Steve" Kodera, received a call from Tokyo informing him of new plans to manufacture picture tubes. Kodera cautioned that a manufacturing operation would be more vulnerable to union penetration and recommended locating it in

---

* According to Morimoto, who was in charge of personnel at the San Diego plant for fifteen years, Sony employed only one in ten applicants and carefully avoided workers who came from unionized plants. Edward Rosiny, Morita's legal mentor in the United States, predicted that unionization was ultimately unavoidable and urged Morita to attempt a sweetheart arrangement with the Teamsters rather than risk infiltration by the more aggressive Longshoremen. Through his Teamsters contacts in New York, Rosiny arranged for union representatives to leaflet cars parked at the plant within weeks of its opening, but employees voted to exclude the union. Later, the International Brotherhood of Electronic Workers also distributed flyers and was unsuccessful. Sony's San Diego plant remains a closed shop.

Irvine, California, but he was overruled. In April 1973, Sony purchased land from the Boston firm of Cabot, Cabot and Lodge, and awarded the construction contract to a giant Japanese builder, the Tajima Company, where Iwama's son happened to be employed. The new plant opened in August, less than a year after Schein had been hired.

Morita had overruled strenuous opposition in mandating a Sony facility in the United States and was desperately anxious that it should succeed: listening to Harvey Schein inveigh against it cannot have been agreeable. Perhaps he shrugged off the experience as part of the cost of working with a real American manager. For reasons of his own, it seems he did not attempt to justify his decision.

However he may have deplored its location, Schein was saddled with the San Diego plant, and he insisted that it should report directly to him and not, as had been contemplated, to Tokyo. Morita consented, but required him to visit headquarters each quarter to confer with Susumu Yoshida, formerly the leader of the Trinitron group and now a company director and head of worldwide television manufacturing and sales. In Schein's recollection, Yoshida was referred to as "the emperor" (actually, he was known as *shōgun*, or generalissimo) and was treated "like a king," receiving managers from all over Sony's growing international empire in a suite at the Takanawa Prince Hotel.

Schein's quarterly meetings with Yoshida, and with other top managers in Tokyo, invariably led to conflict about pricing and profit. He observed, and it drove him to distraction, that some Sony divisions that were losing money in the United States, broadcast equipment, for example, were among the company's most profitable businesses in Japan. Even Trinitron, though it was selling well, was less profitable than Schein believed it ought to be, and was struggling to hold its own against RCA and other American competitors. Schein was convinced that pricing, determined in Tokyo, was the problem. Pointing out that corporate taxes in Japan were much higher than in the States, and that U.S. imports were subject to tariff, he argued that products should be sold more expensively in Japan, diminishing margins and saving corporate taxes, and less expensively in the United States, saving customs duties and increasing profitability. At the end of his presentation, he invariably pressed Yoshida for lower pricing, and Yoshida consistently refused.

Recalling his encounters with Schein, Yoshida explained with a grim smile that Sony's credibility with its Japanese banks was contingent on high profitability in the domestic market; decreasing profits at the parent company to pass savings on to Sony America was considered out of the question. In any event, discounting product to the subsidiary was seen as unproductive because Sony America could be counted on to spend whatever additional dollars it netted on sales and advertising. Yoshida gave his reasons to Schein, who rejected them out of hand as wrong-headed, and the meetings ended in awkward stalemate.

Later, Schein would reflect, with admiration in spite of himself, "Morita was a great con man. He insisted that Sony products should sell high, and he arranged for the world to pay twenty percent more for a Sony product than any other. He always told me, if I could move twice the product he might consider a discount."

Morita of course was aware of Schein's abrasiveness and experienced it himself, but he also appreciated the results Schein was achieving and was apparently willing to give him the room he needed to operate, at least in the beginning. Midway through 1973, he transferred both Inagaki and Sagor to a newly formed Sony subsidiary called Meriton—Inagaki, who chose the name, hoped it might convey a sense of worthy products, "merit-on the product"—and from there, when Meriton failed, to another subsidiary, Aiwa America. Inagaki and Sagor were united in their dislike of Schein, and, though nothing was ever said, it is likely that Morita removed them from the New York office to get them out of Schein's way. Looking back, Schein reflected: "When they liked what I was doing they gave me my head, but as soon as I stepped on their toes, they reined me in."

It wasn't long before Schein began locking horns directly with Morita over strategic and tactical issues. Early in 1975, Sony repurchased the rights to Sony audiotape in the United States from a company called Suprascope. Schein negotiated for Sony with Suprascope's owners, two tough-as-nails Russian émigré brothers named Tushinsky, and retrieved the rights for a modest fee with no royalty obligation. As if to demonstrate to headquarters the proper way to go about creating a new plant, he then concluded a sweetheart deal with Governor George Wallace to help finance a new Sony tape plant in Dothan, Alabama, with tax-free bonds issued by the state. As the plant went into production, Morita instructed Schein to spend $5 million advertising Sony audiotape, and Schein balked: in his view, as he explained to Morita, market conditions did not justify an expenditure of that size on audiotape. Speaking from his bedside phone in Tokyo at 6:00 A.M. Japan time, Morita scolded Schein for "short-term thinking," and ordered him to spend the money. Schein reluctantly complied.

A more dramatic variation of the same conflict occurred as part of the fallout from the Betamax debacle, which played out during Schein's term as president. In September 1975, Sony introduced the United States to Betamax, the first truly portable VCR, the fruit of twenty years of applied ingenuity by Nobutoshi Kihara in response to Ibuka's pleas for a video recorder for home use. The first model, the SL 6300, included a nineteen-inch Trinitron in a console and sold for $2,295. It was too expensive: Christmas season approached and sales remained sluggish. Schein complained to Morita, arguing for a stand-alone Beta deck at a more reasonable price. Morita was reluctant: in his vision, Betamax and television were symbiotically connected. Schein hammered away, and Morita allowed himself to be persuaded. When the stand-alone deck went on sale in February 1976 for $1,295, it flew off the shelf faster than Sony could supply dealers with inventory. As an added bonus, affluent customers were also buying an average of twelve to fifteen half-inch Beta tapes at $16 each.

Early sales were stimulated by a quirkishly humorous Doyle Dane Bernbach ad campaign designed to drive home Morita's notion of "time shift," a phrase he coined during an interview for *Time* magazine. In one of the TV spots, Bela Lugosi as Count Dracula returns to his darkened apartment, switches on his Betamax, and, in heavy Transylvanian tones, enthuses gloomily, "If you work nights like I do, you miss a lot of great TV programs. But I don't miss them anymore, thanks to Sony's Betamax deck, which hooks up to any TV set."

In September 1976, a proposal for a similar advertisement was sent by DDB to Sidney Sheinberg, president of MCA/Universal, for his approval. The top line read: "Now you don't have to miss Kojak because you're watching Colombo (and vice versa). Betamax—it's a Sony." Sheinberg was not amused. Universal owned both these popular TV series, which played opposite each another at 9:00 on Sunday night on CBS and NBC. As he studied the ad, it occurred to him that taping one of his television properties at home might constitute an infringement of copyright. Following a conversation with his lawyers, he felt certain that there were grounds for action. The following week, Scheinberg and his mentor and champion, the Hollywood kingmaker Lew Wasserman, met with Morita and Schein at Sony headquarters in New York. The ostensible reason for the meeting was a discussion about a video playback system, which Universal had spent a lot of money developing. The idea, called DiscoVision, was to prerecord movies belonging to Universal on plastic discs which could be played on a TV screen at home. Sheinberg had already made overtures to Sony about designing and manufacturing the hardware. At some point during dinner in the Sony boardroom, he informed Morita and Schein of Universal's opinion that the sale and use of Betamax constituted an infringement of copyright law, and suggested evenly that the studio would have to seek an injunction if sales continued. Alone with Schein after the meeting, Morita dismissed the threat as empty; people who were discussing a business deal, he assured Schein, were not about to sue each other.

The following month, this time at Chasen's restaurant in Los Angeles, the four men met again, and Sheinberg recapitulated his annoyance and the studio's intention to file for an injunction. Morita remained sanguine, instructing Schein to pursue the DiscoVision conversation actively. Under normal circumstances, it is unlikely that he would have risked cannibalizing Betamax sales by developing a rival system, but apparently he felt confident that a suit would not happen while the DiscoVision dialogue was alive. Harvey Schein did not share his confidence. In America, he tried to explain, people could easily sue, even while they were doing business.

In Japan, lawsuits of any kind were rare. When American lawyers had examined Sony contracts in the process of readying an application to the SEC, they had remarked on the absence of language about sanctions and damages in the event of default. What they found instead was a clause, baffling to them, which specified that both parties would sit down in good faith to discuss any circumstances that caused either party to breach the agreement. Unilateral legal action, in other words, was considered uncivilized and to be avoided at all costs, particularly by organizations doing business together. In light of this convention of social, and correspondingly, of corporate, behavior, it is understandable that Morita might have misread his adversaries. On November 11, 1976, Universal and Disney, whom Sheinberg had by this time involved, filed a suit in federal court against the Sony Corporation, Sony America, and DDB for copyright infringement, and sought an injunction against the sale of Betamax. Schein learned the news at midday on a Friday, which was 2:00 A.M. Saturday morning Tokyo time, and waited until evening in New York to reach Morita on Saturday morning as he was preparing to play golf. His description of what happened when he broke the news is a masterpiece of cultural synesthesia: "There was a pause. Then he gave a moan right out of his *kishkes.* It was like one of those death cries in a Kabuki play."

It was fully two years before round one of the trial began, in January 1979. When Morita had recovered from the initial shock, he dedicated himself to winning the court battle with grim determination, commuting almost weekly from Tokyo to oversee the preparation of thousands of documents and nearly one hundred witnesses. From where he stood, the stakes were high: if Sony should lose and the court were to enforce a penalty against every Betamax unit that had been sold, the financial setback could cripple the company permanently. People close to him at home and at work recall that they had never seen him as physically exhausted or emotionally drained as during this long period of pretrial preparation.

For Harvey Schein, whose eye was ever on the bottom line of the American business, the impending trial was an unsettling prospect, but the national attention that the lawsuit was attracting was cause for quiet exultation. He and Sheinberg appeared on the nightly news with Walter Cronkite for a nationally televised debate in which Sheinberg called him a "highwayman" and Schein asserted Sony's right to "fair use." Jack Valenti, longtime president of the Motion Picture Association of America, was screaming that Betamax was a parasite; Fred Rogers, "Mr. Rogers" to the three million families who faithfully watched his daily television show, announced, and would later testify, that he would consider it an honor if his young friends across the nation saw fit to record his neighborhood to enjoy at a later time. "The publicity was fantastic for business," Schein recalls. "People thought, if Hollywood hates it so much it must be something really good. And there was an incentive to buy now, right away, before it became illegal."

At this juncture, Morita communicated from Tokyo that he wanted substantially more money spent on the Betamax advertising campaign. The advertising budget belonged to Sony America, and Schein angrily refused to make the allocation. He pointed out that Betamax was already enjoying the benefits of nationwide publicity that no money could buy. And he quoted from a poll published earlier that year in *Home Furnishing Daily:* 51 percent of American consumers had heard of Betamax, and 90 percent of those who knew about it declared they were ready to purchase one. At a moment like this, Schein declared, it was irresponsible, not to mention crazy, to throw money away on advertising. Morita retorted angrily that Schein was behaving like the standard American manager unable to see beyond the tip of his nose. The phone call ended inconclusively. Over the weekend, Morita brooded about Betamax and Schein's attitude. Early Monday morning in Tokyo, he phoned again and demanded that Schein spend "several million dollars" over the next two months or be prepared to lose his job. Ray Steiner, eager to get his hands on a fattened marketing budget, asked Tamiya repeatedly whether "Tokyo" knew that Schein was continuing to drag his heels. The question evokes the distance that continued to separate American executives in New York from Tokyo headquarters. Eventually, Schein complied, but when Morita learned that he had decreased advertising budgets in television and audio products to "raise" the funds, he was furious.

Morita's agitation suggests that his emotional investment in Betamax was substantial and highly charged. Sony had already lost its bid to make the system the industry standard in Japan. Matsushita had adopted JVC's Video Home System (VHS) with a longer record-play capacity than Sony's Betamax, and Sharp, Hitachi, Sanyo, and Mitsubishi had followed suit. In Japan, only Toshiba, and later, Sanyo, aligned themselves with Sony and the Betamax. In the United States, largely because of its limited capacity of one hour, the system was faring no better: in March 1977, Matsushita concluded a deal with RCA to manufacture fifty-five thousand VHS machines that year and half a million to one million machines each year for the following three years. In August 1977, RCA announced its SelectaVision VHS, built by Matsushita, with a four-hour capacity for $1,000, $300 less than Betamax. In 1978, GE, Magnavox, Sylvania, and others entered the market with their logos on VHS machines built by Matsushita and priced lower than Betamax with longer record-play capacities. Beginning that year, Sony's market share tilted and began to drop until, by 1980, Betamax was being driven from the home video market.

Sony had staked a lot of time, money, and corporate reputation on the success of Betamax; executives at headquarters had even taken to sporting tiepins and cuff links inscribed with the Betamax logo. For Ibuka and Kihara and the other engineers, it was mortifying to watch a system that they held to be inferior to their own take the lead in the marketplace—particularly when it had been closely modeled on their own Betamax, which they had shown Matsushita and JVC in the developmental stages even while JVC had been working on VHS in secret. This was all painful, but for the engineering group there was at least the comfort of feeling as certain as they did that their technology was superior.

Morita was perhaps less well equipped to buffer his sense of personal defeat at the hands his archrival, Konosuke Matsushita. In his efforts to persuade Matsushita to join Sony, Morita had paid his respects to the old man at his Osaka headquarters several times. Matsushita had listened with his eyes closed and had remained politely noncommittal. On his third visit, Morita, accompanied by Kihara, had walked into Matsushita's office and found a Beta deck and JVC's VHS machine side by side on his desk, their parts and circuitry exposed. The old man pointed to the VHS machine to demonstrate that it had fewer parts and would therefore be cheaper to service and repair. Accordingly, he explained, Matsushita Electric and its subsidiaries would adopt the VHS standard. Kihara recalls that Matsushita accompanied them to the door and stood at the top of the stairs waving "sayonara" as they left. In the car, Morita, red-faced, turned to Kihara and vowed to focus all of Sony's energy on defeating Matsushita in the market.

Now, as Morita spent himself on preparing for a lawsuit that would have terrible ramifications if it went badly, Matsushita was teaming with RCA to overpower Betamax in America. Under these circumstances, it is not hard to imagine why he might have been seized by a sudden need, however unreasonable, to turn up the volume on the Betamax campaign.

From about this time, Harvey Schein began feeling that his future at Sony was uncertain. With RCA's defection to VHS, Morita was desperately in search of an American ally; when Zenith showed signs of interest in Betamax, he put Inagaki and Ray Steiner in charge of the negotiation. Schein interpreted this choice as a clear signal from Morita that his aggressive negotiating skills were no longer valued. The president of Zenith, John Nevin, was among the initiators of a dumping suit against Japanese television manufacturers, including Sony, in 1971, and was known to be anti-Japanese. To Schein's way of thinking, the final agreement was a sweetheart deal served up by Sony in return for Zenith dropping Sony from the class-action suit. "Morita knew I would fight for Sony's rights and not cave in. And so he used Inagaki and Steiner instead. I would sit in my office and see them talking with Nevin through the glass window next door. For me, that was the beginning of the end."

Although Harvey Schein's unalterably American approach to business and his outspokenness made him a thorn in Sony's side, Morita understood that building an American subsidiary would be an exasperating process requiring not only perseverance but a high degree of forbearance. Nevertheless, there was a category of transgression—less strategic than social—which even Morita was not prepared emotionally to tolerate. Finally, it was Harvey Schein's unsentimentality that was beyond enduring.

One festering problem was Schein's attitude toward Kazuo Iwama, which was perceived as disrespectful. When Schein arrived in 1972, Iwama was *pro forma* chairman of Sony America, still resident in New York but having little to do with daily operations. Even so, as a bona fide member of the inner circle of the Sony family, Iwama had the power to countermand Schein's authority if he saw fit. Schein resented this and may have felt threatened by it. Iwama, reticent even by Japanese standards, may have been offended by the other man's style, which he is likely to have perceived as noisy and invasive, and had as little as possible to do with Schein directly. He made no secret of his preference for the affable Ray Steiner, with whom he shared a passion for golf. Schein had no interest in the game, and dismissed Steiner as ingratiating.

In July 1972, Schein's first year in the job, Iwama and his family returned to Tokyo. But before he left he had a number of confrontations with Schein concerning the San Diego plant and hiring and compensation policy in America. First there was Kenji Tamiya's promotion and the raise that Iwama would not approve. Later, Schein wanted an incentive bonus program for his executives and prevailed on Morita to agree. Iwama summoned him to his office and informed him that the program would be installed, but only for subordinates. As the top men in the operation, neither he nor Schein would be eligible. His logic eluded Schein and contributed to Schein's dislike for Iwama.

Even after Iwama returned to Japan, Schein continued, perhaps unwittingly, to offend him. On Friday nights, at the conclusion of quarterly product meetings, Iwama customarily hosted a dinner at his home to which he invited everybody, including the American staffers. Many of the Americans looked forward to these dinners eagerly. Irving Sagor, for example: "Iwama was a wonderful and an unusual individual. The first word that comes to mind in describing him is his warmth. You no sooner met him than you felt a kinship with him. He liked to entertain, and unlike most of the Japanese executives he loved to entertain at his home. He wanted his family to enjoy with him the friendship of foreigners and people who were different. He would always have a Polaroid camera ready, and at one point a picture would be taken of his guests and his children and his wife, and then he would mount it in the album and ask you to inscribe something. Everybody got the same treatment, that was Iwama."

Schein had no time for such bonhomie. There was a Friday-evening flight through Honolulu which would have him smartly home by Saturday afternoon with time to catch up and be at his desk on Monday morning. Schein regularly took that flight, declining Iwama's invitations and almost certainly causing resentment.

For Morita, the last straw may have been a product meeting in Tokyo when Schein lost his temper and shouted abuse at Morita's brother, Masaaki, in front of other Sony executives. The issue, as usual, had to do with pricing for the American market. Schein had requested a remote control tuner for American models of Trinitron. Sony had designed a remote tuner that it was able to produce even less expensively than the standard-issue tuners, but instead of passing the savings along, Masaaki Morita, who was now in charge of TV worldwide, had decided to charge Sony America a premium.

Tuners were in general a sore point for Schein. He had fought with Sony about providing two separate tuners for Betamax, so that one program could be viewed while another was being recorded, and had finally convinced the manufacturing division to give him what he wanted only to be charged for the extra tuner. Now, having come to the meeting straight from the airport after a delayed arrival, tired and jet-lagged, Schein attacked Masaaki Morita for what he labeled an attempt to sabotage the profitability of the American operation.

Later, Schein regretted his outburst, but felt justified in having fought a business decision that he knew would hurt his profit line. Besides, he told himself, he had meant nothing personal. Unfortunately, inside the web of Sony society, or any Japanese subsociety, everything was perceived as personal. Later that same day, Morita called Schein to his office and, without a word about the blowup earlier, suggested it was time for him to get some distance from line management and become chairman of Sony America. "I knew as I listened to him," Schein says, "that he was saying sayonara."

In the summer of 1977, Schein became chairman and, on Iwama's recommendation, Ray Steiner was appointed president of Sony America. Norio Ohga, who had been close to Schein since the CBS-Sony negotiations, tried to mediate a reconciliation at a dinner he arranged in Honolulu. The conversation was heading in a promising direction until Ohga fell asleep at the table. Schein began to complain about Japanese executives indulging in $600 lunches and charging them to

Sony America. Morita advised him unsmilingly not to concern himself with managing details. By the time Ohga revived, a chill had settled in the air.

At the end of 1977, in accordance with his contract, Schein gave six months' notice and triggered his twenty-year term as consultant to the corporation. Morita accepted his "resignation" on the single condition that he remain as a director of the Sony Corporation of America. Iwama, interviewed in an industry newsletter, explained that both sides had viewed the relationship as temporary from the beginning. Schein was impressed with the lengths to which Sony was willing to go to make his departure appear friendly: "They were passionately concerned with form," he told me. "It was the appearance of the box that mattered, not the contents." Schein's summary, at once accurate and misapprehended, evokes the chasm between himself and his erstwhile employers that remained unbridged.

In the spring of 1978, Schein left Sony and went to work for Steve Ross at Warner Communication, but even then, Morita may not have entirely given up on him. The following year, he called to propose a joint venture to develop a video projector system for use in movie theaters; the idea was to decrease the expense of feature film exhibition by replacing film prints with video. Morita had been discussing the idea with Warner and had run into a wall; now it seemed he was approaching Schein directly to go into business with Sony. "He didn't say so explicitly," Schein recalls, "but I had the feeling he was hoping I would go back and help him deal with the Jews in Hollywood who were driving him crazy at the time." Schein began asking the kinds of questions that mattered to him: How much money was Sony prepared to invest? What would happen to Schein if Sony decided to bail? In a matter of days, the dialogue trailed off: it is likely that Schein's habitual directness, the "American-ness" of his approach, had recalled for Morita the difficulty of the relationship over the years. Morita called once more, to ask Schein to accept Masao Morita as an intern and teach him the ropes of American business. Schein met Masao and his American girlfriend and invited them to see *The Sting*. But he was uncomfortable about taking on Masao as an apprentice because he was still new at Warner, and begged off. It was his impression at the time that Morita was angry, and very likely he was, offended by Schein's unsentimental refusal to honor an old, if now broken, connection.

The men had no further dealings. When, in 1989, Sony incurred Steve Ross's wrath by hiring Peter Guber and Jon Peters away from Warner Brothers during the Columbia Pictures acquisition, Schein expected Morita to seek his help with Ross, whom he knew intimately, and was surprised when he was never contacted. Norio Ohga explained: "Harvey had had a falling out with Mr. Morita. We would never have dreamed of asking him for help."

From Sony's point of view, the Harvey Schein experiment was by no means a total failure. Morita would later credit him with introducing powerful features of American management to Sony culture, including budget control and executive compensation. In the process, he had helped to transform Sony America from a mere distribution center to a viable stand-alone subsidiary, and, into the bargain, had nearly tripled the business. Still, Morita came away feeling no less pessimistic than Schein himself about the possibility of mutual understanding between Japan and the West: on the personal level, in the domain of culture, the memory of Harvey Schein's tenure lingered on as an ordeal. It would be fully ten years before Morita would attempt again to realize his vision of a truly global Sony by empowering another American to run the American business. This still bolder experiment in transcultural communication would elevate Michael P. Schulhof to a position of unprecedented power, and would generate its own order of difficulty beyond anything Morita could have imagined.

6

MAESTRO OHGA: THE ART OF PROFIT

Norio Ohga, Morita's chosen successor, is a paradoxical man. Educated in Tokyo and Berlin as a musician, he embodies a combination, rarely found in nature, of shrewd business savvy and an artist's sensitivity. In appearance and demeanor he is as staid as any banker, yet he is a daredevil, a race car driver and a jet pilot who has been known to fly prohibited rolls in his corporate jet. Inside Sony, Ohga is feared by many as a tyrant. There are stories of his hurling prototypes across a room and of scolding terrified subordinates in his booming baritone. To the outside world, he presents an icily superior aloofness. But he is also capable of compassion and warmth: his secretary for thirty-eight years recounts his kindness to her with tears in her eyes, and many others, Japanese and foreign, speak of him as a loyal and passionate friend.

Asked to account for how he sustains deep commitments to art and profit simultaneously, Ohga's answer is always a variation on the same unsatisfying explanation: "I have switches in my head," he recently told me yet again, "and when I throw them, they change me completely. When I fly my jet I'm a pilot and all I think about is flying safely. When I land I remember I must prepare to negotiate a contract. Then I remember I must conduct and I pore over my score."

When Morita prevailed on Ohga to join Sony after pursuing him for years, Ohga resolved to return to his music when he reached the age of sixty, and he has kept his promise to himself. On September 6, 1997, I was in the audience when he conducted the Tokyo Symphony Orchestra in Beethoven's Third and Seventh symphonies at a concert in Tokyo's Suntory Hall. The occasion, a benefit to raise money for the composer's birthplace in Bonn, was cosponsored by Beethoven-House and the Sony Music Foundation, which Ohga controls. Suntory Hall, erected as a monument to himself by Keizo Saji, Suntory Brewery's epicurean owner, seats two thousand people. That night the house was filled with personal guests of Ohga's from the worlds of finance and business. There were surprisingly few foreigners. Tokyo's international community would flock here the following week when Ohga's old friend, Lorin Maazel, conducted, but Sony's chairman was not a draw. The Sony family, on the other hand, was very much in evidence: both of Morita's brothers and their wives; Ibuka's son, Makoto; Ohga's handpicked successor, Nobuyuki Idei, and Idei's wife; and most of the other executives on the Sony board. Clearly, when Ohga conducts, Sony attends.

In Japan, an orchestra does not usually tune on stage. Western concertgoers enjoy the cacophony of scales and fragments of difficult passages as the musicians warm up; in Tokyo, the orchestra appears only when the house is full and settled. The concertmaster sounds his "A," the oboe answers politely, the orchestra briefly confirms pitch a final time, more a gesture than a process, and is silent again, awaiting the maestro's entrance. Are we being spared the disturbance of the random noise of tuning? Or is tuning up, a practice session of sorts, unacceptably personal, tantamount to receiving a visitor in one's private office or taking guests on a tour of the home?

Ohga enters, and moves to the podium with his deliberate, mincing gait. As always, he is impeccable: a perfect tuxedo, gleaming shoes, not a hair out of place (now or at the end of the performance). He acknowledges the applause with a slight bow and turns to the orchestra. I am seated behind the bass viols, looking obliquely across the strings into the audience: the acoustics are frightful, but I have a full view of Ohga's face.

He conducts from memory without a score. His tempi are assured, his phrasing subtle. But he seems even here, as in a business meeting, withdrawn and unapproachable. His eyes closed, his body swaying, he appears to be focused inwardly on the bliss the music brings him. Perhaps, strictly speaking, Ohga's talent alone would not earn him invitations to conduct the Boston Symphony at Tanglewood, the Pittsburgh Symphony, the Israel Philharmonic, or the Metropolitan Opera Orchestra. Perhaps his position as chief of the world's largest sound and music company has gained him access. There are those who suggest that inviting Ohga to conduct advances an orchestra in the global competition for Sony sponsorship on a world tour. Nevertheless, here he is, one of the world's most powerful businessmen, summoning forth Beethoven with his baton from one of Japan's best orchestras, attended by his friends and by the Sony family in Tokyo's most prestigious hall. Before Ohga was twenty he had lost both an elder and a younger brother in freak accidents, and his family used to say about him that he enjoyed their share of luck and happiness in addition to his own. Watching him take his curtain calls, turning to the orchestra to salute the players and back to the audience to acknowledge their applause, one feels that Norio Ohga is indeed a fortunate man. Asked, however, to reflect on his forty-year career at Sony, his reply is unexpected: "What strikes me as unique about my life is the extent to which it took me on a path I had no desire to walk."

Ohga was born in 1930, the fourth of seven children, four girls and three boys, and raised as the darling of the family by his doting elder sisters. His father, a wealthy importer of lumber from Southeast Asia, was away in Hanoi managing his business there until the end of the war. Ohga grew up with his mother and sisters and brothers and the family's servants in a "summer home" near the beach in Senbon Matsubara on Suruga Bay, eighty miles southwest of Tokyo. From an early age he was drawn to art and music. He loved watching his eldest sister paint and practice calligraphy—he would become an accomplished calligrapher himself—and when he was five he sat down at the piano in the Western-style drawing room and taught himself to play. In elementary school he was an A student in all but two subjects, music and physical education. In music, he earned a rarely awarded A^{+} ("outstanding"); in P.E., the best he could do was B. In a recent interview, he recalled that he had always "hated" athletics. Five feet nine inches tall and 180 pounds, Ohga is a large man by Japanese standards, but his robustness seems to have belied, then as now, a delicate sensibility.

In 1943, when he was thirteen, Ohga developed pleurisy and had to stay home from middle school for over a year.* While convalescing, he spent his evenings studying with a man who lived across the street, Ichiro Iwai, the wealthy scion of a merchant family who was living in a large house with his children and three servants after the death of his wife, and who seems to have adopted Ohga as his protégé. Thirty years old at the time, with a degree in electrical engineering from the University of Tokyo, Iwai tutored Ohga in math and science and taught him many other things in the bargain, including how to read electrical diagrams and musical scores. The relationship between the cultivated young widower and the sensitive, eager adolescent who may have been feeling keenly the absence of his father was, for Ohga, formative: "I think Iwai taught me everything he knew, and whatever I understand today about art and beauty and music and science begins with what I learned from him, which was far more than I ever learned in school. In a way, I owe my life to him." When Iwai died of a heart attack at the age of sixty-two, Ohga, in his forties, was bereft.

When the war ended, Ohga returned to middle school vaguely intending to become a mechanical engineer, but he was listless in class, and discouraged by the bleak prospects for engineers in the rubble of the defeat, when MacArthur's occupation was dismantling the country's largest manufacturing organizations. In 1946, at sixteen, he auditioned for the distinguished voice teacher and opera singer Teiichi Nakayama and was accepted as a student. For four years, he commuted to Tokyo once a week, a journey that took three hours each way, for private lessons in German lieder. In March 1949, he passed the entrance examination to the newly founded Tokyo University of the Arts and was one of thirteen students nationwide to be admitted to the first graduating class in the Department of Voice.

Ohga's path might never have led him to Sony had it not been for another of his wealthy neighbors, Kahei Nishida, a man twenty years

---

* By this time, normal classes had in any event been suspended; students from middle school upward were being mobilized to work at munitions factories and other military installations.

his senior whose family money came from textiles and who happened to be related distantly to Ibuka. In the fall of 1950, Ibuka visited Nishida on one of his endless rounds in search of funding, and persuaded him to invest in his fledgling company. Knowing of Ohga's interest in electronics, Nishida told him about Tokyo Telecommunications and suggested he visit Ibuka in Tokyo. Later that year, with a note of introduction in hand, Ohga called on Ibuka at his ramshackle wooden building in Gotenyama Heights. The meeting was brief; when Ibuka took Ohga to the shop floor and demonstrated his brand-new G-Type Tape-corder, Ohga observed that playback was too distorted to be of any use to a musician and excused himself. Later, Ibuka remarked to Morita that he had met a brash student with no manners but an impressive grasp of technology.

Early in 1951, a Tokyo Telecommunications representative named Masao Kurahashi showed up at Tokyo University of the Arts with a G-Type machine. His ostensible purpose was to record the school orchestra as a sound-quality test, but he was secretly hoping to make a sale. Kurahashi was employed by a trading company backed by the Tokugawa family, descendants of the overlords of feudal Japan. He had persuaded the Tokugawas, who were in search of new business opportunities, to purchase fifty Tape-corders from Totsuko at the deeply discounted price of ¥3 million. A check had been delivered into Ibuka's jubilant hands, and now the pressure was on Kurahashi to find customers.

The demonstration ended before it began. In 1951, power was still in short supply in Tokyo, and as evening approached, amperage dropped: when the orchestra was ready and Kurahashi threw the switch, nothing happened. Ohga had been advised of the experiment and was looking on. As a mortified Kurahashi was packing up to leave, Ohga introduced himself and got his business card.

The following week, Ohga went before the faculty to recommend that the school apply to the Education Ministry for funding to purchase a G-Type machine from Ibuka's company. Normally, students did not present to the faculty senate, but Ohga was cocksure, articulate, knowledgeable, and charming. His argument was inventive: dancers relied on mirrors to mark their progress as they worked on perfecting their art; tape recorders would certainly become the musician's mirror. Without the means of playing back a performance, musicians and singers like himself would be deprived of the opportunity to observe and correct themselves.

The machine was priced by Totsuko at ¥160,000, an amount equivalent to annual tuition for 40 percent of the college's freshman students. Nevertheless, Ohga's vision of the Tape-corder as a mirror for musicians proved irresistible, and funding was secured. Though he was only a sophomore, Ohga was given responsibility for overseeing the purchase.

From Ibuka's point of view, an outspoken college student had transformed himself into a premium customer: this time he dispatched his secondhand Datsun truck to pick up Ohga and convey him to Gotenyama in style. When he arrived, Ohga handed Akira Higuchi a list of modifications he wanted made, with an accompanying sketch and wiring diagram. He was asking for changes in the stabilization motor to reduce flutter and wow, and for a different kind of microphone and input connection. "I could see that he was right," Higuchi recalls, "and as an engineer I wanted to make those changes because they were definitely improvements." When Higuchi showed Ohga's specifications to Ibuka and Morita, they were astonished by the level of his technical understanding.

Before long, Ohga had become a familiar figure at Totsuko. He was invited to attend technical meetings, impressing everyone with his encyclopedic knowledge of tape recorders in particular, and engaged Ibuka and his engineers intensely in conversation about the future of audio technology. Eventually, Ibuka directed that no prototype would be put into production before Ohga had had an opportunity to test it and render an opinion. Often he was critical. Recording a set of tuning forks, he would point out the distortion in pitch and tonal quality as he played back the paper tape and would reiterate his judgment that the machine was still too crude to be of use to a musician. Recently, recalling these early days, Ohga reflected on Ibuka and Morita's tolerance of criticism from a college music student: "They actually treated me as an equal, and that was an attitude you would never have found in any normal Japanese business executive."

The day he graduated in 1953, Tokyo Telecommunications threw a party for Ohga at the Gotenyama office. After breaking open a keg of Morita sake to toast his success—celebrations at Sony still begin with the dramatic stoving in of a Morita keg—Morita called Ohga aside and invited him to join the company "informally."* He proposed paying him a full starting salary, ¥3,000 a month, and in return asked only that he stay in touch and continue to share his thoughts about the future of the technology. Ohga protested that he was determined to pursue a career as a soloist and had no intention of becoming a businessman, but Morita insisted that he accept the offer in the spirit it was being made, with no obligation to himself. Ohga promised to think it over and, in his excitement, left his diploma behind and had to retrieve it the following morning.

By any conventional standards of Japanese business protocol, Morita's offer was unthinkable. Nor was the additional cost of Ohga's "salary" a trivial matter at a time when the company was ramping up for transistors and funding was in short supply. But Morita in particular had detected in Ohga what he judged to be the gifts Sony would need to grow, and he was determined not to lose him. This would not be the last time he would act on impulse to lure people to Sony from other careers. But Ohga, who would prove to be his biggest catch, was the hardest to land: it was six years before he finally persuaded him to join the company formally in 1959. Once he did come to work, Ohga met and abundantly exceeded Morita's expectations. When he took over as president in 1982, the business stood at $15 billion. When he turned over daily control in 1995 to Nobuyuki Idei, revenue had grown to $45 billion.

In the year that followed, until he left for Europe in the summer of 1954, Ohga continued to study with Professor Nakayama as one of a handful of students who had been invited to remain in a makeshift "Masters" program, and embarked on what promised to be a successful career as an operatic baritone. He sang a number of roles in Wagnerian

---

* The term Morita used, *shokutaku keiyaku*, means a "part-time contract," but his intention seems better expressed in English by the notion of an "informal agreement."

and Italian operas—Wolfram in *Tannhäuser*, the Commendatore in *Don Giovanni*, Count Almaviva in *The Marriage of Figaro*—which were produced as radio broadcasts. He also appeared in recital, performing Schubert's *Winter Travels* and other German lieder. His accompanist on these occasions was the woman he would marry on his return to Japan from Europe in 1958, the pianist Midori Matsubara.

Meanwhile, he continued to participate in technical meetings at Totsuko, and even poached talent for the company. This practice was highly irregular for someone who was a student and not formally an employee, but Ohga's capacity for disregarding protocol when he perceived an advantage was always a match for Morita's aggressiveness. From Tokyo's finest hi-fi store, where he was a regular, Nishikawa Electric, Ohga shanghaied two engineers at once, Susumu Yoshida, the "general," and Masahiko Morizono, the "father" of Sony's commercial video camera business. In conversation one day, Ohga suggested to Yoshida that he was wasting his time in his job and offered to introduce him and Morizono to a company that was dedicated to technology and where a person could grow as fast as his talent took him. "He was very smart," Yoshida recalled, "and he was sensitive but without the darkness that sometimes goes along with being an artist. He was sunny and bright and sensitive, it was hard to imagine that he was just a student." Ohga arranged a meeting with Ibuka at a small restaurant near Shinagawa Station, at the bottom of Gotenyama Heights. Yoshida recalls with a smile Ibuka's assertion that the company employed three to four hundred people, and his subsequent surprise, on joining Sony, when he was assigned employee number 195. Morizono, who knew and respected Ibuka but says he had no idea who Ohga was, signed up at the same time.

During his senior year at university, Ohga had met the eminent German singer Gerhard Hüsch and had been invited to become his student in Berlin. In June 1954, he departed Yokohama on a Japanese freighter bound for Manila, Singapore, Bombay, and points west. At Manila, because anti-Japanese sentiment was still running high in the Philippines, passengers were advised to go ashore at their own risk; Ohga was one of the handful of travelers who took the opportunity to tour the city hurriedly. In Bombay, he received a letter from Morita informing him of the birth of Morita's second son, Masao.* From that point on, long letters scrawled in what Ohga describes as "Morita's hideous hand" were waiting for him at every port of call. That he found the time to write given the frantic pace of his business life is further evidence of his intense interest in Ohga.

Ohga spent his first year studying in Munich while he waited for permission to enter Berlin, still an occupied zone administered by the Allies. Once there, he enrolled as a regular student at Kunst Universität and spent three years earning a second degree, graduating in 1957 at the head of his class. The cost of his stay in Europe was paid for by his family, but Totsuko helped him with foreign exchange through the Mitsui Bank, and continued to deposit his monthly salary in a Mitsui Bank account in Tokyo. Ohga in return sent detailed reports of the latest developments in the German tape recorder industry, which, as he was at pains to point out, was far ahead of Japan's. He provided diagrams of the newest Telefunken machines, and urged Totsuko's engineers to work on faster revolutions and acetate-based tape of the kind made in Germany by BASF to improve the quality of playback.

Morita kept closely in touch with weekly letters, and sent Ohga one of the first model TR-55 radios to come off the production line. Ohga was excited by the new product, and promptly showed it to Michiko Tanaka, the resident doyenne of Japanese high society in Germany.†

---

* Morita wrote that he had chosen for "Masao" a Chinese character composed of two identical elements top and bottom, in hopes of raising "a child who will be entirely straightforward and honest with nothing to hide." Morita's two sons grew up referring to Ohga as "Uncle Ohga," and when Masao came to work for Sony, Ohga would say that he had known him from the time he was in his mother's womb.

† Michiko Tanaka was the daughter of a famous Japanese painter who was sent to Vienna to study art in the early thirties. Her charm and beauty quickly gained her access to Viennese society. At age nineteen, she married a wealthy businessman three times her age named Jurius Meinl, known across Europe as "the coffee king" for his chain of cafes, and became the reigning social queen of Vienna. In a twist worthy of an Italian opera, the aging Meinl, recognizing that he had grown too old to satisfy his still youthful wife, introduced her to a German superstar of screen and stage, Viktor de Kowa, and arranged their marriage, providing Michiko with a large dowry and a mansion in Berlin, where she was living in splendor when Ohga met her.

Ohga had met her while he was still in Munich, and she had helped him with his visa to Berlin and given him use of a chauffeured car while he was getting settled. Tanaka borrowed the TR-55 and demonstrated it to everyone in her broad acquaintance in Berlin, creating what may have been the first interest in Sony in Europe.

When Ohga returned to Japan late in 1957, he married Midori Matsubara and joined an opera company. The couple intended to spend their lives together as musicians. Morita persisted, urging Ohga to spend one day a week at the company, doing "nothing in particular." While in Europe, Ohga had become acquainted with the head of the Japan Musical Instrument Company, soon to become known as Yamaha, and Morita had heard rumors that he was also being actively courted by Yamaha's president. Ohga did resume his visits to Totsuko, but continued to resist pressure to go to work; he was aided in his resolve by his wife's determination that he should continue his career as a musician. In 1958, as a preliminary to building a Sony sales network in Europe, Morita asked Ohga to reconnoiter the European market and consult with him on strategy. At the end of a month of traveling, Ohga was joined by Morita in London and they returned to Japan together via New York. Morita assumed they would fly, but Ohga was set on crossing the Atlantic on the USS *United States:* "It was a seven-hundred-thousand-ton vessel, and could travel at thirty-five knots!" he explained to me. "That made it the fastest passenger ship in the world. I had to see for myself what that was like!"

Their Atlantic crossing threw Morita and Ohga together in close quarters for, by Ohga's measure, "four days and just over ten hours" with nothing to do but eat, sleep, and talk. According to Ohga, he did most of the talking, and his insistent theme was the necessity of bringing the company into the modern age with product planning and more stylish product design and marketing. By the time they reached New York, Morita was even more certain that Sony needed Ohga, and renewed his efforts to recruit him. Ohga held out for another six months, continuing to receive a salary, until Morita invited him to dinner with his wife, a schoolmate of Yoshiko Morita's and inclined to respect her, and proposed that he wear, in the Japanese idiom, "two pairs of sandals," continuing his singing career while working for Sony full-time. Ohga says that Morita assured him that night that if he came to work he could count on being president of Sony one day.

In April 1959, Ohga entered Sony formally at the top rung of the middle-management ladder, as *buchō*, general manager, a position that normally, then as now, would take an enterprising employee twenty years to achieve. As head of the Second Manufacturing Group, he was responsible for manufacturing and sales of Sony's broadcast equipment and three hundred employees. By night, after long days in Sony's manufacturing facility at Atsugi and traveling the country selling to broadcasters, Ohga pursued his career as a singer. But the moment came when he was obliged to recognize that his double life was no longer sustainable. He was singing the role of Count Almaviva in *The Marriage of Figaro* and had arrived at the theater in Hiroshima at the end of a day he had devoted to a futile attempt to sell Sony tape recorders to a local affiliate. With just enough time to get into costume and makeup, Ohga went on stage and sang his way without incident through Acts One, Two, and Three, the most challenging of all, in which the count is on stage from beginning to end. With the demanding portion of the performance behind him, he changed costume and sat down on a trunk backstage to wait for the count's final entrance at the end of Act Four. Exhausted, he dozed off, was awakened by the orchestra playing the cue for his entrance, and realized he was sitting on the wrong side of the stage. With no choice, he appeared on stage from an unexpected direction—"Here you are, my sweet Susanna"—and threw Figaro and Susanna off their timing (Susanna was being sung by Kyoko Ito, who would become a distinguished soprano). That night, he resolved to give up his musical career. Asked what had impelled him to choose Sony over music, he replies that it was his sense of responsibility to the people working for him. By that time, as head of the entire tape recorder division, Ohga had a staff of more than one thousand employees.

The speed of Ohga's ascent within Sony was unprecedented and remains unique. Within two years of starting, in addition to managing the tape recorder division, he had requested and received responsibility for product planning, industrial design, and advertising for the whole company. "I changed Sony's image," he told me. "When I came on the scene, it wasn't really such a modern company at all. And I'd been telling Morita for years that what we needed to do was create products that looked smart, stylish, international, and start advertising stylishly, and that's what I undertook to do, and it's amazing that Morita let me do it all, as young as I was."

Japanese executives do not meet visitors in their own offices, which are held to be private, but in "guest-reception rooms," which become increasingly spacious and ornate as the floors in the building ascend toward top management. I am interviewing Ohga in the "parlor" that adjoins his office on the seventh and top floor of Sony headquarters in Gotenyama and is reserved for his personal use, a stiffly formal room furnished with a long, low table and eight plush chairs positioned too far apart to allow comfortable conversation. Abruptly he stands, motions to me that I am to follow him, and strides out of the room without a glance at the head of Corporate Communications and the two freshmen in attendance to tape-record our conversation (with two tape recorders as insurance against the possibility of mechanical failure). I hurry after him, down the hall past the large Monet that he and Mickey Schulhof purchased in New York on the occasion of Sony's fiftieth anniversary in 1996, to the executive elevator. No word of explanation. Emerging on the ground floor, Ohga crosses the lobby to the entrance of the Sony products museum, heedless of the bows in his direction from people halted in their tracks in gestures of respect, moving as if through his private space, and enters the softly lit museum. Here, in long showcases that hug the contour of a gently arcing wall, Sony's entire product line is on display in chronological order. Ohga approaches the point on the arc corresponding to the year 1960, and says, taking my arm, "You can see the difference before and after I got involved." There is no arguing this point: the tape recorders, radios, and TV sets spotlighted behind the glass become noticeably sleeker and more modish upstream of where we stand. For thirty years, Ohga minutely managed what consumers around the world identify today as the Sony look: the heft and feel of the product, the matte black of the metal, the placement, color, and even the size of the Sony logo.

But Ohga's interest was never limited to style; he insisted that Sony tape recorders should have automatic level adjustment circuitry, which Sony pioneered in the Japanese market, and argued successfully for another first, the built-in microphone. Over the years, his review of Sony prototypes at the Product Planning Center, created by him and staffed by designers he has hired and trained, were always tense occasions. Morita was generous with compliments and encouragement; Ohga looked for faulty design. In February 1990, David Sanger of the *New York Times* witnessed Ohga in action at the center:

> He stops at the display of a new television set for introduction in 1991, clearly unhappy: all the peripherals connected to it—the stereo, the compact disc player, the videodisc, the CD-ROM player—require separate remote controls.
>
> Ohga launches into a lecture. The television, he tells the silent room, is "the center of the home." Just as Sony transformed the TV business once with Trinitron, it must do it again, making the television set a single, simple control center, so that you point the remote control only at the screen rather than at each individual machine.
>
> Briefly, Ohga is challenged by a tall young woman. The television is her project, and she raises some problems with Ohga's approach. He listens carefully, but gives no ground.
>
> "I want direct control of everything, all through the television," he tells her. She retreats.

During his early years as a full-time employee, Ohga also recreated single-handedly the company's domestic advertising. When he arrived in 1959, Sony ads featured a line drawing of a small boy named Attchan, a well-known comic strip character that Sony had licensed from the cartoonist for its exclusive use as, according to the borrowed English term in the contract, an *aikyatcha*, or "eye-catcher." Ohga was appalled, and in short order had persuaded Morita that a stick figure with a sappy smile was not an appropriate logo for a company with international aspirations. "Mr. Ibuka and Mr. Morita were great visionaries and entrepreneurs," he explained to me circumspectly, "but they didn't necessarily have much sense of style." By 1961, "the Sony boy" had disappeared and Ohga was personally writing the leads for all of Sony's domestic advertising. He also managed layout, choosing fonts and sizes, and insisted that print ads for newspapers be positioned where readers were most likely to see them, just beneath the TV program listings or on the op-ed page.

In the mid-sixties, Ohga began to manifest an aptitude for dealmaking and a perspicacity about the future which rivaled Morita's own. In 1965, he met with Wisse Dekker, then head of the Asian Division inside Philips Electronics N.V., to discuss Philips's newly developed tape cassettes, which would soon replace reel-to-reel tape. Philips was competing with Telefunken and Grundig in a race to establish its cassette tape as the worldwide standard, and it wanted support from Japanese electronics manufacturers. In Ohga's judgment, the Grundig cassette was slightly more sophisticated than the Philips version; nonetheless, he had already decided that Sony should align with Philips because he viewed the company as a more vibrant brand worldwide than Grundig, with a stronger organization and more to offer Sony in potential for future growth. Speaking in German, a language he commanded better than English, he informed Dekker that he was in conversation about adopting the Grundig cassette with Max Grundig personally. If Sony were to align with Grundig, in view of the market share the two companies controlled in Europe, the United States, and Japan, Philips would certainly lose the opportunity to control the worldwide standard. On the other hand, if Sony joined Philips, the Dutch company would be assured of worldwide control, and Sony was ready to sign if Philips would waive royalties altogether. Dekker agreed.

The following year, 1966, leveraging his connection with Dekker, Ohga concluded a free cross-licensing agreement between Sony and Philips. Now both companies were licensed to make use of the other's technologies or manufacturing processes free of charge. For Sony, this deal was a major triumph for two reasons. As it developed products, the company had been obliged to pay licensing fees on countless Philips patents in the past, and that obstacle had been removed in perpetuity. At the same time, Philips, a company vastly larger and wealthier than Sony, a household name around the world, was acknowledging the probability that Sony would be innovating on a level that would make the exchange worthwhile.

Morita was certain that the venerable Dutch company that had impressed him so vividly on his first trip to Europe in 1953 would not agree to view Sony as an equal and assured Ohga that he was wasting his time. When the agreement was concluded, he was both surprised and proud. According to Ohga, Philips in general was opposed and would never have come to the table at all if it hadn't been for Dekker, who lobbied executives throughout the organization, persuading them of his certainty that Sony would deliver the goods over time. In years to come, Ohga's relationship with Dekker, who was soon to become president and later chairman of Philips, would yield a variety of benefits to Sony. In the early eighties, Ohga and Dekker would lead the collaboration that produced and brought to market the compact disc.

In view of Ohga's success with Philips, it is not surprising that Morita entrusted him the following year with a negotiation that would prove to be one of the most important in Sony history: a joint venture with CBS Records. Over lunch with Harvey Schein in October 1967, Morita had decided to enter into a deal, and that same day had turned over responsibility for handling the negotiation to Ohga, thirty-seven at the time. Harvey Schein was known to be a cunning and aggressive deal-maker who left no money on the table, but Morita was confident that Ohga would hold his own.

The principals agreed at once that the new company would be capitalized at $2 million, each parent to invest $1 million. The first issue was a name for the venture. With his characteristic emphasis on building the Sony brand, Morita wanted "Sony-CBS Records," but Schein insisted that CBS and its Columbia record label were household words around the world and must come first. Ohga yielded, then exacted an "equivalent" concession from CBS when the discussion turned to the issue of royalties. Schein wanted to charge the venture an "all-in price" for each record master provided by CBS Records. "All-in" meant that CBS's markup on the actual cost of acquiring rights to the master and producing it would be folded invisibly into the bill it presented to the joint venture. Ohga refused. He had already proposed that the parent companies take 1 percent each of annual gross "off the top" (irrespective of profit or loss), a mechanism to help ensure that their original investment would be repaid, and Schein had eagerly agreed. Now Ohga pointed out that marking up the cost of record masters in addition to taking 1 percent from gross receipts would amount to "double dipping." The joint venture, he declared, would pay CBS its actual costs per master and not a penny more. As Schein faltered, Ohga pressed home his advantage by reminding him that Sony had compromised on the issue of the name, which was more important than profit to a small company building its presence around the world. Schein felt trapped and finally conceded a point that, in the ensuing twenty years, would cost CBS hundreds of millions of dollars. "Ohga suckered me," he recalled with a smile. "He gave up on the name because he knew he could more than make it back when we got around to royalties and profit."

CBS/Sony Records, Inc., was signed into being in Tokyo on March 1, 1968. The English documents were vetted by a young lawyer named Walter Yetnikoff whom Schein had recruited from the same law firm he had left for CBS.* Yetnikoff remembers driving through a pouring rain in a Sony car to deliver the document book for signing to Morita's house, where he was sick in bed with the flu, and waiting in the foyer while Yoshiko Morita shuttled back and forth to the bedroom with the documents. Yetnikoff would later become head of CBS Records and would play, as an intimate of Ohga's, a catalytic if abrasive role in Sony's acquisition of both the record company and Columbia Pictures.

On paper, Morita was president of the new company, but on the first day of business he informed Ohga that managing it would be up to him. Ohga claims that this announcement came as a complete surprise and reflects wryly that he might have reconsidered the 2 percent top-off payment to the parent companies had he known that the ensuing problems would be his own to manage.

---

* Rosenman, Colin, Freund, Lewis, and Cohen. In 1972, Morita's mentor in legal matters, Ed Rosiny, joined the firm as a partner and brought Sony's business with him. Located on Madison Avenue and 56th Street, opposite Sony headquarters in New York, the firm continues to play a role in the company's U.S. legal affairs: Paul Burak, another partner, sits on both the Sony America board and the compensation committee.

The business began badly. It was common practice in the trade for record companies to make the rounds of the retailers once a month to collect what they were owed in cash. Ohga pronounced this "feudal," and insisted that money be remitted to a CBS/Sony account. He also rejected the conventional returns policy and required retailers to purchase a percentage of the records they ordered. When the merchants protested and Ohga refused to yield, the Japan Record Dealers Association declared a boycott against CBS/Sony Records which lasted throughout what is remembered at Sony Music as the "long hot summer of 1968." That autumn, the impasse was broken when Simon and Garfunkel's "The Sounds of Silence" arrived in Japan on the Columbia label and swept the country off its feet: as a condition of distributing the hit single, Ohga required dealers to sign a new contract and they had no choice but to agree.

By the end of the second year of Ohga's leadership, the joint venture was returning dividends to both parents equal to 100 percent of their initial investment, and continued year after year to repay the initial investment in addition to royalties due on individual albums. On the fifth anniversary of the business, Ohga was able to use profits to build CBS/Sony its own headquarters building, in front of Ichigaya Station in Tokyo, without borrowing a penny. A celebration dinner for the entire staff was held at the Pacific Hotel, and Walter Yetnikoff, Harvey Schein, and Goddard Lieberson, president of CBS Records, flew from New York to attend. Ohga bid the crowd welcome and thanked them for their hard work. In the middle of his speech, his feelings overwhelmed him and he paused, fighting back tears.

Ohga's strategy was to stay just ahead of the constantly shifting tastes in popular music. Judging that classical music was, for the time being, out of vogue, he limited it to less than 10 percent of his investment. He also ignored *enka*, an entire genre of native ballads sung by its own galaxy of star performers and conveying a particularly Japanese brand of melodrama. Instead, he developed and marketed a series of artists including Momoe Yamaguchi and the Candies, who were created to mirror the fantasies of teenagers and adults in their early twenties. The sensibility he identified was naïve and odorless: astonishingly, given his parallel life as a classical musician, Ohga created what he describes as a "cutie-pie boom." CBS also fed the joint venture many profitable acts from its own Columbia and Epic labels, beginning with the Simon and Garfunkel hit.

The unlikeliness of Norio Ohga as pop music czar is breathtaking. "Beethoven," Ohga explained when I asked him about the paradox, "will always be Beethoven. And then there is business. I have a switch I can throw as I move back and forth between worlds." In the early seventies, the place to be seen in Tokyo after midnight was a disco in Akasaka called Byblos. Morita, not surprisingly, enjoyed dancing the night away in the private part of the club upstairs; Ohga hated the noise, and was observed on more than one occasion as he backed toward an exit and then bolted to freedom. When Morita called to let him know that Andy Williams was in town, he wanted to know if Mr. Williams was a professional golfer. He accompanied Morita and his son, Hideo, to a Williams concert and declared unsmilingly when it was over, "This man cannot sing!"

Though he may not have admired some of the music he packaged, Ohga was so successful at marketing and selling it in Japan that the record company under his leadership grew to become Sony's, and CBS's, most profitable division. The profits generated became a source of ongoing contention. CBS's share quickly climbed above $10 million annually: the succession of CBS presidents under the founder and chairman William Paley eyed the retained earnings hungrily and were continually demanding that larger portions of the pie be "repatriated" in dividends. Ohga politely and immovably rejected these demands (Sony had the contractual right to determine distributions). How would it look to outsiders, he would remonstrate, if the parent companies were to withdraw tens of millions of dollars above and beyond the 100 percent annual return on original investment, not to mention royalties! Year after year, CBS Records executives made the pilgrimage to Tokyo to petition Ohga and returned empty-handed. Finally Paley made the trip himself but fared no better than his subordinates. Sometimes Ohga would commiserate, complaining that his hands were tied because "Mr. A. Morita" refused to release the funds. Sometimes he would grow angry and throw an intimidating tantrum. "I was the retained earnings watchdog," he told me. "When CBS reached out its hand, I would lunge and growl."

In the course of managing CBS/Sony, Ohga became friendly with both Harvey Schein and Walter Yetnikoff. With Schein, he shared a sense of having created a successful company, and this feeling was a bond. After 1972, when Morita lured Schein to Sony from CBS, Ohga had less to do with him professionally, but the men remained close and continued to socialize with their wives, in Tokyo and New York. In 1976, Ohga tried unsuccessfully to mediate between Morita and Schein in the last days of their relationship, hoping to save Schein's position inside Sony. Then, on the twentieth anniversary of the founding of CBS/Sony, Ohga and Schein had their own falling out over what appears at first glance to have been a trivial matter but should probably be ascribed to deep cultural differences.

Sony had invited Schein and his wife to the anniversary party in Tokyo, and had sent them first-class air tickets. Schein wanted to exchange the tickets for economy-class seats around the world for himself and his wife and their two grown children, and asked Ohga to make the necessary arrangements through the Sony travel bureau. Ohga said he would try, but week after week the reissued tickets failed to arrive. Shortly before the date of the party, Ohga came for dinner to Schein's apartment in New York, and Schein confronted him about the tickets. Sony had sent him first-class tickets because it wished to honor him, Ohga explained, his face clouding with anger; it was disrespectful and unseemly for Schein to sell his tickets and ride in the back. Schein retorted heatedly that Ohga had no business interfering in his family travel plans and threw the tickets in his face. Ohga rushed out of the apartment. Schein's wife pursued him and brought him back, but the damage had been done: the men did not meet again.

Ohga's relationship with Walter Yetnikoff ran deeper and was more complex. If Morita and his first American mentor, Adolph Gross, were an unlikely partnership, Ohga and Yetnikoff are not easily conjured in the same room together. Ohga tends to be restrained and distant; Yetnikoff, though said to be much mellower than in his days as a bellicose drinker, remains contentious, incorrigibly and gleefully invasive.

Nevertheless, it seems that they were close friends for over twenty years. By way of explanation, Ohga told me repeatedly that Yetnikoff was a very serious, quiet, "even somewhat timid" corporate lawyer when they met: "He had a charming wife named June, and I thought they were an ideal couple. The only unfortunate thing was, she couldn't cook. It would be hard to find someone as bad at cooking as she was. You know the expression 'tone-deaf'? Well, she was cooking-deaf! Still, they were a wonderful couple."

In the early years of the joint venture, while he was still working as Schein's in-house counsel, Yetnikoff served as consigliere to Ohga, helping him to organize a mail-order record club in Japan and advising him on budgeting costs for developing new artists and producing their records. Notwithstanding Ohga's memory of Yetnikoff as gentle and timid, he was known to others even then as a man committed to outrageousness. Ohga himself referred to Yetnikoff affectionately as "the funny American." Yetnikoff called Ohga "Herr Ohga" or "The Ohga." Several times a year he traveled to Tokyo, and the men were often in each other's company. They took long walks and engaged in intimate conversations about the purpose of life. The Ohgas were childless, and Ohga confided in Yetnikoff his sense of resignation and loss. Over time, Ohga's wife, Midori, and Yetnikoff's first wife, June, became friendly, and the couples socialized at one another's homes. The Ohgas lived in Tokyo with a chow chow called Martin. Ohga would cook, preparing steak and roast beef and his speciality, crêpes suzettes, and serve French cheese and rare wines. When he was in America, he would visit Yetnikoff at his house in Wilton, Connecticut, and, later, Great Neck, swim in his pool, which he always complained was too cold, and fall asleep on the couch with his glasses on.

Yetnikoff recalls that Ohga could be "both condescending and kind." His personal secretary was having trouble with a boyfriend, and Ohga was very solicitous, inviting her home to dinner and in general looking after her. In matters of business, then as now, Ohga's temper was volcanic. Yetnikoff remembers an incident from the mid-seventies in the New York office of Arthur Taylor, then president of CBS. Taylor, who had taught Renaissance history at Brown, was a student of Japan who prided himself on understanding the culture. When he tried explaining to Ohga his perception of Sony's problem with distributions, Ohga began screaming at him, waving his arms—"You think you know Japan better than I. The joint venture is *over!*"—and stormed out of the office. Yetnikoff went after him and tried to calm him down, but that night, over lobster, one of his favorite foods, Ohga was still insisting that he intended to cancel the joint venture. According to Yetnikoff, he remained agitated until the conversation turned to investing CBS/Sony money in a plane for the purpose of importing lobster to Japan. Yetnikoff had been working on Ohga to compromise on the dividend issue by putting some of the retained earnings to work in profitable investments. For a brief period they were serious about lobsters, until they learned that shellfish could not be imported to Japan. Later they went shopping together for citrus groves in California, flying around in a chartered helicopter, and purchased for the joint venture two thousand acres of lemon trees north of Los Angeles near Oxnard.

In 1972, when Schein left CBS for Sony, Yetnikoff succeeded him as president of the International Records Division. From this point on he was Ohga's opposite, representing CBS's equity position in CBS/Sony. Twice a year he was in Tokyo trying to extract increased dividends from Ohga and failing, but the men remained fast friends. One day they were out on a walk and happened to pass a Porsche dealer. Yetnikoff stepped inside and wrote a check for a new car as a gift to Ohga from CBS. Business, after all, was booming under his leadership, and he had yet to receive a bonus. Ohga was pleased, but declined to accept the car until Morita had been given the opportunity to approve the bonus, which he did in a memo that afternoon. When the Porsche arrived, Ohga discovered it was an automatic shift and gave it to his wife: a race car driver, he was not about to drive a "hydromatic." Subsequently, Yetnikoff paid Ohga additional bonuses.

In the business arena, the men had other conflicts. The dividend issue was a constant irritation. Then there was Guenther Breest, a record producer in Hamburg championed by Ohga who, in Yetnikoff's view, was "ruining the company" with recordings of Broadway musicals with inferior casts. Another blowup occurred late in 1989, after Sony had purchased CBS Records, when Ohga approved a $20 million purchase of rights to record Herbert von Karajan's complete opera performances on video laser disc. Yetnikoff was furious, screaming into the telephone that he was not about to do "this lousy junk" that would not earn the company a penny. When Ohga insisted, Yetnikoff took his complaint directly to Morita, who brushed it aside. Today Ohga says that the project's passionate advocate was not himself but Sony's current president, Nobuyuki Idei, who was responsible for developing the videodisc. In any event, Yetnikoff's protest notwithstanding, the project was completed and lost money.

In 1975, Yetnikoff became president and chief executive of CBS Records, Inc., worldwide, in his own words, "führer of records." It was then, Ohga said, his voice heavy with judgment, that Yetnikoff began to transform into an entirely different human being: "He left his pretty wife and became a womanizer, and he was rude and arrogant and would scream into the telephone and then slam the receiver down. . . . watching Walter made me realize how suddenly and unexpectedly a man's life can take a sharp turn." Yetnikoff listens to what Ohga has said about him and smiles grimly: "I'd say he was the one who changed, certainly in his attitude toward me, for no reason I could see . . . I am very disappointed in Norio Ohga."

If Ibuka was the muse and champion of the transistor age, and Morita's principal marketing triumph the Walkman, Ohga rightfully lays claim to having taken Sony into the digital age with the compact disc. The technology at the heart of the CD as it was to evolve, the "general junction laser," was developed in the early sixties at MIT's Lincoln Labs and improved at Bell Laboratories later in the decade. By 1974, Philips had incorporated the laser in an early video laser disc, and Sony was developing its own version based on the Philips model. Independently, a team of audio engineers led by Heitaro Nakajima were at work on digital sound. By early in 1974, Nakajima had a machine capable of digital recording, but it was the size of a refrigerator and weighed several hundred pounds. Subsequently, his team devised a processor that permitted digital signals to be recorded on tape and played back on a U-Matic three-quarter-inch player. This system was still unwieldy and had its own technical problems, but when it was performing optimally, it delivered a clarity of sound that could not be retrieved from analog signals.

Sometime in 1975, the optical and audio teams came together at the Central Research Laboratory and went to work on recording digital audio information directly onto a laser disc. Ohga was already enamored of digital recording, which he had likened to "removing a heavy winter coat from the sound." When he learned that an "audio laser disc" was in development, he ordered it brought to completion as a top priority no matter the cost. To reduce the chance that Ibuka might wander in on his endless rounds, the work was housed in a building at a remove from headquarters on Gotenyama Heights. Ibuka had grown up in the analog age and was immovably convinced that analog sound was the genuine article. In this conviction he was not alone: with the exception of Nakajima and his small team, Senri Miyaoka and Toshitada Doi, Sony's audio engineers were united in their opposition to digital technology. Doi reflects that Nakajima championed and drove the project forward in the face of ongoing skepticism from Sony's analog engineers, who continued to predict that there was no future in digital sound. Later, listening to the system when it had been refined, Ibuka acknowledged its superiority, but he never focused on the technology.

In the spring of 1976, the team proudly presented Ohga with an audio laser disc thirty centimeters across, the size of an LP record, a musical "platter" with a capacity of thirteen hours and twenty minutes of digital sound. For their pains, they received a withering lecture on the folly of engineering for its own sake and the importance of developing business sense. A principal reason for Ohga's excitement about creating a new standard in the recording business was the opportunity it would create to combine new hardware with the rich "software" holdings of CBS/Sony Records. But at current rates for acquiring and recording music, in the neighborhood of $75,000 an hour, each and every audio disc with this capacity would cost the record company over $1 million to produce. Ohga still uses the story to ratify his assertion that engineers left to their own devices would run any business into the ground because business is never what excites them. "I love technology," he says, "and I love technical details, but inside Sony I was always first and foremost a businessman."

Meanwhile, in Eindhoven, Holland, Philips's audio division was at work on its own variation of the optical laser disc. In the spring of 1979, Philips's chief audio engineer, L. F. Ottens, traveled to Japan to demonstrate the device to Sony and other electronics manufacturers. Ohga was in the hospital. On his way to a Sony plant on March 16, his helicopter had crashed—"we had a lousy pilot," he told me in English—and he had broken his back. In July, when he was able to travel, he visited Philips's headquarters in Eindhoven and was shown a first-stage prototype of an audio laser disc that was 11.5 centimeters across and could hold one hour of digital information. One faction inside the Dutch company had been inclined to choose Matsushita as a partner, but Wisse Dekker, who was now president, and the head of Philips's audio group, Johan Van Tilberg, were Sony admirers and cast the decisive votes in Sony's favor. Ohga was impressed by what he heard and promised to secure Morita's approval for a collaboration.

From August 1979 until June 1980, when they presented their standards to the Digital Audio Disc Conference, Sony and Philips teams, physicists and audio engineers, alternated visits to one another's laboratories in Tokyo and Eindhoven. Ohga and Van Tilberg applied terrible pressure, asking originally that the project be completed by the end of October. According to the Sony team leader, Toshitada Doi, the collaboration accomplished three years of work in ten months (the teams continued to refine their work together until April 1981). As technical problems surfaced, each team addressed them separately. The solutions were tested and compared at the next meeting. The atmosphere was competitive—which side would end up with more patents in the finished product?—and there were heated arguments. Doi recalled an October morning in Eindhoven when the unusually clear and sunny sky abruptly darkened in the middle of a dispute and a storm began. A junior member of the Philips team suggested quietly that the thunder and lightning signified Ohga's and Van Tilberg's anger at time being wasted in arguments.

There was also by all accounts a goodly measure of bonhomie. The

Dutch team introduced the Japanese to the pleasures of cold, salted herring eaten whole from a cart on the street and gulped down with *jenever*, the potent Dutch gin. Doi went so far as to prevail on Philips to serve the herring, considered a street food, in its executive dining room. In Tokyo, the Philips team learned to enjoy sake hot and cold, and joined the Japanese for evenings of drink and song. On at least one occasion, Ohga himself put in an appearance and translated the Japanese lyrics into German for the Dutch engineers.

As the optical disc evolved, debate focused on two issues. One was the number of bits, the units of digital memory which ultimately determined the quality of the sound; Philips maintained that fourteen bits would be sufficient, but Doi argued that sixteen bits, though more costly and complex, would help distinguish their product from competing systems in development. Sony's audio engineers continued to drag their heels, insisting that fourteen bits, which enabled a dynamic range far superior to analog sound, would suffice, but Nakajima supported Doi's position and would not budge. The second issue related to size and capacity: Philips proposed an 11.5-centimeter disc, which would fit into an audio car system in the European market and would allow a recording capacity of sixty minutes. Ohga was adamantly opposed on grounds that a sixty-minute limit was "unmusical": at that length, he pointed out, a single disc could not accommodate all of Beethoven's Ninth Symphony and would require interrupting many of the major operas before the end of the First Act. But seventy-five minutes would accommodate most important pieces of music, at least to a place where it made musical sense to cut them. At sixteen bits, the disc would have to be twelve centimeters to accommodate seventy-five minutes. In the end, Philips agreed to Sony's specifications.

At the final session of the first round, held in Tokyo in March 1980, the two teams tested one another's error correction systems on discs that had been scratched, marked with fingerprints, even dusted with chalk. The Philips system proved inadequate to these extreme conditions, and Sony was judged the winner. There were protests from the Philips team that the test conditions were extreme, but their manager, J. P. Sinjou, agreed that the test had been fair, and the Sony error-correction mechanism was adopted. Later, Nakajima and Ohga decided to agree that the CD technology should be considered a fifty-fifty contribution between Sony and Philips. "It was a very Japanese conclusion," Doi says, "and some of us objected, but Mr. Nakajima was a man of great integrity, and we went along with him." The name chosen for the product was Compact Disc Digital Audio System.

As early as September 1977, twenty-nine manufacturers around the world had formed the Digital Audio Disc (DAD) Conference with an eye to achieving standardization. In June 1980, in Salzburg, DAD reviewed three systems, the optical system proposed by Sony and Philips, a mechanical system developed by Telefunken, and an electrostatic system that belonged to JVC, Japan Victor Corporation. In April 1981, DAD announced that it was endorsing both the Sony-Philips and the JVC systems. JVC, a subsidiary of Matsushita and a hated rival inside Sony, had developed the VHS system that had effectively driven Sony's Betamax out of the market by 1980. Fifteen years later, Sony would lose again, this time to Toshiba and the Toshiba camp, in the standards war that raged around the product now known as DVD, digital video disc. In the case of the compact disc, the superiority of the Sony-Philips approach—the laser read the information on the disc without physical contact with its surface—and Sony's privileged access to music through CBS/Sony Records generated a competitive advantage for Sony. Before long, most electronics manufacturers around the world had joined the Sony-Philips camp and were licensing their technology.

The CD lacked a player equal to the disc's potential, and creating an acceptable CD player at an affordable price proved to be a hellish job. Ohga had declared an immovable deadline of October 1982 for bringing the CD and player to market in Japan, and it seemed impossible to all concerned that components could be ready for mass production in time. The lens to focus the laser beam, for example, had to be of microscope quality. Today's lenses are stamped from a plastic mold; in 1981, three four-millimeter lenses were assembled in a copper mounting. It took a day to grind fifty lenses, and two hours for the assembly, which tended to crack when soldered, at a cost of $800 each. Sharp licensed its laser to Sony, but there was no guiding mechanism for moving the beam and the lens across the spinning disc in unison in order to read the two billion bits on a twelve-centimeter disk. Because the scale was so much smaller than in the videodisc player, none of the previous solutions to this scanner could be applied. Sony also faced the challenge of converting the analog signal to digital and back again. Finally, the player's circuitry required miniaturizing five hundred integrated circuits into three LSI circuits. Once again, for perhaps the final time in Sony history, audio engineers brought bedding to their labs so they could work around the clock. By mid-1981, Sony had a prototype player, which held the disc upright so it was visible spinning through the glass at the front of the machine. It was nicknamed *goronta*, a neologism that conveyed the notion of bulk and unwieldiness. Philips was developing its own CD player independently.

While Sony continued to work on refining the player, Ohga went on the road with his associates at Philips to prepare the record industry for the advent of CD technology. By this time, he felt certain that CDs would replace records sooner or later and was expecting resistance from the record industry. But he was not prepared for the unanimity and the violence of the opposition. The most dramatic moment occurred at the International Music Industry Conference sponsored by *Billboard* magazine in Athens in May 1981. Ohga, accompanied by his new right hand, Michael Schulhof, arrived with Sony's prototype player and a digital recording of performances by the Berlin Philharmonic conducted by Herbert von Karajan. Morita had met the maestro, Hitler's conductor of choice, in Vienna on his first trip to Europe in 1953, and the two had evolved a personal relationship over the years. Morita was frequently von Karajan's guest at his house in Salzburg, and when a concert brought Von Karajan to Japan, he invariably visited Morita (Yoshiko remembers her husband and the conductor swimming nude laps in the indoor pool in the basement of the Moritas' home). In April 1980, Morita announced he had something he wanted the maestro to hear and surprised him with a digital recording on three-quarter-inch videotape of a von Karajan rehearsal of *Die Walküre*. When the conductor pronounced the sound "superior to anything achievable with analog technology," Morita asked if he would allow Sony to make digital recordings of his performances for use in demonstrating the CD, and von Karajan agreed. Thereafter, he became an impassioned and ultimately effective advocate of Sony and Philips in the face of record industry opposition.

In Athens, however, record executives from around the world were not impressed. On the contrary, they were furious: the CD was the ugly brainchild of people in hardware who knew nothing about the software business, they asserted. The new format would require an enormous capital investment, and all it promised to do was threaten LP records, the heart of the business! Hans Timmer, later the Philips chairman, was also there, and recalls, "Ohga must have been shaken, but he didn't show it. He was calm, and he kept explaining that CDs would never scratch and that the sound was superior. But they shouted him down." Jerry Moss, then president of A & M records, screamed that CDs would kill the industry because the perfect digital master would invite and facilitate piracy. At some point, executives stood up in the auditorium and began to chant a slogan that sounded like a Madison Avenue nightmare, "The truth is in the groove! The truth is in the groove!" Even Timmer, a thickset bull of a man, was anxious: "I think we almost—we barely escaped physical violence," he told me.

Nostalgically, Ohga recalls joining the others that night, including Schulhof and his wife, Paola, for a "defeat banquet." Paola, who is Greek, chose a fish restaurant on the docks in the Piraeus and impressed Ohga indelibly when she led the party to the kitchen and inspected their fish for freshness by examining the flesh through the gills. "We sat down to console ourselves with some delicious seafood. We had certainly lost the day, but I wasn't despondent by any means. I sat there and resolved to go *my own way* [Ohga used the English phrase]. You might say I came back triumphant in defeat and resolved to go it alone."

The stakes for Sony, and for Ohga, were high. With the rout of Betamax now in full play, Sony earnings were dropping—1984 would be the worst year in its history to date—and the 8-millimeter video camcorder was still several years away from market. The company desperately needed a breakthrough product, and its fortunes would be tied to the success or failure of the CD.

Ohga had already committed the company to CDs with a $30 million mastering plant built in 1981 in Shizuoka Prefecture and financed entirely with profits from CBS/Sony Records. For fifteen years he had been retaining earnings despite CBS's protests on the principal grounds that it was unseemly to siphon excessive profits from the joint venture. Perhaps he truly felt that this was so, but he was also waiting for the right moment to use the money to fund a new business. Now the opportunity was at hand. In the space of three years, Ohga would use record company profits to build two additional CD plants, one in Salzburg and another in Terre Haute, Indiana. CBS was reluctant to approve the expenditures, but Ohga persuaded Yetnikoff's boss, CBS president Thomas Wyman, that he was being handed a brand-new business free of charge. Wyman signed off on the plants, but CBS, uneasy about becoming the first major record company to endorse the compact disc, asked Ohga to delay the launch for a year. Ohga was disgusted with what he called CBS's "stodgy conservatism" and, as chairman of CBS/Sony Records, decreed that the launch would proceed on schedule. Philips agreed but was a year late in bringing its own CD player to market.

On October 1, 1982, Sony introduced to the Japanese market its first CD player, the CDP 101. At the same time, CBS/Sony Records released the world's first fifty CD titles. There was some jazz and some pop, including Billy Joel's *52nd Street*, but the catalog was weighted toward masterworks from the classical canon, reflecting Ohga's judgment that classical music fans would better appreciate the benefits of digital recording. Ohga had hoped to release the CD in 1977, the centennial anniversary of Thomas Edison's invention of the phonograph, and to this day speaks regretfully of missing the target by five years. The importance he attaches to a historical connection to a great American entrepreneur reveals the postwar Japanese view of American achievement as the definitive yardstick of its own progress.

The launch generated excitement, but the first CD player was priced prohibitively at ¥168,000, roughly $700: once audiophiles willing to pay any price had been skimmed from the top of the market, sales began to lag. By November 1984, the second anniversary of the CD, Sony had developed a new player, the D-50, which was half the size and one-third the price of the original, and the market came roaring back to life. At the end of that year, LP records produced in

Japan outnumbered CDs by ten to one. By 1986, CDs had climbed to forty-five million titles annually, overtaking records to become the principal recording format. By 1988, production grew to one hundred million CDs, and by 1992 had tripled to three hundred million CD titles a year.

But opposition to the technology was deep-rooted and persistent, particularly among musicians, who formed a group called MAD—"Musicians Against Digital"—and the recording engineers who controlled the sound studios. Detractors maintained that digital perfection was artificial and empty. The singer Neil Young put it as well as anyone: "The mind has been tricked, but the heart is sad."*

Ohga handed responsibility for overcoming professional resistance to the man who had led the Sony team during the Philips collaboration, Dr. Toshitada Doi. A jazz saxophonist who idolizes Charlie Parker and a charmingly eccentric, original man, Doi was given his own business unit in 1980, staffed by thirty engineers from the Audio Division, and charged with developing and marketing a Sony line of digital recording and mastering equipment for studio use. At the time, a twenty-four-track analog recorder cost between $20,000 and $30,000; when Doi's team developed a digital recorder priced at $150,000, he went to Hollywood in search of a first customer and found Stevie Wonder, who was delighted with the quality of the machine and bought one for his Wonderland Studio.† Impressed that

---

* Dissatisfaction with digital audio is like an underground river that rises to the surface from time to time. In a May 7, 1998, article, the *New York Times* reported that audiophiles in swelling numbers are replacing their CDs with records because the sound is much closer to live music. Asked to comment, Ohga is dismissive: "Anyone who cannot hear that CDs are incomparably superior to records, 9.7 toward a perfect 10, has a tin ear and no business listening to music."

† An awkward moment occurred when Doi brought Wonder an improved model of the recorder-mixer, which Sony had ingeniously "coffee-proofed." Recording artists and engineers often work through the night and habitually consume large quantities of caffeine. Coffee spills are a commonplace and could shut down an expensive session if the equipment were not impervious. On the improved model, the mixing board had been covered with a thin sheet of plastic and LED indicators had been added to allow the mixer to see what he was doing, rendering the device unusable by Wonder. With Band-Aids, which allowed him to feel positions of the toggle switches, the problem was hastily corrected.

Wonder had gone digital, others tried the recorder and followed suit, including the jazz pianist Herbie Hancock. Both Wonder and Hancock joined von Karajan as active advocates of digital technology and Sony equipment, endorsing Sony products and even accompanying Doi to demonstrations and audio fairs. Before long, five or six of the top forty hits in *Billboard* had been recorded on Sony's nine-channel model PCM 3324 digital recorder, and by 1984, Doi's business unit had secured a 90 percent share of the commercial digital market.

As CD production around the world began to climb, the CD mastering plants that Ohga had built with joint-venture money became lucrative profit centers. In 1984, as Betamax was disappearing from the American market, the Terre Haute plant in Indiana generated Sony America's entire profit for that year. Ohga took pleasure in reporting the profitable numbers to CBS in person.

In 1986, shortly before he lost his job to Larry Tisch, then CBS president Thomas Wyman, the man who had reluctantly approved the plants, asked Morita and Ohga to visit him in New York on their way home from a Sony conference in Spain. In his office, Wyman explained that CBS profits were down and asked whether Sony would consider purchasing the plants from CBS/Sony and distributing the proceeds fifty-fifty to the parent companies. "We arranged for CBS/Sony to sell us those plants at cost," Ohga chuckles, relishing the memory, "and they've remained highly profitable down to this day."

By the time Ohga oversaw the CD launch in 1982, his power inside the organization was comprehensive and, effectively, supreme. Approval from Morita was still required before major decisions became final, but Ohga controlled budget allocations from product development through marketing, and was by now the principal architect of product planning and business strategy.

The steepness of his climb to power remains unique in Sony history. In June 1964, only five years after he entered the company, Morita had seated him on the board, making him at age thirty-five Sony's youngest director. Ohga relinquished his directorship temporarily when he moved to CBS/Sony Records in 1968 and began to manage the record business, first as senior vice president nominally reporting to

Morita, and from 1970, as president and representative director of the CBS/Sony group. For five years, he belonged to the record company and disappeared from view in the rest of Sony. In 1972, with CBS/Sony in orbit and generating robust profits, Morita called him back, promoting him to managing director (*jomu torishimariyaku*) of Sony and chairman of the record company. From this point on, he continued to oversee the record business, spending one morning a week in the new headquarters building in Ichigaya, and resumed control of budgeting, design, and market planning for the Sony Corporation. In 1974, he was promoted again, to senior managing director (*semmu torishimariyaku*), and was invited to join the company's supreme governing body, the Executive Committee.

In January 1976, as Sony was preparing to launch Betamax, management responsibility at the top of the company shifted: Ibuka ceded the office of chairman to Morita and became honorary chairman; and Morita appointed his brother-in-law, Kazuo Iwama, president and made Ohga deputy president and representative director. Having two classes of board membership—director and representative director—is one of a number of unique features of Japanese corporate governance. As a representative director, Ohga was empowered under the Japanese Commercial Code not only to vote on issues brought before the board, but to bind the corporation legally with his signature. In 1976, there were twenty-six directors on the Sony board, all company executives, but only eight representative directors. Ohga's promotion amounted to formal recognition that he had been admitted to the innermost circle of power in the organization. The fact that he had been designated Sony's first executive deputy president (*fuku shachō*),* is additional evidence that Morita intended to make good on his promise that he should one day lead the company.

The outside world saw this move as a major realignment, and the media speculated about its significance and cause. According to Ohga, the impetus was provided by Morita, who, though he had been president for only five years, was feeling bored and exasperated by the

* The literal meaning of *fuku shachō* is "vice president," but the actual rank is much higher than VP would suggest. NEC and the other companies use "senior executive vice president."

quotidian demands on his attention. By this time, Morita was already sitting on the boards of several American companies, in general augmenting the international activity that would claim his time increasingly as he became the country's most visible corporate statesman. Not that Morita intended to relinquish control of the big picture: beginning in 1981, he made himself Sony's first chief executive officer, becoming chairman, representative director, and CEO until he passed the new title to Ohga in his third year as president in 1985.

The choice of Kazuo Iwama for president was Morita's. He had admired, even idolized, Iwama for his scientific brilliance and discipline since the days of their childhood together as neighbors in Nagoya. Two years his senior, Iwama turned sixty in February 1976. If he was ever to be president of Sony, something Morita wanted for him, the time had come. Hiroko Onoyama, Morita's secretary in New York for twenty years, was visiting the Morita family in Hawaii for New Year's 1976 and remembers "Chairman Akio" telling her that two of the happiest days in his life were Iwama's marriage to his sister and the board's recent approval of his appointment to president.

If Morita was impatient with daily management, Iwama had no larger appetite for managing the details of what had now grown into a billion-dollar business. By 1976, he was preoccupied with the challenge of the "charged couple device" (CCD), which would enable the video camcorders that Sony pioneered in the eighties. The CCD, a tiny electric eye, was a semiconductor that could "see" color and convert it to electrical impulses. Along with so many other breakthrough technologies, it had come from the Bell Labs in the late sixties. In 1972, Iwama had organized a team inside Sony's Central Research Laboratory to develop a CCD for use in an electronic camera. The work was funded and carried out in secret: integrated circuitry was another of the new technologies that Ibuka had declined to understand. According to Dr. Makoto Kikuchi, who came to Sony in 1974 to head the lab, Iwama was increasingly alarmed that Ibuka's intransigence would impede the company's ability to stay on the leading edge. Even Morita, Iwama worried, was becoming so involved in international business that he was falling behind the technology. These concerns made the CCD project all the more appealing, just the kind of formidable challenge that Iwama felt Sony needed to keep its innovators on their toes: instructing his team to fix their sights on Kodak, the world leader in photographic imaging, he dedicated himself to the project with all his energy and passion.

The result, according to Ohga, was that beginning in January 1976, Sony had two chairmen and no president. He was given to understand, in the implicit way of the Japanese family, that he was to run the daily business. "I would discuss things with Mr. Iwama, but he would say, 'You handle it.' Or he would come to me, and say, 'I'll need ten or twenty or however many million dollars this quarter. Make sure you have that much on hand.' He wouldn't explain anything, and I'd of course do as he asked."

At the time, and to this day, decisions of major importance were made at regular Tuesday-morning meetings of the Executive Committee. Membership was limited to the chairman, the president, and several other executives chosen by the chairman from the ranks of senior managing director or executive deputy president. After January 1976, Ibuka attended when the spirit moved him. The committee was controlled by Morita, and regular members at the time included Iwama; Ohga; Susumu Yoshida, the "general" of Trinitron days; Masahiko Morizono, the engineer credited with developing Sony's broadcast products business; and the accountant dispatched by Morita's father in 1951 to oversee the family investment, now a senior managing director and unofficial CFO, Mitsuzo Narita, who was also recording secretary at the meetings. Later, Morita's brother Masaaki would also be invited to join. From time to time, as necessary, other executives were asked to present to the committee, but were normally asked to leave the room before decisions were reached.

The dynamics of decision-making inside the committee were, and remain, arcane. Ken Iwaki, rumored for a time to be Ohga's choice for the next Sony president, understood the process as well as anyone: "Ibuka-san rarely appeared, and when he did he only commented on money matters. The committee was in Morita's hands. When Chairman Akio said, 'Yes, that sounds like a good idea,' that was an end to it, it was decided. When he said, 'Not a good idea,' no one brought it up again. And Narita would write down the result. But when Morita would say, 'That's really very interesting' but not 'yes' explicitly, Narita would record in his book, 'Unresolved due to lack of approval from chairman.' Sometimes I'd negotiate with Narita after a meeting. Since the chairman had said, 'Hmm, that's interesting,' I'd urge him to change 'unresolved' to 'Chairman interested, suggests we investigate further.' Sometimes he'd agree."

"When Ohga came in as deputy president, he did a lot of the talking, mostly about product development. Our Betamax was being driven out of the market by the VHS camp, and we were looking hard for new products to come back with. Ohga was particularly strong in the new-product area, and Chairman Akio was just as passionate about products. Iwama-san was never much of a product man, so he'd mostly just listen, and others would chime in, but when Akio-san and Ohga-san were in agreement, and Akio-san said, 'That's really interesting,' Iwama would jump in and say, 'Sounds right to me,' and that was it. We spent about half our time with Ohga leading the discussion about new products. He was pushing CDs in those days, and then eight-millimeter video cameras."

At just the moment Iwaki describes, before the CD was ready for market, when the company was searching desperately for a new product that might reverse the losses occasioned by Betamax, Morita conjured the Walkman out of thin air. He had long argued that creative marketing was as important as technology and innovation. The Walkman is the prime example of his prodigious skill as a market pioneer.

The product originated in a personal request from Ibuka late in February 1979 for a highly portable player that would allow him to listen to stereo recordings on long international flights. Ohga, in charge of Sony audio at the time, conveyed the request to Kozo Ohsone, general manager of the Tape Recorder Division, reminding him that Ibuka was scheduled to travel to the United States in March. In just four days, Ohsone and a team of engineers modified a small, monoaural tape recorder called the Pressman, which Sony had designed for use by journalists, replacing the recording mechanism and speaker with a stereo amplifier and stereo circuitry. When the altered player was connected to headphones, the high-quality stereo sound it delivered was a surprise to everyone.

The player was brought to Ohga for approval when he was in the hospital with his broken back; he remembers phoning CBS/Sony Records from his hospital bed to arrange for a selection of classical music tapes to be presented to Ibuka before his trip. Ibuka was delighted with the sound. Returning to Tokyo, he showed the device to Morita, who took it home to try it over the weekend. Saturday morning, he carried it with him to the golf course, and that evening he passed it around the dinner table, laughing with pleasure at the look of surprise on his guests' faces at the unexpected richness of the stereo effect.

All his life, Sony's new products were Morita's favorite toys. By the time a prototype was ready for the production line, Ibuka's volatile imagination might have leaped beyond it to the next challenge, but Morita invariably laid claim to the first working model for himself, and bore it home excitedly to explore its features and demonstrate to his family and friends. When he was pleased, which was often, he was lavish with compliments to all concerned. When he discovered what he considered faulty design from the consumer's point of view, he took it personally and communicated his displeasure. Sony's first portable home-video camera was an example. As with the precursor to the Walkman, Morita had taken the prototype home on a Friday. The following Tuesday, at the weekly lunch attended by technical division heads, he asked for the name of the camera's designer and had him summoned to the dining room. Had the man tried using the camera, he inquired. Certainly, he had. But had he taken it skiing? No, because the product had not been publicly announced and could not be taken outside where it might be seen by the competition. Morita, who had taken the camera with him on a ski trip, now pointed out that the start switch was too small for someone wearing ski gloves.

This time, at the regular Tuesday-morning meeting of the Executive Committee, Morita appeared with Ibuka's special-order toy in hand and declared that Sony should bring it to market. Because the product would be aimed at young consumers, Morita felt strongly that it should go on sale before students went on summer vacation on June 21, in less than four months. Ohga remembers about the meeting only that Morita was emphatic, as certain as he had ever seen him. Based on Ken Iwaki's description of the dynamic in the room, it is likely that the others went along with "Chairman Akio" whatever doubts they may have had.

In fact, from the moment Ohsone's engineers began to design an approach to producing the player in mid-February until the Walkman was launched, Morita was alone with his enthusiasm. No one believed that a player without the capacity to record would catch on. According to Ohsone, even Ibuka had his doubts. Morita was immovable. He had watched teenagers on vacation in Japan and the United States lug their radios with them to the beach or into the mountains. How could they resist the opportunity to immerse themselves in their music while they played, exercised, or simply walked down the street?

Ohsone remembers the winter of 1979 as he rushed the Walkman into production as the most frantic period in his Sony career. There was no time for designing from scratch: the tape transport and stereo circuitry were lifted from other Sony players. Morita had asked for two additional modifications: twin outputs to allow two people to listen to the same tape, and a button to depress the volume and enable conversation with the headphones in place. These refinements had been suggested by Yoshiko Morita, who had not enjoyed her first experience of headphone isolation. The first iteration of the Walkman had his-and-hers headphone outputs and a fader button, which Sony advertised as the "hotline function."

Morita had also objected to the headphones that had come with Ibuka's set, which were heavy and as large as the player itself. As it happened, the Research Laboratory was already developing lightweight headphones designed for outdoor use (the H-AIR model), and these were added to the package.

The Walkman project was founded on Morita's certainty and determination; there was no conventional development process, and no market testing. From the outset, Morita insisted that the product must be affordable to teenagers. Asked to name a price, he declared ¥33,000 ($125 at the time), explaining with a smile that 1979 was the company's thirty-third anniversary. At this low figure—the Pressman sold for $400—design engineers cautioned that thirty thousand units would have to be produced to achieve economy of scale. At a time when the company's most popular tape recorder was selling only fifteen thousand units a month, a first run for the Walkman of thirty thousand units was beyond imagining. Nevertheless, Morita ordered thirty thousand units. When he learned that the sales force was expressing dismay, he announced to the entire Audio Division that he would resign as chairman if thirty thousand units failed to sell.

Morita orchestrated every aspect of the Walkman campaign, beginning with the press conference to introduce the product. Normally, product introductions were held indoors. When journalists arrived at the Sony building in the Ginza on June 22, they were escorted onto buses and taken to Yoyogi Park, where each was handed a Walkman and asked to push the play button. While reporters stood under the trees listening to a recorded pitch with background music in stereo, Sony staffers and models demonstrated how to enjoy the Walkman on roller skates, skateboards, or riding a tandem bicycle on a date.

Morita also had a hand in the early print ads, which were aimed at the young and active and emphasized speed and mobility: a girl with long blond hair bent low on her racing bike, a blond roller skater in summer shorts, a roller-skating couple, hand in hand. The Sony Walkman was positioned as a passkey to youth and sportiness. One poster placed a Walkman alongside three pairs of shoes. The banner read: "Why man learned to walk."

Less overtly, the early campaigns emphasized the lightness of the headphones and the stylishness of the earpieces compared to the bulky, earmuff look of conventional earphones. In the first poster for the Tokyo launch, a tall, leggy American girl in leotard and heels grooves to the music coursing through her Walkman headphones, her left arm thrown exuberantly upward. Just behind her stands an elderly Japanese Buddhist monk in summer kimono. On his shaved head, he wears clunky old-fashioned phones, and observes the young lady with a look of admiration and envy. Other ads featured two comely models in evening dress and headphones on the deck of a ship in New York Harbor, or a blond Cleopatra in a sheer bathing costume reclining at the edge of a swimming pool at dusk. In every case, the images reinforced the notion that the Walkman and its stylish headphones were a fashion statement.

The emphasis on fashion was partly Morita's response to general uneasiness inside Sony about a product that required headphones. "We were all skeptical," Ohsone recalls, "because at the time, in Japan, anything you put in your ears to hear with, including headphones, was associated with impaired hearing, and deafness was a taboo subject. Even the word had been disallowed for use in newspapers and magazines." Morita had anticipated this objection and was ready with an answer: Sony would create a new fashion, a "headphone culture."

The Walkman was launched in Japan on July 17, 1979, more than three weeks behind Morita's deadline, and for a stomach-churning month, nothing happened. Then, from mid-August on, dealers had trouble keeping their shelves stocked. The first thirty thousand units were gone by the middle of September, and for the rest of the year, production capacity had to be doubled and tripled every month. The first wave of buyers were music fans in their mid-twenties, but by the fall, the Walkman had become the fashionable new way for teenagers to listen to music, and Morita's prescience was confirmed.

The original marketing plan called for introducing the Walkman to the United States and Europe in September, but exporting was out of the question at a time when the production organization was scarcely able, and by December unable, to keep up with domestic demand. Although foreign sales did not begin until February 1980, advertising had been placed in the fall. Aware that "Walkman" was Japanese English, Morita had approved a suggestion from the advertising department that the product be sold under different names in different countries. The player was to be called "Sound-About" in the United States and "Stowaway" in Britain, where "Sound-About" was already registered. In Sweden, where the lawlessness conveyed by "Stowaway" was deemed objectionable, it was named "Freestyle."

In November, Morita telephoned Sony from Paris and told Ohsone that he had decided that the product should be sold as Walkman everywhere. All summer and fall, tourists and airline flight crews had been bringing the player home, and a Walkman had been given as a gift to the musicians in the Berlin and the New York Philharmonic orchestras. The result was that Morita had been besieged by friends in New York and London and Paris asking how they could lay hands on a Walkman for their children, and he realized that the name had appeal whether or not it was proper English. When Ohsone pointed out that advertising using other names had begun, Morita replied that his next call was to marketing at Park Ridge, New Jersey, and hung up.

Between 1979 and 1990, under Ohsone's supervision, Sony developed and launched eighty different models of the increasingly portable stereo player. In the process, Walkman became a household word around the world and, as Morita had predicted, established headphone music on the move as a singular feature of international youth culture. The phenomenon has always been closely associated with Morita personally. When he was knighted at the British embassy in Tokyo in October 1992, two English tabloids headlined the story, "Arise, Sir Sony Walkman."

Ohga by his own admission never paid any attention to the Walkman. Reflecting on his attitude as cumulative sales approached 250 million units at the end of 1998, he reaffirmed his admiration for Morita's market sense: "I could never be bothered because it had no technical interest. When they showed it to me in the hospital I was preoccupied with CDs and optical laser technology, which was much more difficult and more interesting. Frankly, I couldn't see why Sony should make a product that was boring technically. And that is the major difference between me and Mr. Morita. He had the merchant's intuition that allowed him to see what it would become. If it had been up to me, it would never have happened."

In August 1982, Kazuo Iwama died of colon cancer at the age of sixty-three. People close to him worried that the stress of the Betamax debacle had contributed to his untimely death. The grieving company organized a formal corporate funeral, the first in its history, at which Ibuka and Morita presided. Ohga took his place in the procession with the immediate family as they entered the hall bearing Iwama's ashes in an urn.

Ohga had been close to Iwama. As a demonstration of their affection and respect, they had purchased adjoining funeral plots for their families in a temple cemetery in Kamakura not long before Iwama had fallen ill, a practice not uncommon among Japanese colleagues and less macabre in their social tradition than it somehow feels in ours. Ohga, in deference to his superior, had chosen his plot just below Iwama's, on the slope of a hill. "I never dreamed," he told me, "that his ashes would be installed there so soon after we had arranged for the space." Iwama died just before the CCD chip he had labored for ten years to develop at a cost of $100 million was to be put into mass production. When the first chips came off the line several months after his death, Ohga took one to the cemetery in Kamakura and affixed it to his gravestone with rubber cement. "I said to him, 'Iwama-san! Here's the CCD you made for us. Look through it and see how beautiful the world appears!" In 1990, Ohga led a group of engineers to Kamakura to pay their respects and to report to Iwama that cumulative production of CCDs had reached ten million units. In the tradition of his mentors and champions, Sony's founders, Norio Ohga is a sentimental man.

In September 1982, one month after Iwama's death, Ohga was appointed president of the Sony Corporation, a position he had been guaranteed since even before he joined the company formally in 1959. His precipitate climb to power was of course enabled by Morita with Ibuka's approval; and the process of his investiture points toward a contradiction, possibly a paradox, at the heart of the organization. By the late seventies, Sony was a multibillion-dollar transnational company with a board of directors—albeit composed of insiders with fealty to the founders—and multiple levels of management. In principle, decisions were made by consensus, which reflected the Japanese emphasis on the subordination of individual needs to the good of the social group. But, in fact, as illustrated by Ohga's ineluctable progress toward an autocracy of his own, the animating power of the organization had always resided, and would continue to reside, with the founder-owners and those few others chosen by them for inclusion in the "family." Iwama was the first, and though he happened to be related to Morita by marriage, the literal family connection appears not to have guaranteed admittance to the inner circle: Morita's brother, Masaaki, for example, notwithstanding his distinguished career at Sony, was not granted access to the highest level of privilege and power. The next to be anointed, the adoptive son, was Ohga. There are those who distinguish the nature and quality of Ohga's power from the founders' own, seeing it as less elemental and sweeping, functional only in reference to, or as a derivation or echoing of, Morita's desires and intentions. Perhaps there is merit in this view; for although Ohga was known and feared as a tyrant in the heyday of his own power, it was then and remains unthinkable now that he would act in a manner contrary to Morita's wishes, any more than Morita was ever capable of betraying Ibuka. In any event, to fathom Sony's behavior as an organization, its acquisition, for example, of Columbia Pictures in 1989, it is necessary to look beneath the rational exterior that the company presents to the world and into the web of family connections at its emotional center.

7

EXTENDING THE FAMILY: THE RISE OF MICKEY SCHULHOF

Michael P. "Mickey" Schulhof was adopted into the Sony family by Morita and Ohga, much as Ohga had been adopted by the founders, and his career at Sony, like Ohga's own, was protected and privileged. Since the early sixties, when he founded Sony America, Morita had been searching for an American capable of assuming the burden of dealing with other Americans even as he grew the business in the United States. Harvey Schein had shown promise and had delivered, building a solid management infrastructure and introducing the company to budget and cost control, but Schein had not been able to modulate his abrasive American style, and eventually had worn down Morita's tolerance.

Gradually, over time, Schulhof filled the vacuum created by Harvey Schein's departure. Morita and Ohga immediately perceived him as the smartest person around, ambitious and entrepreneurial, and he turned out to be masterly precisely where Schein had been inept, at tuning in and accommodating the Japanese sensibility. Beyond that, there was a personal chemistry between Schulhof and, in particular, Norio Ohga, which forged a powerful bond between them. As a result of this connection, Schulhof was able to influence the company's behavior profoundly during the late eighties and early nineties. Acting as Ohga's representative, if not his alter ego, he was a principal figure in Sony's acquisition in 1987 of CBS Records and, two years later, Columbia TriStar Pictures. Subsequently, as CEO and chairman of all the American companies, he managed a $15-billion piece of Sony's global business with a degree of authority never accessed by any foreigner before or since at Sony or any other Japanese business organization. History is likely to reflect that Mickey Schulhof represents Sony's boldest, and possibly most vexed, experiment in acculturation.

Schulhof was born in New York City on November 30, 1942. His parents—his Czechoslovakian father was educated in Prague, his mother was a Berliner—had met and married in Brussels in 1939 and had fled the threat of Hitler that same year. In Belgium, Mickey's father had worked in his wife's family business as a publisher of greeting cards. In New York he started a similar business, Reproducta, Inc., which he grew into the largest manufacturer of Catholic greeting cards and Mass announcements in the country, and he became a wealthy man in the process.* In 1946, the Schulhofs moved to the largely Jewish community of Great Neck, Long Island—Mickey's mother had been informed that the best school systems in the state were in Scarsdale and Great Neck—and Mickey grew up there in affluence with his two younger brothers and sister. Schulhof's parents were widely read and had cultivated tastes in classical music and art: his father was a trustee of the Guggenheim and the Modern Museum of Jerusalem; his mother was for many years a member of the Acquisition Committee at the Whitney Museum. The Schulhofs were also patrons of Jasper Johns, among others, and had built a sizable collection of modern American painting and sculpture, which included Cy Twombly and Alexander Calder. Mickey's mother, Hannalore, was a violinist who played in the Great Neck Orchestra and who took her children to concerts and museums in New York City. Mickey developed a broad

* Rudolph Schulhof's business strategy had the virtue of entrepreneurial simplicity. He made the rounds of Catholic parishes and persuaded churches to give Reproducta their congregation mailing lists in return for the promise of an annual donation. Marketing greeting cards directly to parishioners was profitable, and the churches got an annual piece of the action. Schulhof dismisses the possibility that growing up in a Jewish family business that catered to Catholics had in any way prepared him to work for a Japanese company.

knowledge of music and art and a discriminating ear and eye, refinements that his Japanese champions later perceived and appreciated.

From Great Neck High School, Schulhof majored in physics at Grinnell College and learned to fly a plane, an accomplishment that would figure importantly in his relationship with Ohga the pilot and lifelong student of aviation. After Grinnell, Schulhof completed a master's degree in physics at Cornell and a doctorate in applied physics at Brandeis. His final year in graduate school, 1969, coordinating a research project between a physicist at Brandeis and Brookhaven National Lab, he commuted from Boston to Upton, Long Island, in a rented Cessna he piloted himself. The following year, he received a postdoctoral research fellowship from Brookhaven and spent a year in residence, publishing papers on "magnetic phase transitions" in new materials, and lecturing at the Max Planck Institute in Munich and elsewhere. It was during this academic year, 1970 to 1971, that he decided to leave science to look for a career in business. He characterizes the decision as painful but unavoidable given his growing sense that life as a research physicist would not be sufficiently adventurous to satisfy him. He may also have wanted a larger income than physics was likely to provide. In the summer of 1967, while vacationing with his father at the princely Hotel du Cap on the French Riviera, Schulhof had met the daughter of a wealthy Greek businessman who had been born in Thessaloníki and raised in Paris, and they were married in April 1969. Life together as graduate students was one thing, but Schulhof's salary at Brookhaven may have indicated that a career in science was unlikely to support his own, not to mention his young wife's, material expectations.

In December 1971, the younger Schulhofs were vacationing in Jamaica, at the Round Hill Resort in Montego Bay, when they encountered Clive Davis on the tennis court. Davis was the lawyer who had been hired by Harvey Schein and would soon replace him as head of the International Records Division at CBS Records. When Davis heard that Schulhof was considering a career in business, he invited him to CBS, an invitation that Schulhof dismissed as pleasantry and ignored, but which Davis followed with phone calls after the holidays. In April 1972, Schulhof went to work as assistant to the vice president of operations at CBS Records. In September of that year, Harvey Schein left his job at CBS to become president of Sony America. Two years later, when he was looking for a personal assistant, a friend in the record-jacket business, Paul Shore, recommended Schulhof. By this time, Schulhof had been asking himself whether a record company had any reason to appreciate a physicist—when Schein called, in April 1974, he jumped at the opportunity to move on.

At a time of frantic preparations for the Betamax launch in the United States, Schulhof's arrival at 9 West 57th Street might have gone unnoticed at Tokyo headquarters had not Schein boasted to Morita that he was hiring a Ph.D. in physics to work as his assistant. Applied physics happened to be Morita's field: he was sufficiently intrigued to introduce himself to Schulhof as a "fellow physicist and colleague" on his next trip to New York. Evidently, he was impressed: before he left the office, he suggested that Schein assign to Schulhof the job of organizing an audio speaker assembly plant that Schein had been promoting to Tokyo as a cost-saving measure. When Schulhof protested that he knew nothing about speakers, Morita summoned him to Japan to observe a Sony assembly operation. On his return, he set up and briefly managed a new speaker assembly facility south of Hazleton, Pennsylvania. Schulhof says he learned two important things from this experience: that business suited him, and that Akio Morita was an autocrat who could make anything he desired happen at Sony.

Between 1974 and 1980, building on a growing reputation as a quick study with a keen, entrepreneurial mind, Schulhof established a niche for himself as a corporate troubleshooter to be assigned by Schein or Morita directly to faltering Sony businesses in the United States. In 1975, he took over the foundering Business Machines Division, which sold dictating equipment and one of the earliest electronic calculators, SOBAX, a product that the company was unable to market successfully and eventually discontinued. Schulhof was given financial control and responsibility, his first actual experience in running a Sony business, and claims to have grown U.S. market share from 7 to 20 percent. Shortly after he had taken over the division, he recalls that Morita walked into his office and told him he was giving up his seat on IBM's International Trade Board because IBM was a competitor in business machines and he wanted to avoid any conflict of interest that might complicate Schulhof's job. The following year, 1976, when Schein and Edward Rosiny terminated Sony's distribution agreement with Suprascope, Schulhof was put in charge of Sony sound tape and hi-fi equipment, which was losing market share to Panasonic. Schein also sent him to Dothan, Alabama, where he had negotiated a sweetheart deal for land and buildings with Governor George Wallace, to organize a tape manufacturing plant. In 1977, he was asked to fix the company's service organization; Sony was acquiring a reputation in the United States for selling products at premium prices and providing inferior after-sales service. Schulhof was promoted and became, at thirty-five, the youngest foreign vice president in the company.

In January 1978, Sony America was rocked to its foundations by the death in Tokyo of its recently installed president, Ray Steiner. The previous summer, a possibly angry and certainly exasperated Morita had removed Harvey Schein from the front line and moved him upstairs and out of the way into the office of the chairman, a "promotion" that Schein had correctly interpreted as the beginning of the end of his Sony career. On Iwama's recommendation, Ray Steiner had been appointed president. A Midwesterner from South Bend, Indiana, with a degree from Notre Dame, Steiner was an unlikely candidate for the job, strikingly different from the other Americans at Sony. His protégé, Kenji Tamiya, speaking Japanese inlaid with English, paints a vivid picture: "I was six when the war was over, and I grew up watching movies from the United States, and to me, Steiner was *a good old American hero* right out of an American movie, big and tall, and cheerful, and honest." Steiner was solid, but there is nothing to suggest that he was a dazzler, and despite his reputation among the Japanese in the office for being easy to deal with, in dramatic contrast to Harvey Schein, he was continually frustrated by the difficulty of communication: "In my day," he told Kunitake Ando, who had been assigned to the New York office, and who tended to agree with him, "Americans and Japanese will never understand one another."

On January 18, 1978, Steiner arrived in Tokyo with Tamiya and Schulhof to attend a series of product meetings. He spent the following day at headquarters, and had an early dinner with Tamiya that night at the Pacific Hotel, where they were staying. By 8:00 the men had parted and returned to their rooms, intending to get a good night's sleep. That day, Norio Ohga had traveled north to Sendai. When he got back to his office at 8:30 in the evening, he found a new putter on his desk with a note from Steiner, and telephoned to thank him. "Ray picked up the phone in his room and he was gasping and said he had chest pains and thought he was having a heart attack. Then he hung up!" Ohga called Schulhof, and told him to go to Steiner's room with a bellhop in case he was unable to open the door. Schulhof rushed in to find Steiner unconscious on the floor. A doctor and an ambulance arrived to take him to a hospital. Ohga drove to the hospital, picking up Schulhof and Tamiya at the hotel on the way. When they arrived they were informed that Steiner had died and was downstairs in the morgue.

Ohga recalls a detail that seems like a macabre variation on the theme of cultural disparity. Steiner's body had to be embalmed before it could be returned to the United States for burial, but as most Japanese funerals are Buddhist and involve cremation, no one knew where to find an embalming facility in Tokyo. Morita, who was always quick to assume leadership at a moment of crisis, determined that the American Air Force base at Yokota had a facility that had been set up to process GI casualties during the Vietnam War, and he arranged for Steiner's body to be sent there.

Morita called an emergency meeting to discuss what to do next at Sony America. The meeting was scheduled for February 6, a Monday, in New York, but a blizzard struck the Northeast during the previous night and buried the city under snow. Morita's secretary at Son-Am, Hiroko Onoyama, phoned Morita in Tokyo at two in the morning New York time to alert him to the storm, and at the last possible moment the meeting was shifted to the Hotel New Otani in downtown Los Angeles. Morita arrived from Tokyo with Iwama; Kenji Tamiya, now the senior man at the New York office, flew in from New York with Morton Fink, vice president of corporate communications; Joe Lagore, recruited by Steiner from Sylvania in sales and marketing;

and Bob Dillon, the chief financial officer under Schein and Steiner, a man the Japanese called the "Boy Scout."*

By all accounts, including Ohga's, Morita was "invigorated" by a crisis. He embraced it with excitement, if not eagerness, as an opportunity for positive change, and became more decisive and commanding than ever. Now, certainly, a crisis was at hand, for though the choice of Steiner for president may have been a temporary measure until someone with more promise appeared, there was no one around to replace him. To make things much worse, Harvey Schein was known to be looking for another job (that same week of record snowfall, he was sitting down with Steve Ross to discuss his employment at Warner). In fact, the situation was more dire than Morita knew: less than two weeks later, on February 18, his longtime friend and mentor, Edward Rosiny, would also die.

Before the meeting was over that day, Morita had decided to divide Sony America into two divisions, Consumer Products, to be run by Tamiya, and Professional and Broadcast Products, to be headed by Koichi "Mike" Tsunoda. Tamiya was also made responsible for overseeing the entire operation as deputy president.

Schulhof was not invited to the Los Angeles meeting, perhaps because Morita considered him too new to the company, but it seems that Schulhof was on his mind: later that week, in New York, Morita asked him to serve as co-head with Tsunoda of the Professional and Broadcast Products Division. According to Schulhof, he asked for smaller but sole responsibility for an entire division, whereupon Morita, to accommodate him, created a third division to include tape, business machines, and other products that did not belong to the other two. As a name for the business he was now to run, Schulhof chose Sony Industries. Part of his new assignment was to create a components business. Until then, ever since he had declined the Bulova order in 1955 for a hundred thousand transistor radios because Bulova wanted its own brand on the product, Morita had been unwilling to sell com-

* Ohga's characterization of Robert Dillon is vivid: "He was the kind of guy who sent you a Christmas card every year with his family gathered for a color photo on the front, about a dozen children, all dressed up, and his grandchildren, too."

ponents based on proprietary Sony technology. Now, with the acceleration of standardized platforms, and with technology migrating across company lines, he conceded that it was time to do business as an original equipment manufacturer, or OEM, selling unbranded components to competitors for use inside their products. Schulhof hired a man named Kevin Finn from TRW and charged him with building a sales organization in the United States to sell Sony semiconductors and other components. When the business began to grow, it was passed to Sony headquarters in Japan.

There is a sharp division of opinion inside the organization about the quality of Schulhof's business performance. His detractors, and they are legion, suggest that Morita and Ohga's esteem for him was unfounded or, at best, exaggerated. Many go so far as to imply that Schulhof took advantage of their reliance on him to further personal interests at a cost to the company in the billions of dollars. The harshness of such a judgment is certainly colored by envy and by lingering resentment at Schulhof for having ignored the chain of command and reported to Morita and Ohga alone.

Other executives who knew him and worked with him closely, particularly during the seventies, retain the impression that Schulhof was a gifted strategist who executed his diverse missions impressively. Harvey Schein himself acknowledges as much, and so does Kenji Tamiya, who was Schulhof's boss beginning in 1978. If Schulhof had a manifest weakness, it was as a manager: the parts and service business he took over briefly failed to recover under his command. According to Kunitake Ando, who was sent to Sony America in 1976 and replaced Schulhof as head of service operations, Schulhof was a hands-off manager who disliked spending time at parts and service shops: "We had maybe five hundred fifty people in service across the country, and maybe one or two of them only had gone to junior college. All the rest were high school graduates. Someone like Mickey Schulhof could never communicate with those guys, and he just couldn't make that business work." It seems likely that Schulhof was, as Ando implies, an elitist.

On his frequent trips to Japan, Schulhof met and impressed Sony's important scientists and technicians with the reach of his own technical understanding. Ibuka himself pronounced him a fascinating young man and looked forward to talking shop with him, as did Senri Miyaoka, Makoto Kikuchi, and many others. Iwama, despite the fact that he was a fellow physicist, seems to have had his reservations. Miyaoka remembers being scolded by Iwama for discussing his work on Trinitron picture tubes with Schulhof. "I was stunned," he told me. "He was a Sony man, and a physicist at that. But Iwama said I wasn't to reveal any proprietary information because there was no telling when he might decide to leave the company!" What Iwama observed in Schulhof that made him doubt his loyalty may have been ambition. Personal ambition is a force in human nature perfectly familiar to the Japanese. But in a society that continues to resist the notion of individuality as a virtue, ambition is still considered shameful and is generally cloaked. Schulhof seems to have worn his ambition on his sleeve. As Ando recalled, "He was very clear that he would only speak to someone who was beneficial to him. When he shook hands with you, he might be looking at someone else in another direction—very ambitious."

Schulhof's relationship with Norio Ohga originated in their shared passion for airplanes and aviation. When Schein requested something special to commemorate the company's thirtieth anniversary in 1976, Morita decided to purchase a corporate jet for Sony America. Schulhof's first memory of Ohga is their meeting in the spring of that year, when Ohga walked into his small office in New York and asked him to help Sony find an airplane. The men shopped together, kids in a candy store, took test flights together, and spent hours afterward comparing notes. Eventually, they agreed that "the Porsche of corporate jets" was the Falcon 10, a nine-passenger airplane made in France.* Schulhof negotiated what he considered to be an excellent deal on Sony's behalf, $1.3 million including two licenses for Sony pilots, then asked Morita for permission, in view of the discount he had arranged, to purchase an additional license for himself. Morita agreed, and nine months later, when Schulhof was licensed, asked him to fly the plane to Japan so that

* Later, Sony would become the sales agent for Falcon in Japan, and for Aérospatiale, Morita's preferred helicopter manufacturer.

he could show it off to the Sony family. Schulhof remembers having made the crossing in three days, with stops to refuel at Edmonton, Anchorage, Adak Island in the Aleutians, and Kashiro in northeastern Hokkaido. In Ohga's recollection, Schulhof was at the controls as far as Adak but then turned back twice when he encountered headwinds on the final leg of the journey and ultimately left the Falcon to the Sony pilot and arrived in Tokyo on a commercial liner. What is clear in any event is that the purchase of the Falcon and its arrival in Tokyo constituted an auspicious moment, and that Schulhof was very much a part of the celebration. As it happened, the jet plane's N number, or identifying number, was N 30th; and November 30, Ohga reminded me with a broad grin as he told me the story, was Mickey Schulhof's birthday.

Throughout the eighties and into the nineties, the Sony trio—Morita, Ohga, and Schulhof—jumped at every opportunity to fly off in their Falcon 10 or, later, the larger Falcon 50 or 100, to escapades around the world. Jetting off became a thematic pastime, an emblem of their privilege and a basis of their bond. Their journey to Nome, Alaska, was typical. Since 1978, Morita had decided that the annual management conference should be convened in Anchorage because it was an equal distance across time zones from Japan and Europe. The day before the annual meeting, Ohga proposed that they fly themselves to Nome, which he had been longing to see since he was ten years old in 1940, when Japan's first solo flight across the North Pacific had landed there. This was just the sort of historical marginalia that Ohga and Schulhof delighted in and stored abundantly in memory, and Schulhof was immediately enthusiastic. Morita's curiosity was unbounded and ignited at the hint of any adventure. Landing in Nome, Ohga paused to reflect that this was the landing strip where the Japanese monoplane had touched down. In town, on the only street, all three were surprised to find that everyone appeared to be Japanese—Morita orchestrated photographs with the Eskimo residents. From Nome they flew to Point Barrow because they knew it was the farthest point north on the American continent, and they received a second surprise: there was no town at all, only an airstrip with a few houses nearby. Ohga recalls experiencing with awe the vastness of America. They had also planned to fly to an offshore oil deck, but time was running out and they returned to Anchorage to rejoin the meeting.

In 1979, Schulhof's younger brother died in an airplane crash coming in for a landing in Westchester. It was a freak accident, a midair collision. Ohga gave me a pilot's demonstration of how it happened, his hands converging, palms down, at just the angle to produce a blind spot between two aircraft on the same approach. The loss of his brother left his father and youngest brother struggling to manage the greeting card business. The following summer, at an executive retreat in Hawaii, Schulhof drew Morita aside and confided that his family was in trouble and that he felt obliged to resign from Sony to help them. Judging from Schulhof's account, Morita's response is evidence of his high regard for Schulhof and recalls the offer he made Ohga just before the latter left to pursue his studies in Europe: "'Don't do that,' he said. 'I'm an eldest son, you're an eldest son—I understand your responsibilities. Go and help your family but don't leave Sony!' I said that was very generous of him but I didn't know how to do that. And he said, 'It's simple. You keep your Sony office and your Sony secretary, and we run a phone line from our switchboard directly to your business. The only request I make of you is that if I want to reach you or if someone wants to reach you on Sony business, they won't have to know where you are physically.'"

Schulhof agreed to give it a try: for six years, until he returned to the company formally at the time of the CBS Records acquisition in 1987, he worked at Reproducta and doubled as a Sony agent without portfolio. He invented a title for himself: chairman of strategic planning. At Morita's urging, he kept his seat on the board of Sony America and maintained his office at 9 West, next door to Tamiya, who became president in 1981. Tamiya remembers him dropping in at the end of the day to inquire about Sony business. In his first year away, Schulhof estimates that he spent 95 percent of his time on greeting cards and only 5 percent on Sony. Gradually, over the next three years, the ratio inverted.

It was at this point that Ohga began relying on Schulhof as his right hand in the United States. In 1980, he brought him to Japan for a crash course on CD error-correction algorithms from Toshitada Doi, then dispatched him to Eindhoven to negotiate format standards with Ottens at Philips. That year and the next, Ohga and Schulhof lobbied the record industry on behalf of the CD, flying together in the Falcon around the world. It was Schulhof who urged Ohga to attend the Athens conference in May 1981, and Schulhof who made the presentation from the stage and took the heat directly.

By early 1982, Ohga judged that the time had arrived to build a CD plant in the United States, and entrusted Schulhof with making it happen. The first step was securing approval from CBS. Walter Yetnikoff, who was now the president of CBS Records, had resisted accepting the CD as a new media standard, despite his long-standing friendship with Ohga, and was opposed to spending $35 million of joint-venture money on a new plant. It was not until Schulhof proposed purchasing a building that CBS owned in Terre Haute, Indiana, and converting it, that the president of CBS, Tom Wyman, took the carrot and approved the investment, overriding Yetnikoff's reluctance.

Schulhof oversaw every phase of the Terre Haute project, from architect's plans to personnel. Initially, the factory was designed to produce a hundred thousand discs per month, but demand for CDs in the United States climbed steeply and volume had to increase continually. On a regular basis, using Sony's jet, Schulhof would pilot himself to Terre Haute and instruct the plant manager, whom he had hired, to increase capacity and output in accordance with his own gauge of the growing market. Throughout, he was acting as an independent agent who reported to Ohga alone and stood, unthinkably in the context of a Japanese organization, outside the consensus. Even at a company as proudly renegade as Sony, this degree of independence represented an assault on the normative architecture of business society, and the community of Sony executives found it difficult to endure. Schulhof himself in retrospect acknowledges the discord he created and deems it reasonable: "No one was sent to Japan and there were no formal presentations. I was acting as an entrepreneur and I guess that's how I was able to keep the factory ahead of the market and maintain Sony's position as market leader in CDs. On the other hand, there is the fair criticism coming out of Japan, 'There's Schulhof running off again, making decisions on his own, clearing it only with Ohga and not taking the rest of us in the organization through the process.'"

Clearly, the source of Mickey Schulhof's power was a decision made by Morita and Ohga, and a desire, to empower him. The question remains: What was it about Schulhof that attracted them? There is no explaining away the mystery of human relationship. Nonetheless, there are congruities that account for something. Schulhof, like Morita, was the eldest son in a traditional family and, furthermore, a physicist; Schulhof had lost a brother and so had Ohga, in a boating accident. Beyond these coincidences, Schulhof's capacity, his style, his assumptions about life, powerfully impressed both Morita and Ohga. To this day, four years after he was finally persuaded by his inner circle of advisers to fire Schulhof, Ohga maintains that he was by far the brightest and most able American executive who ever worked for Sony. When asked to explain, he invariably talks about quickness. Time and again, using virtually the same language, he has explained to me: "If you told Mickey one thing, he'd come back with ten more. With anyone else, you'd have to explain everything, but he'd get it all at once and then go out and explore all the angles and come back with options." Clearly, Ohga also admired the breadth of Schulhof's knowledge and his command of detail: "Whatever you discussed with him, he could answer you like a specialist, a professional. When you try to discuss music with most people, or airplanes with most people, the conversation doesn't go very far. With Mickey, if we spoke of airplanes he was on the level of an airline pilot, and since he was a real learner, if I asked him to look into something he'd come back in a flash with the most detailed report you can imagine. No matter what the subject he'd be there and master it." Finally, and very important, Mickey Schulhof, like Ohga, was a cultivated man. Ohga had never encountered an American businessman with whom he could share his own passion for classical music and art, and the pleasure he took in this aspect of their association is evident in his reminiscences. It was Schulhof who arranged for Ohga to conduct the Metropolitan Opera Orchestra in 1993, his debut concert in the United States, and in 1995 the men went shopping together for a work of art to commemorate the company's fiftieth anniversary. Through family contacts, Schulhof located a large

Monet, two by two meters, which had been placed quietly on the market by a private collector in New York, and negotiated a bargain price for the painting, a well-known work from the artist's late period at Giverny that the collector had acquired from Monet's son Philippe in 1946. The Monet now hangs outside Ohga's office on the seventh floor of the Sony headquarters building in Tokyo. To avoid requests for a loan from museums, it was agreed that not only the purchase price but the identity of its new owners would not be disclosed: according to Ohga, the painting is considered to have "disappeared into thin air" and is being hunted by curators and Monet scholars around the world.

Ohga tells the story of this secret purchase of a Monet masterpiece with relish. In the telling he conveys his feeling that this variety of epicurean experience became possible in the company of Mickey Schulhof—was, in a sense, definitive of their relationship, and could be enjoyed with Schulhof alone. In essence, it seems likely that their bond had ultimately to do with the fact that Ohga saw in the younger man a reflection of himself, of his own image of himself, agile of mind, broadly knowledgeable, cultivated, and that this recognition led him to treat him as, in his own words, "a younger brother." He once told me, "What made Mickey so special was his inherent savvy and his good breeding. . . . He was arrogant, but considering how smart he was, his arrogance was understandable." It was hard to avoid the impression that Ohga was also talking about himself.

Schulhof's ascent toward the zenith of his power began in 1986, when, together with Walter Yetnikoff, he managed the acquisition of CBS Records. The $2 billion purchase is a vivid and dramatic illustration of the importance of personal relationships in Sony's process of decision-making: had Schulhof and Yetnikoff not been involved, each with his own strong claim on Ohga's attention, reliance, and loyalty, it is likely that Sony's acquisition of the world's largest record company might not have happened.

By 1986, Yetnikoff, securely in command of CBS Records as president and CEO, with solid personal connections to most of the division's superstars, including Michael Jackson, Bruce Springsteen, and Barbra Streisand, had often reflected that a sale of the business to anyone at all was likely to augment his own power and wealth. By the summer of 1986, a controlling interest in CBS, Inc., had been acquired by Lawrence Tisch, a no-nonsense businessman who had grown his family's summer camp business in New Jersey into a conglomeration of parking lots, movie theaters, and hotels called Loews Corporation, becoming a billionaire in the process. Yetnikoff was well aware of Tisch's reputation as a man who would sell anything for the right price, and had engaged him in "theoretical" discussions about spinning off the record business from the moment he took over as president of the parent company. There was no love lost between these men: Tisch considered himself a serious Jew and was not amused by Yetnikoff's profanity and irreverence. Yetnikoff thought Tisch a sanctimonious hypocrite, referred to him as the "evil dwarf" and even the "kike upstairs," and relished opportunities to demonstrate that his knowledge of Judaism was superior to Tisch's, who boasted that he studied with a rabbi. "I may not be that religious," Yetnikoff thundered, "but nobody out-Jews me!"

Despite their mutual hostility, Yetnikoff believed that Tisch would not move in the direction of a sale without consulting him, and was flabbergasted to learn from a Wall Street acquaintance in October 1986 that Tisch, less than a month after he had taken over, was considering an offer from the food-business tycoon Nelson Peltz, who knew nothing about broadcasting or music. When Yetnikoff barged into Tisch's office spewing abuse, Tisch informed him that he was welcome to purchase CBS Records himself if he could produce $1.25 billion in cash, the price Tisch had quoted Peltz, by the end of the year. Yetnikoff began working the phone, trying to raise money for a buyout, and quickly found that he had no hope of making Tisch's price without including CBS-Sony Records, a multihundred-million-dollar business, in the deal. This meant that Sony had to be involved because Ohga had included in his original negotiation with Harvey Schein a block on selling the joint venture without approval by both parties. A few days before the November board meeting, Yetnikoff placed a call to Schulhof at Sony headquarters in New York. While Yetnikoff and Schulhof had shared seats on the Sony America board for a number of years, it is surprising, given Walter's long-standing friendship with Ohga, that he chose to approach Schulhof first. Yetnikoff claims that his choice was merely a matter of proximity, that Schulhof was, after all, just down the street: possibly he was aware of Schulhof's influence on Ohga and wanted him involved in the pitch from the outset.

Schulhof was just flying the Falcon back from a Sony mission in California and took Yetnikoff's call at the Teterboro Airport in New Jersey, where the plane was hangared. He told Yetnikoff that he would call him back and phoned Ohga at his home in Tokyo, where it was early morning (phone conversations between Schulhof in New York and Ohga in Tokyo at dawn would play a critical role in the acquisition process of both CBS and Columbia). Ohga had been longing for an opportunity to lose CBS as a partner in the record business. For eighteen years, as he grew CBS/Sony Records into the largest and most profitable subsidiary in the conglomerate, he had been fighting with CBS about retained earnings and dividends. In that space of time, CBS chairman William Paley had hired and fired eight presidents—someone waggishly described them as "Paley's disposable lighters"—and with each new regime the dispute had reheated. Now, speaking to Schulhof from his bedside in Tokyo, Ohga said he was strongly in favor of making an offer and instructed him to phone Morita for his approval. Morita was willing to go along, and told Schulhof over the phone to find an investment banker and to arrange a meeting with Tisch. Within twenty minutes of receiving Yetnikoff's call, Schulhof called him back to say that Sony was seriously interested in acquiring CBS Records. He then phoned Paul Burak, the lawyer from the Rosenman firm who had taken Ed Rosiny's place as American legal adviser to Morita, and Stephen Schwarzman, Peter Peterson's partner in the Blackstone Group and a personal friend, and, without disclosing what he wanted to discuss, called a meeting at the Mayfair Regis for 7:30 the following morning.

Over breakfast, Schulhof described the situation and what he considered his verbal authorization to proceed. From the hotel, the men went directly to Yetnikoff's apartment on 56th Street. According to Schulhof, "[They] woke Walter up at 10:00 A.M." Yetnikoff remembers that his wife at the time, Cynthia Slamar, shook him awake at 8:30 to convey the doorman's report that "three guys from Sony" were waiting downstairs. By his own account, with his drinking problem spinning out of control, he was accustomed to crawling out of bed each day with a pounding hangover, and, this exciting morning being no exception, he asked his wife to prepare strong coffee. At their meeting, which Yetnikoff attended in his bathrobe, Schulhof asked if he could deliver CBS Records without serious defections by management and, more important, recording artists, and Yetnikoff assured him he could and would if the deal was right. He then asked for a fee of $50 million for himself and "the *mishpocheh*," the family of managers and producers all over the world who reported to him. Schulhof agreed at once—an early foreshadowing of the degree of power and latitude that he would receive increasingly from Ohga. Yetnikoff told me that he had bungled the moment: "I knew when Mickey said yes without hesitating that I could have asked for much more, maybe twice as much, and gotten it." Yetnikoff proceeded to reserve as a personal fee from the $50 million an amount he declines to mention, but when I suggested $10 million, he shook his head, smiled, and said, "More, much more!" According to an unconfirmed rumor, he asked for and secured from Schulhof a promise of a $10 million bonus in advance and an additional $10 million in equity and other compensation.

From Yetnikoff's apartment, Schulhof phoned Larry Tisch to confirm the asking price of $1.25 billion and to place Sony's bid on record. Tisch had bad news: he had been polling the directors of the company and was finding resistance to the notion of selling. In his view, it was not unlikely that the board would reject the proposal at the meeting the following day.

Tisch's fear was confirmed; Paley had argued forcefully against losing the record business and had persuaded the board to decline the offer. Yetnikoff was devastated at the loss of power and wealth that he had glimpsed. Sony was also disappointed: in an interview in the *New York Times Magazine*, Morita wanted to know how the board of a company could say no to a deal that had been approved by its president. What he did not say, but was certainly the source of his bafflement, was that no Japanese director would ever oppose a decision made by the head of the company. More to the point, no company president in Japan, and certainly not at Sony, would have made a decision of such importance without prior agreement from the chairman.

Masaru Ibuka (left) and Akio Morita (center) inspect an early "production line" at Sony's tape-recorder plant in Gotenyama Heights in 1960. The makeshift operation in the postwar years is a far cry from the company's streamlined sleekness today. *Courtesy Sony Corp.*

Akio Morita and his family at Haneda Airport on their way to live in New York in July 1963. Relocating his family to a foreign city was a bold and disruptive move, but Morita was determined to establish a foothold for Sony in the U.S. market. *Courtesy Yoshiko Morita*

A mock battle between Akio Morita (left) and Masaru Ibuka, Sony's cofounders, circa 1970. No one ever witnessed them in true conflict. *Courtesy Sony Corp.*

Harvey Schein and Kazuo Iwama on the forty-third floor of 9 West 57th Street, the new headquarters Schein chose for Sony America in 1973. The bright smiles belie the antagonism that characterized Schein's relationship with Morita's brother-in-law. *Courtesy Harvey Schein*

Akio Morita with his mentor and consigliere in New York, Edward Rosiny, in October 1975. When Rosiny died in 1978, Morita was inconsolable. *Courtesy Sony Corp.*

Norio Ohga remembers Walter Yetnikoff in the early years of their relationship as a "shy, sincere young man with perfect manners." Later, after Yetnikoff had become head of CBS Records in 1975, Ohga would wonder "how anyone could change so much." *Courtesy Norio Ohga*

Henry and Nancy Kissinger join the Morita family in October 1979 to celebrate Yoshiko Morita's birthday at Maxim's de Paris, the restaurant that Morita brought to Tokyo and used as a private club (Sony insiders referred to it as "the executive dining room"). Morita stands behind the Kissingers and Yoshiko with his two sons, Hideo (right) and Masao, and his daughter, Naoko. *Courtesy Yoshiko Morita*

Yoshiko and Akio Morita on the pirate ship that was the set for *Hook*, January 1991. Left to right: TriStar captain Mike Medavoy, chairman of Columbia Pictures Peter Guber, and director Steven Spielberg. Guber arranged an opportunity for Morita to hobnob with Hollywood royalty whenever he was in town. *Courtesy Peter Guber*

From left: Michael Schulhof and Norio Ohga with Seiji Ozawa, music director of the Boston Symphony Orchestra, in 1992. Ozawa's wife, Vera, and their son, Yukiyoshi, look on. Ozawa has been a close friend of the Sony family since 1978 when Herbert von Karajan urged him to ask Morita for a demonstration of digital recording. *Courtesy Michael Schulhof*

Akio and Yoshiko Morita with Bill and Hillary Clinton in 1993 following a cello recital at the White House by Mstislav Rostropovich. Yoshiko remembers that Clinton approached Morita after the concert to say hello — they had met once before at the American embassy in Tokyo — and that Hillary insisted on a photo. It arrived from the White House when Morita was in intensive care after the stroke that ended his career. *Courtesy Yoshiko Morita*

Norio Ohga (left) and Michael Schulhof with Cyndi Lauper in 1995 after her command performance for twelve hundred Sony dealers from around the world at the company's annual blowout in Hawaii. In 1988, following the CBS Records acquisition, Lauper had posed for a *Time* magazine cover photo with an exuberant Akio Morita and was considered a charter member of the Sony Music family. *Courtesy Michael Schulhof*

Ohga conducts the Tokyo Metropolitan Symphony Orchestra in an all-Dvořák program at Suntory Hall in January 1995. When he finally succumbed in 1959 to Morita's campaign to lure him into the company, Ohga promised himself that he would return to music when he reached the age of sixty, and he has kept his promise. *Courtesy Sony Corp.*

In the presence of Norio Ohga, his champion and mentor, Nobuyuki Idei (left) appears self-conscious, at once deferential and impatient. Ohga, unconcerned, strikes his usual pose. April 1997. *Courtesy Sony Corp.*

When Bill Gates accepted Nobuyuki Idei's invitation to play a round of golf in June 1998, Idei sent a Sony helicopter for the twenty-minute trip to the Glen Oaks Country Club near Tokyo. Though he does not say so, parity with America's digerati is important to Idei. *Courtesy Sony Corp.*

Akio Morita with his eldest son, Hideo, at his oceanside estate near Honolulu in the summer of 1996. Hideo's relationship with his father was stormy, but since taking over management of the family's affairs, he flies from Tokyo to spend weekends with Morita in Hawaii several times a month. *Courtesy Yoshiko Morita*

Larry Tisch had not abandoned the idea of divesting CBS Records, and he began to explore opportunities to sell it off in a public offering. At the same time, he left the door open for a private acquisition. When Disney extended a feeler in his direction, Tisch had visions of a bidding war and contacted Schulhof to ask if Sony were still interested. The price, he was careful to mention, to compensate for the recent increase in capital gains tax, now stood at $2 billion. Schulhof went back to Ohga and Morita and once again, even at a price that had fattened by 60 percent in only nine months, they agreed to move ahead.

Sony's—Ohga's and, more decisively, Morita's—analysis of the cost-to-value ratio where CBS Records was concerned is revealing, and is moreover an important key to understanding their subsequent behavior during the acquisition of Columbia Pictures. The Betamax debacle of the seventies had confirmed Morita in his vision of hardware and software as interdependent: looking back on Betamax, he had often reflected that Sony would not have lost the battle to VHS if he had controlled a movie and television library at the time. Now, in 1987, with Sony about to launch a new digital audiotape (DAT) and players, and in early development of a recordable version of the CD, called the MiniDisc, CBS Records, which included the Columbia and Epic labels and had deep catalogs in every music category, was a tantalizing asset.* Second, CBS/Sony Records was by this time Sony's most profitable subsidiary, an unstoppable money machine, and partnership in that venture with Nelson Peltz, the owner of Snapple, was not a development that either Morita or Ohga was prepared to consider. The asking price of $2 billion, even at a time when the Japanese economy was cash rich and the yen stood at an all-time high of ¥80 to the dollar, was for Sony in 1987 a very substantial sum of money. But Morita was never loath to invest heavily when he perceived the likelihood of long-range

* Goddard Lieberson had attracted important popular and classical artists to the Columbia label, including Simon and Garfunkel, Tony Bennett, Miles Davis, Johnny Cash, Johnny Mathis, Pete Seeger, Barbra Streisand, Vladimir Horowitz, Glenn Gould, Arnold Schoenberg, Bruno Walter, and many others, not to mention numerous all-time best-selling albums, including the original cast recording of *My Fair Lady*. Later, on the Epic label, Walter Yetnikoff managed superstars such as Bob Dylan, Bruce Springsteen, Cyndi Lauper, the Byrds, and Michael Jackson.

return. According to Mickey Schulhof, who more than anyone had reason to know, "Morita's attitude was, 'I can always recover money. I can't recover people or companies and in the long-term scheme of things it doesn't matter whether we pay a little more or a little less.'"

If Morita did have reservations, they had to do with Sony's long-term ability to manage and grow the business. According to Ohga, at a meeting of the Executive Committee, he framed his concern by asking what Sony would do if Ohga were to die after the company had laid out $2 billion for a worldwide record business. Morita's emphasis on the critical importance of skilled management, coupled with his relative indifference to the size of an initial investment, would provide an important key to understanding the company's behavior during the Columbia Pictures acquisition.

Following the CBS board meeting in September 1987, Tisch called Schulhof to report that the board had deferred the decision on the record company and was calling in an outside consultant to evaluate the business. Yetnikoff, who had been asked to leave the room when the vote was taken, was shocked. Sony was also stunned, and publicly silent. Rumor had it that Tisch was once again considering the option of spinning off the business in a public offering.

On October 19, 1987, "Black Monday," the stock market crashed, and Tisch's certainty in the brightness of his financial future seems to have been badly shaken. The next morning he was on the phone to Schulhof once again: if Sony were still interested in the record company, he would make certain that Bill Paley would go along. Notwithstanding the market tumble, the price stood at $2 billion.

For the third time Schulhof contacted his bosses in Tokyo. Characteristically, Morita was untroubled by the price: "If it was worth $2 billion last month, it's still worth $2 billion," he told Schulhof. What he did insist on was a letter signed by Tisch and Paley guaranteeing they would recommend the sale to the board jointly. Within a day the signed letter was delivered to Schulhof.

If Morita was cavalier about cost, Ohga invariably argued for pennies: as he pitted his will against Tisch's across the table in the final days of negotiation, the deal nearly came apart again. At issue from Ohga's point of view was some $40 million in overaccrued pension benefits. CBS had overinvested in its pension fund, but was insisting that only the amount required to fund the pension be transferred to Sony at the time of sale. Ohga wanted the full amount reflected in the balance sheet transferred. At a breakfast meeting in his office just before the closing, with Schulhof, Paul Burak, and Steve Schwarzman also at the table, Tisch turned to Ohga and suggested with a smile that congratulations were now in order as all issues had been resolved. Ohga shook his head unsmilingly and suggested that one issue "remained open." Without a word Tisch stood up, threw his napkin on the table, and left the room. According to Schulhof, Ohga appeared dumbfounded and asked him for an explanation of what had just happened. "He's negotiating," Schulhof observed, "and he's just as hard as you." In fact, Tisch was more than a match for Ohga: that day, November 17, he had CBS prepare a press release for the next morning announcing that the deal was off. Morita continued to insist that money was not an issue, and Ohga yielded. On November 18, Schulhof went to Larry Tisch's office and accepted his terms on Sony's behalf.

Yetnikoff claims that the $40 million was not the only money Sony "handed back to Larry Tisch on a silver platter." The record division's fiscal year began on November 1. Harvey Schein had made this arrangement while he was still at CBS, and it had the effect of moving the decisive Christmas season forward so that prospects for the coming year were clear by January 1. Inasmuch as the deal was closing late in November, Yetnikoff maintained that the November and December revenue belonged to Sony. Tisch rejected this assertion with characteristic abuse, and, according to Yetnikoff, despite his advice to Schulhof to hang tough, Sony left the money on the table.

The acquisition was formalized two months later, in February 1988. So that both boards could approve the contract simultaneously, meetings were scheduled at the same time, 7:00 P.M. in New York, 9:00 A.M. the following morning in Tokyo. Schulhof and the Sony team had assembled in the CBS boardroom when a CBS lawyer pointed out that Schulhof was not a director of the parent company, Sony Corporation, and needed a letter authorizing him to execute a $2 billion contract. Schulhof called Morita at home and Morita asked for a fax number; five minutes later, he received a handwritten note: "I authorize Michael Schulhof to sign the purchase contract for two billion dollars on behalf of Sony Corporation for CBS Records, Inc. Signed, Akio Morita, Chairman of the Board." According to Schulhof, this note was the only written communication generated during the process of negotiation for the record company. Later, during and after the more complex acquisition of Columbia Pictures, the lack of documentation would confuse and dismay many of the American businessmen involved in Sony's U.S. affairs. Schulhof himself was not concerned. Sony was, after all, a family business, and Schulhof assuredly was family.

Yetnikoff and Schulhof were given credit for having managed successfully what was at the time the largest acquisition of an American company by a Japanese organization, at a price that represented five times earnings and would prove to be a bargain. Yetnikoff had made good on his promise to deliver the business to Sony without defections by management or major talent. Ohga immediately installed him as president and CEO of the new company, which Ohga himself later renamed Sony Music Entertainment.

Schulhof rejoined Sony America formally at this time, as executive vice president of operations. On paper, he reported to a man named Neil Vander Dussen, an engineer recruited to manage Sony's broadcast division, who, after several years of effective stewardship, had been appointed president of Sony America in December 1986 on the recommendation of Kenji Tamiya. By Vander Dussen's own account, though he was theoretically Schulhof's boss, the new executive vice president paid scarcely any attention to him. In view of the events about to occur, that is not surprising: the ink on the CBS Records deal was barely dry when Sony set out to acquire a major film studio in Hollywood, a prospect that ignited the imaginations of both Yetnikoff and Schulhof and set them to dreaming heady dreams of status, wealth, and power. The CBS Records acquisition prefigured and set the stage for Hollywood. Certainly the personal dynamic that drove the process would be identical. Ohga would be in control but strongly influenced by his two closest American friends, each with his own agendum. Morita, increasingly behind the scenes as he stepped into his role as international business statesman, would retain the force of a prevailing wind. Clearly, the nature of this four-way relationship and the balance of power it determined are critical to understanding the process that was to launch Sony on its boldest experiment in assimilating a foreign property. In the course of that process, Walter Yetnikoff would be unseated, and Mickey Schulhof would briefly ascend to heights that no other foreigner has ever achieved in a Japanese organization.

8

ONE FOR CHAIRMAN AKIO: THE COLUMBIA PICTURES ACQUISITION

On September 25, 1989, Sony announced its offer to pay $3.2 billion and to assume an additional $1.6 billion in debt to acquire Columbia Pictures Entertainment, a company that included Columbia Pictures and Columbia Pictures Television, TriStar Films, and the Loews Theatre chain owned by TriStar, 820 screens in 180 locations nationwide. At a time when Columbia's fortunes were at low ebb and its asset value was estimated at no higher than $22 a share—the stock had traded at $12 at the beginning of the year and was now at $21—Sony agreed to pay $27 a share, a figure that analysts characterized as "a wild multiple" and that would earn millions of dollars for Columbia's exultant shareholders.

Three days later, on September 28, to induce the producers Jon Peters and Peter Guber to run the studio, Sony announced its intention to purchase the Guber-Peters Entertainment Company (GPEC) for $200 million, a price some 40 percent in excess of its estimated value at the time. The purchase was concluded in full knowledge of a recent contract binding Guber and Peters to Warner Brothers for five years and in spite of the near certainty that Warner would file suit as a result. In mid-October, as expected, this came to pass. Sony agreed in short order to a settlement out of court which cost an additional several hundred million dollars. All told, expenses related to acquiring the

Hollywood property and its management came to just under $6 billion.

The Columbia Pictures acquisition, at the time the largest purchase of an American company by a Japanese organization, was received in the United States as evidence that the Japanese were plotting in an inscrutable and treacherous way to acquire America. In truth, national paranoia aside, the unparalleled climb of the yen against the dollar between 1985 and 1990 was driving a Japanese shopping spree deep into U.S. territory. Prime real estate in Florida, Texas, and California, small and medium-size businesses in the manufacturing sector, hotels, country clubs, and even banks, were being annexed and developed as fast as paper could be signed. In the process, landmark American monuments were falling into Japanese hands—Firestone Tire Company, the Tiffany building, and that same year, the purchase by the Mitsubishi Real Estate Company of New York City's Rockefeller Center. Now the Japanese had reached all the way into Hollywood, profane and sacred bastion of some of the country's most cherished fantasies of the American dream. The week of the purchase, *Newsweek*'s cover dressed the statuesque lady in the Columbia logo in Japanese kimono, and declared in a banner, "Japan Invades Hollywood!"

If the Japanese acquisition was grounds for resentment and uneasiness, the spectacle of Japan's apparently most sophisticated company being eaten alive by Hollywood and, in particular, by two of that community's mythic tricksters, Jon Peters and Peter Guber, was reported with uncontainable glee. Hollywood observers watched in disbelief as Sony appeared to be willfully relinquishing its buyer's leverage at every turn in the negotiation: when the final reckoning was announced, the derisive conclusion headlined across the country was that Pearl Harbor had been avenged. Subsequently, observers with a taste for schadenfreude were treated to five years of pleasure, which culminated in mid-1994 when, after repeated attempts to curb spending and bring the business under control had failed, Sony took a colossal write-off of $3.4 billion.

As a response to the largely negative, and distinctly anti-Japanese, publicity occasioned by the acquisition, Sony chose a policy of silence which it maintains to this day. The rationale was twofold. Any response was likely to further energize what Sony perceived as a negative "media frenzy." More concrete, there was the threat of litigation. Teruo "Ted" Masaki, currently deputy president of the Sony Corporation of America and a paragon of circumspection, explained the official company position in his flawless English: "It's an area that is so sensitive . . . not in the sense of what actually transpired . . . I can certainly, if I chose to, fully explain. I was there during that period of our negotiation. But we've been through a number of litigations surrounding the entire acquisition, not based on any one particular event, but these litigations, whether they are well founded or otherwise, place a tremendous burden on us, particularly because of the issues involved. It takes up a lot of executives' time, and whenever you have litigation, particularly in the U.S., you have juries, and all sorts of factors go into assessing what all of it means to Sony, and it's a major, major risk and a major, major burden. It's getting to a point now where new litigation based on the acquisition is unlikely, but there are potential and ongoing legal disputes about that era. All this could be discoverable: any evaluation, assessment, description, I may or may not make all have the potential of being used, and that from a legal perspective leads us to be cautious. We're just not comfortable in waiving what would otherwise be a privileged area of information which we shall have to assert should we be exposed to new litigation."

Masaki declines to be specific about "the series of litigations [Sony] has been exposed to" already, but these are likely to have been "10-D5" class-action suits of the variety that shareholder groups almost automatically file at the time of any major acquisition. He also neglects to mention that the severance agreements drawn between Sony and Walter Yetnikoff in 1990 and Mickey Schulhof in 1995 contain mutual-confidentiality and "no-attack" clauses: anything revealed by Sony which might be construed as implying criticism of either Yetnikoff or Schulhof, principal agents in the acquisition, could expose Sony to legal action.

In any event, there is little room for doubt that what is, or would be, at issue in any litigation was the question of Sony's objectivity during the acquisition process: did every important decision reflect careful consideration of the shareholders' interests, or was the company neglecting its fiduciary responsibility? In particular, there were two critical moments, which observers found inexplicable and which seemed to lead to the conclusion that Sony was either self-destructive or foolhardy beyond belief. The first of these was Mickey Schulhof's announcement early in September 1989 that Sony would increase its bid from $15 to $27 dollars a share, a price that analysts agreed had no basis either in current performance or potential. The second, probably related, moment, which followed on the heels of the first, occurred when Sony yielded to pressure from Guber-Peters's lawyers and agreed not only to purchase the Guber-Peters Company without a prior release from Warner, but to indemnify the principals, Jon Peters and Peter Guber, in the event of the lawsuit that was certain to result.

While no one involved has been entirely forthcoming, a picture nonetheless emerges, blurred in places and sketchy in others, suggesting that the process which resulted in Sony's acquisition of Columbia and the Guber-Peters Company was profoundly influenced by the web of personal relationships among the principals. There is nothing to suggest that anything unethical occurred. What does seem true is that the decisive role played by personal relationships was, if not distinctively Japanese, then certainly characteristic of Sony historically and unlikely to be observed inside any American business organization engaged in a decision of similar magnitude.

In the spring of 1988, shortly after the purchase of CBS Records, Mickey Schulhof called Walter Yetnikoff to report that he had just returned from Tokyo with instructions to begin shopping for a film studio, and to suggest that Yetnikoff accompany him to California for a look at MGM. Since the mid-seventies, Morita had felt that Sony could have turned the defeat of Betamax into victory if it had owned a library of films and tapes to release in Beta format. Now, given the recent successful introduction to the U.S. market of both its video Walkman and a new line of popular 8-millimeter video cameras and players, the prospect of controlling a software bank of its own was more than ever appealing. The successful acquisition of the record company at a price that was already looking like a bargain may also have emboldened Morita to consider a new business that Sony knew nothing about.

Yetnikoff had been urging Morita and Ohga to move Sony into the movies for years, arguing that a rich harvest was waiting to be reaped from a synergy between music and film. He had tasted the excitement of filmmaking himself when in 1986 he had been one of three executive producers on the Disney film *Ruthless People.** He had also observed, with envy that he readily admits today, the power wielded as the head of Warner Brothers by Steve Ross, a man whom he admired and wished to emulate. If Ross could manage both film and music at Warner, what was to stop Walter Yetnikoff from doing the same at Sony?

As he was not known in Hollywood as a player, Yetnikoff asked Michael Ovitz to help broker Sony's search. Not surprisingly, Ovitz agreed at once: it would have been clear to him that an association with Sony as it acquired a Hollywood studio was likely to hold abundant possibilities for personal advancement, including the top job at the studio, a position that Ovitz himself was not loath to consider.

During the summer of 1988, Ovitz introduced Sony to top management at Paramount, MCA/Universal, MGM, and Columbia. Schulhof and Yetnikoff attended most of the meetings together. On his own, Yetnikoff set up a meeting with Ohga and Michael Eisner and Jeffrey Katzenberg at Disney.

Morita had long admired MCA/Universal, but Lew Wasserman was not yet in a mood to sell. At Paramount, Martin Davis, chairman and CEO of Gulf and Western, the parent company, was amenable to a sale, but proposed to structure a board of directors that he would continue to control after the acquisition. Morita's friend and adviser Peter Peterson considered this idea preposterous: "I said, 'Marty, my God, they're buying the company. What are you talking about?'" On Peterson's advice, Sony backed away.

---

* In his 1984 contract with CBS, Yetnikoff had negotiated the right to produce movies on the side; this was the first time the company had permitted one of its division heads to accept free-lance work.

By the fall, it was clear that only MGM and Columbia were possibilities. At the time, MGM was controlled by Kirk Kerkorian; Morita met with him personally and may have discussed a price of $1 billion. But Kerkorian advanced his own outrageous demands, including retention of the MGM logo, the roaring lion, for use as a trademark for his hotel and casino chain. According to Ohga, he disqualified MGM when he learned that the studio had sold the bulk of its film library to Ted Turner. Subsequently, just weeks before Sony announced that it would buy Columbia, the Qintex Group of Australia acquired MGM/UA for $1.45 billion, igniting anxiety in Hollywood about foreign domination which was substantially deepened by the Sony purchase.

Columbia alone among the major studios had retained rights to its entire library of films, 2,700 titles, one of its principal appeals, for it was software that Sony wanted. The studio had been founded in 1924 by Harry and Jack Cohen from New York's Lower East Side. The brothers had kept the small studio profitable through the Depression by cranking out B Westerns and Three Stooges movies. Over the years, the Cohens had also attracted to their lot a number of major directors who had created some masterpieces for Columbia: Frank Capra's *It Happened One Night*, Elia Kazan's *On the Waterfront*, Robert Rossen's *All the King's Men*, David Lean's *Lawrence of Arabia* and *The Bridge on the River Kwai*, Fred Zinnemann's *From Here to Eternity*, Edward Dmytryk's *Caine Mutiny*, Sydney Pollack's *Tootsie*, Ivan Reitman's *Ghostbusters*, and Steven Spielberg's *Close Encounters of the Third Kind*.

Though Columbia had twelve Oscar-winning films for Best Picture of the Year, its film library was not esteemed as one of Hollywood's finest. In television, on the other hand, it was the industry leader, with 23,000 episodes of 260 series in syndication, including "I Dream of Jeannie," "Bewitched," "Gidget," "The Monkees," and "Dennis the Menace." At the time of the acquisition, the Columbia TriStar Television Group owned "Days of Our Lives" and the two most popular game shows in TV history, acquired from Merv Griffin Enterprises in 1986, "Wheel of Fortune" and "Jeopardy!," and was producing several of the biggest hit series in prime time, ABC's "Who's the Boss?," the Fox network's "Married. . . with Children," and CBS's "Designing Women." It was estimated that these three series alone would eventually bring the studio $500 million in syndication rights.

Ohga says his interest in Columbia had to do above all with its TV production capacity: "Movies are a studio's marquee," he told me, "but the money is made in television. I was interested in how many programs they could turn out each month."*

In November 1988, Yetnikoff and Schulhof sat down in New York with Victor Kaufman, chairman and CEO of Columbia, and Lew Korman, president and COO, to discuss a deal. When Kaufman had determined that Sony was in earnest, he passed Schulhof to the man who would represent the seller, the multibillionaire investment banker Herbert Allen, Jr. The scion of a vastly wealthy family, Allen was an eccentric loner with a lethal business mind. In 1973, against the recommendation of his father and uncle, the cofounders of the investment house of Allen and Company, he had acquired some Columbia stock because he found the movie business "more interesting than heavy metals." By 1981, he owned 6.7 percent of the company and was chairman of the board, controlling the business with his friend and associate, the producer Ray Stark. In 1982, he cashed in his original holdings at a very substantial profit—he had paid on the order of $3 a share—by selling the company to Coca-Cola for $70 a share when the stock was trading at $40, but retained his influence at the studio, becoming a member of the Coca-Cola board and in essence overseeing the business for Coca-Cola. During the summer of 1988, when word was out that Sony was seriously interested but before negotiations began, he is said to have quadrupled his holdings in Columbia in preparation for a second Columbia bonanza when the acquisition occurred.

Initially, Morita and Ohga were interested in acquiring a controlling interest in Columbia and had instructed Schulhof to discuss Coca-Cola's 49 percent and a little more. But a partial sale would not have

* Ohga's emphasis has been validated over time: since the acquisition, the television group has frequently been the driver of profits at the studio. In 1998, "Seinfeld" was sold into syndication for a record-breaking sum.

maximized Herbert Allen's personal holdings, and before long he had maneuvered Sony into negotiating for the entire business. His asking price was $35 a share. At a time when Columbia stock was trading between $12 and $21, this was an impressive demonstration of *chutzpah*. It was also evidence that Allen had read Sony intuitively as hungry, cash rich, and naïve about the movie business. Sony entered the negotiation hoping to pay somewhere between $15 and $20 a share.

The accuracy of Sony's due diligence remains controversial in and outside the organization. As Allen well knew, no one inside the company was equipped to evaluate a Hollywood film studio, not Ohga, certainly not Schulhof, not even Walter Yetnikoff, who was a music man. On Yetnikoff's recommendation, responsibility for estimating the value of Columbia's assets was given to Michael Ovitz. For a month or two, in return for a fee of $11 million, Ovitz worked at quantifying the value of the film and TV holdings and a number of film projects still in development, including *Flatliners*, and a film to star Kevin Costner called *Revenge*. He then handed his numbers to Sony's investment brokerage of choice, the Blackstone Group, where they were plugged into financial models projected into the future. Blackstone's projections, high, average, and low, were in turn passed to Ohga via a team of financial analysts inside Sony.

Blackstone's founder and chairman was Morita's longtime friend and adviser Peter Peterson, the former secretary of commerce under Nixon; Peterson's partner, Stephen Schwarzman, was a close friend of Mickey Schulhof's. The company had first become involved with Sony in formal business dealings at the time of the CBS Records acquisition, when Schulhof had invited Blackstone into the process with Morita's enthusiastic approval. The purchase of CBS, a far simpler transaction, had generated handsome fees for Blackstone. This time it seemed only natural, from Schulhof's point of view, and very likely from Morita's as well, that Peterson once again should be involved.

Others, including Norio Ohga, were not so sure. In 1992, at Morita's insistence, Peterson would be invited to become the first foreign outsider to sit on the Sony board—Mickey Schulhof and Jack Schmuckli, head of Sony Europe, had joined in 1990—and, at this writing, is the only foreigner to retain his seat. As chairman of the board, it would be improper for Ohga to go on record with anything negative about a director, but in conversation with me he did express a reservation: "Peter Peterson is a company director and we must not criticize him. And Pete is a huge Sony fan. He has gone out of his way to help us. At the same time, he has Blackstone. When we're not doing business together he gives us plenty of sharp advice, but the minute a deal is involved the edge goes off him a bit. . . . Blackstone is a very ambitious place." Ohga's veiled implication is that Peterson may have been caught in a conflict of interest because consummation of the purchase would—and in fact did—deliver very large fees to Blackstone. A corollary concern is whether Blackstone's final evaluation was "accurate" or skewed on the high side. Ohga never voiced this doubt to me, but there are those inside Sony who maintain that he was critical of Blackstone's performance. Peterson, who is outspokenly appalled at the "informality" with which the acquisition was handled by Sony generally, dismisses out of hand the possibility that Blackstone was in any way responsible for what many believe to have been an overvaluation: "I told Akio, 'You ought to get a film expert in here to evaluate these libraries and to look at the new films,' and Schulhof or Yetnikoff came up with Ovitz. He provided every number in the financial analysis, and we put it in the models—on the libraries, the prospective sales, and so forth. He was paid $11 million to do that, probably did a month's work."

Schulhof is critical of what he interprets as Peterson's refusal to accept any measure of responsibility: "Ohga had a corporate group run by a fellow named Tokunaka who was modeling the price, taking the Blackstone numbers, which is why Ohga is still angry at Blackstone, modeling the kinds of returns we could expect . . . Pete is half right when he says we got the information from Ovitz, but Ovitz's contribution was minimal. Whether I was directly responsible for running the picture company or not, it was there on my watch and therefore I take responsibility for the actions. Blackstone can't shirk its responsibility either. It put its name on the projections. If it didn't agree with the projections, it shouldn't have put its name on them."

During and after the negotiation, observers laughed at Sony for seating Mickey Schulhof across the table from Herbert Allen. Ohga himself acknowledged to me that Schulhof was "no match" for a man whom he characterized, in English, as "the shrewdest, the most shrewd" businessman in America. "But remember," he added, "Walter Yetnikoff was there all along and working very hard to keep the price down; and Mickey was on the phone to me every single day at six-thirty in the morning for instructions on how to proceed. I approved every move." While it isn't clear that Yetnikoff was effective in holding down the price as Ohga suggests, there is evidence that Schulhof was not empowered to bargain independently. In March 1989, Ohga suffered a mild heart attack while at home and spent a month recuperating. The illness was not reported, and Allen and his team were confused and exasperated when the negotiations were put on hold. According to Yetnikoff, Ohga had suffered "several" other coronary episodes that had been concealed, including one in Salzburg while he was attending a performance of *Don Giovanni* with Yetnikoff. In July, while in Cologne with Morita and Schulhof, Ohga had a more serious attack and had to have an artery dilated in an emergency angioplasty. Once again, while Ohga recovered, Sony suspended negotiations without a word of explanation.

In August 1989, a discussion took place in Tokyo at a meeting of the Executive Committee which would have a decisive impact on the end-game to be played out during the frantic month of September. By this time, with Ibuka rarely in attendance, the meetings were dominated by a running dialogue between "Chairman Akio," as his top executives still refer to him, and Ohga. Nothing was "determined" or "concluded" until it had been voiced by Morita explicitly. Ohga, who was running the company, would imply his own feelings and desires but was careful not to assert an opinion in Morita's presence unless invited to do so. What resulted was a subtle dance of signals and responses between Sony's founder and his protégé and chief executive. The other members of the committee were welcome to comment, and did from time to time express their own feelings openly, but in general the tone and direction of the debate were established by the Morita-Ohga dialogue.

There were six regular members of the committee in addition to Morita and Ohga, including Morita's brother, Masaaki, about to be dispatched to New York as chairman of the Sony Corporation of America, where Mickey Schulhof would ignore him; Executive Deputy President Masahiko Morizono, the engineer who created Betacam and built Sony's broadcast and commercial video business; Senior Managing Director Tsunao Hashimoto, one of Ohga's closest advisers; and Executive Deputy President Ken Iwaki, rumored at the time to be the front-runner to succeed Ohga as president.

Iwaki, who was serving as secretary, recalls that Chairman Akio opened with a question: With Ohga in fragile health, was it wise for Sony to be entering a business it knew nothing about? Until this moment, Ohga had strongly advocated purchasing Columbia because of its film library in particular. His confidence in his own health badly shaken, he now accepted Morita's invitation to express misgivings. The price remained unreasonably high; and it appeared that Herbert Allen did not intend to budge from his asking price of $30 per share. Besides, while he understood the music business, he knew nothing about Hollywood. And there was the question of his health . . . Ohga's tentative remarks were followed by a pause. Then Morita spoke: under the circumstances, it seemed that the prudent thing to do was to abort the Columbia acquisition, which he now proposed that Sony do. In his minutes book, Iwaki wrote: "per chairman, Columbia acquisition abandoned."

According to Iwaki, that might have been the end of it but for a chance remark by Morita that night at dinner with members of the committee. At a pause in the conversation, Morita put down his cup of green tea, and said softly, "It's really too bad. I've always dreamed of owning a Hollywood film studio." No one spoke. But when the committee reconvened, Ohga announced that after careful consideration he now felt strongly, assuming the chairman concurred, that Sony should move ahead after all. Morita agreed, enabling a conclusion. The previous entry was amended to read that Sony, per the chairman, would proceed with the acquisition. Iwaki recalls: "We decided to proceed, based on what Chairman Akio had said the night before, and we also agreed that if we wanted the studio we would have to pay the price. Since we had no real means of evaluation, we wouldn't really be able to bargain. It would be a sort of take-it-or-leave-it proposition." If

Iwaki's recollection is accurate, it seems that Morita's long-standing dream of owning a Hollywood studio, or, more properly, the regret that losing sight of his dream inspired, was the basis, or certainly a basis, for Sony's commitment to acquiring Columbia at more or less the asking price. If that is true, it is additional, dramatic evidence that the environment in which the Sony Corporation has historically conducted its affairs is less public than personal: less rational than sentimental.

The commitment to acquiring the studio in principle did not mean abandoning the search for someone to manage the business; Morita continued to insist that Sony must not venture into Hollywood unless superior management was in place. Victor Kaufman, who had been moved by Coca-Cola from TriStar to Columbia in 1987, was the most recent in a procession of studio executives who had paraded through Columbia during the eighties, leaving debt on the books, few if any hits, and the business in disarray. According to Ted Masaki, Sony viewed Kaufman and his team as interim stewards installed by Coca-Cola to keep the hedges trimmed until it could sell the studio, but it was far from clear who might be available and capable of replacing him.

Michael Ovitz had been waiting for the right moment to make his own pitch for the top job at Columbia. Ovitz was head of the talent agency he had founded, Creative Artists Agency (CAA), and was widely acknowledged as one of the most powerful players in the entertainment industry, a man who could single-handedly "package" a movie deal that included top stars and top directors, many of whom he represented and advised. Apparently, he had decided it would take a rich reward to make it worth his while to move away from the talent to the management side of the table: when he approached Sony late in August, the terms and conditions he presented to Schulhof are said to have been exorbitant beyond imagination, including a multimillion-dollar salary and bonus, equity, control of the board, and the freedom and funding to acquire other businesses. Peter Peterson, who was included in the review, recalled the moment with wry dismay: "The stuff with Ovitz was just wild. I said, 'This isn't even your company. If you put up $2 billion, Mike, maybe you'd have a seat at the table.' I mean, the demands were outrageous. I've never seen anything like it."

Schulhof conferred with Ohga in Tokyo, and Sony rejected Ovitz's proposal. According to Yetnikoff, Ohga was more offended by Ovitz's demands for control than for remuneration. From this point on, Ovitz was no longer involved in the negotiations. Publicly, he represented that Sony had asked him to accept the job and that he had intentionally created terms he knew would be unacceptable as a means of avoiding the awkwardness of turning the company down. In 1996, when Nobuyuki Idei was looking for his own man to run the studio, Ovitz surfaced again and once again presented impossible terms.

While Ohga was unable to accept Ovitz's proposal, he was disappointed to lose him. On Yetnikoff's recommendation, he had come to Los Angeles to meet Ovitz in the glass and marble CAA building designed by I. M. Pei and had been duly impressed with his power and presence in the Hollywood community. Now the search had to begin again.

In mid-August, the pressure began to build. Allen phoned Schulhof to inquire about Sony's intentions and to let him know that another buyer had appeared. In truth, Rupert Murdoch, who had folded 20th Century–Fox into his News Corporation, briefly considered a merger with Columbia and suggested a price of $15 per share, but Allen was still confident that he could maneuver Sony into paying more. On August 13, as Allen was applying pressure to Schulhof, Walter Yetnikoff, on orders from his doctor, entered the drug and alcohol rehabilitation program at the Hazelden Institute in Minnesota, and took himself out of action for a month. Yetnikoff remembers taking phone calls from Schulhof during his confinement: "Imagine: there was one public phone for about twenty guys, including some whopping big ones with their own serious problems, and I'm half dead myself. Mickey tells me we're going ahead with the Columbia acquisition, and I ask him, 'Who's going to run it?' and he tells me, 'We're Sony, we'll find someone!' and I remember saying to him, dazed as I was, 'It won't be that easy—remember, moviemaking is just about as people-intensive as prostitution!'"

Yetnikoff emerged from rehab on September 10, and says he had not been in his office for five minutes when Schulhof called "in a state of high anxiety" about locating a manager to run the studio. Jeffrey Katzenberg had been on his short list, but Katzenberg had just renewed his contract with Disney. After several days of considering and then rejecting Frank Price, a deliberate executive who had run Columbia briefly in the past and who would eventually return, Yetnikoff placed a call to an old acquaintance in California, the producer Peter Guber, to say that he had an exciting opportunity to discuss and would like to see him in New York. However chastened and quieted he may have been since Hazelden, Yetnikoff, by his own admission, still had his eye on the power seat at the top of what would become the Sony entertainment conglomerate—possibly he reasoned that a smart move would be to win acceptance as head of the film operation for someone he knew well and might control.

In 1980, with money from Polygram, the entertainment arm of Philips Electronics N.V., Peter Guber, thirty-eight at the time, had gone into business with Jon Peters, a man three years his junior who seems to have regarded him as an elder brother. In 1981, they had coproduced a film that lost money, *An American Werewolf in London.* The following year, they had installed themselves in the community as "winners" with *Batman* for Warner Brothers, the biggest box-office hit since Steven Spielberg's *Jaws.* They had followed *Batman* with Spielberg's *The Color Purple*, *Flashdance*, and *The Witches of Eastwick*, a film based on the novel by John Updike and starring Peters's friend Jack Nicholson. When Yetnikoff reached Guber in September 1989, he and his partner were still basking in their triumph at the Academy Awards in April when their most recent film, *Rain Man*, starring Dustin Hoffman as Tom Cruise's autistic brother, had won Best Picture for 1988.

At dinner in New York on Sunday night, September 17, as he listened to Yetnikoff describe Sony's determination to purchase Columbia if the right management could be found, Guber was both elated and very cautious. Yetnikoff's eagerness was clear, and his invitation to run a major studio was flattering. The Sony connection in particular was titillating. Guber admired what he calls "the design magic" of Sony products and was awed by the company's global reach and power. But there was a downside, and it was steep. Guber and Jon Peters were partners in a public company, recently renamed Guber-Peters Entertainment; a major if not the principal asset of the company was a lucrative five-year production contract with Warner Brothers which had only recently been signed. Even assuming they could get free of their Warner contract, and Guber had reason to think they might, how would the shareholders of Guber-Peters react to losing the major revenue streams that flowed from the Warner deal? Then there was Columbia itself: to Guber's mind, the history of takeovers, beginning with the Banc de Paris, and the continuous changes in management had left "little nicks and cuts and scars" in the company's walls, and he was leery of its ability to recover.

At dinner that Sunday night, Guber expressed his reservations, emphasizing his contract with Warner, but Yetnikoff seemed undeterred. What he did want to know was whether Guber, assuming he were interested, would want to include Jon Peters in the deal. This question was loaded: though Guber and Peters referred to themselves as "Karmic twins," drove identical green Range Rovers, and were generally inseparable, they were dramatically different people. Guber, the son of a junkyard owner in Somerville, Massachusetts, had a B.A. from Syracuse University and a law degree from NYU. He was polished, smooth, a fabulous talker, and very tactical with a keen, strategic mind. In 1968, he had moved to Hollywood with his wife and baby daughter to work in business administration at Columbia, and in five years had climbed to vice president of worldwide production under David Begelman. Three years later, he left the studio and went into business for himself to produce, in 1977, *The Deep*, with Jacqueline Bisset, and, in partnership with Neil Bogart, the "Disco King," *Midnight Express*, before joining Jon Peters.

Peters was the shadow side of Guber's dazzle. He had grown up angry in Encino, California, hating his stepfather, and had a violent temper. After four turbulent years in and out of high school, he had trained as a beautician while working as a bouncer and discovered that women loved to have him do their hair. By 1973, he had his own salon on Rodeo Drive in Beverly Hills, and a growing clientele of female stars who were drawn by his brooding good looks and somehow dangerous charm. After a brief marriage to Lesley Ann Warren, he became hairdresser and then consort to Barbra Streisand, who recreated him as a film producer. In 1974, Streisand showed up at the office of Warner Brothers studio chief John Calley with Peters in tow. Calley had been working on Streisand to accept the lead in a remake of *A Star Is Born*, which he had been developing with John Foreman and the writers Joan Didion and John Gregory Dunne. Streisand had been reluctant. Now she informed Calley that she would take the part on condition that Peters direct. Calley was appalled—Peters was a hairdresser with no movie experience—but he agreed, confident that Streisand would never allow herself to be compromised at the hands of an incompetent. Three weeks into the shoot, she told Calley that she needed a rewrite and suggested bringing on a director to shape the story as a film, leaving Jon Peters free to "oversee" the creative process. Calley hired Frank Pierson to rewrite and direct the film, which John Foreman produced. *A Star Is Born* opened to scathing reviews—the *Village Voice* proclaimed, "A Bore Is Starred"—but the film was a box office success, grossing $90 million, and the sound track album was at the top of *Billboard*'s charts for many weeks. Peters took credit as executive producer and was transformed overnight by the mysterious alchemy practiced in Hollywood into a "movie man." His next move was *Eyes of Laura Mars*, a stylish thriller starring Faye Dunaway, which he produced for Columbia. By this time, Guber had left Columbia, and had installed himself in an office that happened to adjoin Peters's on Hollywood Way, just outside the Warner Brothers lot in Burbank. The men became friends and partners.

Yetnikoff had become acquainted with Guber through Peters. After *A Star Is Born*, as he built a career as a producer, Peters also managed Barbra Streisand and had negotiated her new contract at CBS Records with Yetnikoff. But Peters was not a tame animal, and, despite his lethal charm under the right circumstances, Yetnikoff was apprehensive about the impression he was likely to make on Mickey Schulhof, not to mention Morita and Ohga. Accordingly, he may have been hoping that Guber would consider the top position at the studio as a solo for himself. Guber was emphatic: he and Jon Peters were copartners in their public company; there could be no question of his moving without Jon. In any event, he cautioned Yetnikoff, he was not at all certain that he should move at all.

Despite his misgivings, Guber was sufficiently intrigued to accommodate Yetnikoff's request that he meet Mickey Schulhof the following morning, and they spent the better part of the day with Schulhof in his office at 9 West 57th Street. To Schulhof, Guber would have seemed an exotic and fascinating creature, a man who wore his hair in a ponytail and dressed in jeans and a polo shirt yet had accumulated impressive wealth and power. He might also have been surprised to discover, contrary to type and his own expectations, that Guber, like himself, had been married to the same woman for twenty-five years and had two children in their early twenties (briefly, Schulhof's son and Guber's daughter would become engaged). Though Guber denies that he had any attention of setting a hook, that seems to be precisely what he accomplished over the course of the day. As he recalls the meeting, the more he protested his uncertainty about the offer, the more eagerly Schulhof pressed him to consider. He claims that he shared with Schulhof his view of Columbia as a damaged old car being polished for sale, and told him he had no desire to become caretaker to a business on its way down. Schulhof assured him that Sony was committed for the long term. Guber alluded to his contract with Warner, but added that securing a release from his friends there, Terry Semel and Bob Daly, was not likely to be a problem when and if the time came. If Schulhof was alarmed by this news, or by Guber's negative appraisal of Columbia's health, he has no memory of it. Possibly, he was too relieved at the appearance of someone who might be qualified to satisfy Morita's insistence on superior management to worry about details. Schulhof denies, however, that Guber swept him off his feet as later was reported: "It was a mixed bag. I didn't know him. This was the first time we'd ever met. I had done a little bit of background checking, I talked to Herb Allen, who told me that Guber was a very capable filmmaker. When Ohga asked my opinion, I said, 'Look, I don't know either one, but we're going to need for the first three years for somebody to reinvigorate the studio because the one missing piece in all this is that there is no current production of any value. And having a filmmaker at the top makes sense to me. And Walter knows this guy, so let's try to negotiate something with them."

To Guber, Schulhof conveyed only eagerness, which both tickled and, he claims, unsettled him. As he was preparing to return to California at the end of the day, Schulhof said, "We're going to want you to do this."

At home in Los Angeles, Guber took his wife, Lynda, to a local sushi bar and they laughed together about the deal he was being offered. Lynda was opposed to going forward: they didn't need the money and it seemed like trouble. Guber recalls that he was disposed to agree with her: "At thirty-seven, I would have had an orgasm at being invited to the conversation. Now I was more sober. I knew intrapsychically there would be problems built into the walls. I loved being invited to the party, but I wasn't sure I wanted to go, let alone run it."

The next day, Guber conferred with Jon Peters, his lawyer, Paul Schaeffer, and the chairman of Guber-Peters Entertainment, a billionaire Australian shopping mall developer named Frank Lowy. The previous year, Guber had flown to Australia and persuaded Lowy to invest $45 million in the company, making him the principal stockholder. When Lowy heard the story, he reminded Guber that he was already a principal in a public company, and warned him that "jumping ship" was not an option. Instead, he urged Guber to make the deal contingent on Sony's acquisition of Guber-Peters Entertainment at an inflated price. Guber represents that such a strategy had not occurred to him until Lowy spelled it out. Perhaps, although imagining Peter Guber in the role of ingenue is not easy. In any event, from this moment on, Guber would position Sony's acquisition of his company as a condition of his agreement to run the studio.

The next day, Guber flew back to New York on Frank Lowy's private 727. At a meeting with Yetnikoff and Schulhof to which Peter Peterson and Steve Schwarzman had also been invited, Guber introduced the notion of Sony acquiring Guber-Peters Entertainment, emphasizing the importance of treating his shareholders fairly. Afterward, he spent thirty minutes alone with Peterson, who questioned him "with an icy face" about his career, a subject that surely elicited from Guber a beguiling account. Later that day, Schulhof informed him that all parties were agreed they wanted him as chairman of the board and CEO of Columbia Pictures. Guber says he was not surprised. The rhetoric of his explanation is typical: "I had the credentials, I was twenty-five years in the business, married for twenty-five years, produced the greatest films in the last ten years in Hollywood, five times for best picture, fifty Academy Award nominations, had been at Columbia for six years, had been the CEO of Polygram for six years, produced music and television shows, had four degrees, all these degrees. Why wouldn't I be a candidate!"

Schulhof explained that the final decision could only be made by Akio Morita. As Morita happened to be staying in California at Pebble Beach, he hoped that Guber would agree to fly to San Francisco to meet with him at once. In the colorful embroidery he stitched for me, Guber represents himself as having been dragged by the heels into a process that was beginning, as the circle of people who knew about it widened, to make him deeply anxious. That may have been the case. Perhaps he had also glimpsed the possibility of a sale of Guber-Peters that would deliver a large profit to himself and his partner and was feigning hesitation as part of a brilliant play to make it happen. In any case, a meeting with Akio Morita was too exciting to turn down, as he put it to me, "an experience for my book of memories!" Schulhof was duly impressed by "Guber's" 727, and asked if would take Sony's man in New York at the time, Ken Hoshikawa, along. "I told him there were plenty of seats and I'd take anyone he liked, including the Illinois state marching band!"

At the Lodge at Pebble Beach, Morita had a stomachache and received Guber and Hoshikawa in the bedroom. Yoshiko Morita was also there. In Guber's recollection, Morita, propped up in bed, put on his glasses as they came into the room, smiled his radiant smile, and asked, "Why should we buy Columbia Pictures?" Guber protested that he hadn't the foggiest notion. "Then why are you here?" Morita asked. Ken Hoshikawa began speaking in Japanese and Morita silenced him with "just a silken wave of his hand, the slightest movement you could imagine." It was his understanding, Guber explained, that Sony was considering the purchase of Columbia and was also considering offering him the opportunity to run the studio. Morita nodded, and was silent for "ten to fifteen seconds that felt like three and a half years." Finally he asked Guber to tell him a little about himself, explaining that he wanted to hear it, "from [his] own mouth," a perversion of the English idiom which stuck in Guber's memory.

Morita dismissed Hoshikawa, and Guber delivered himself of what he claims was a measured account of his career. At that point the atmosphere in the room abruptly lightened, and the Moritas, Yoshiko included, chatted with him for an hour about life in California and his family, playing golf, ham radio, electronic imaging, and a number of other things unrelated to the job. Yoshiko remembers having seen at once that Morita would enjoy spending time with Guber and quietly canceling a dinner engagement to extend the meeting. As Guber was leaving, Morita asked him to fly to Tokyo to meet Ohga, explaining that he would make the decision whether to buy the company and whether to hire Guber to manage it. Guber claims that he demurred politely, suggesting that he would decide for himself whether he wanted to run the studio, whereupon Yoshiko Morita exclaimed laughingly, pointing to her husband, "No, no, he makes the decisions about everything, including you." From his bedside, Morita placed a call to Tokyo, and, while Yoshiko chatted with Guber in English, he spoke for ten minutes with Ohga in Japanese. On his way out, Guber asked Morita to sign a copy of his memoir, *Made in Japan*, which he had read on the plane from New York, and asked Yoshiko to take a picture of them together. Not a word had been spoken about the job. Outside, Guber mentioned to Hoshikawa how relieved he was to be on his way home. "What do you mean?" said Hoshikawa. "Mr. Morita told Mr. Ohga that you would be in Tokyo tomorrow!" "I didn't agree to go," Guber objected. "You didn't say you wouldn't."

Guber returned home to Los Angeles that night, and flew to Tokyo the following day for his first meeting with Ohga. Having been cautioned by his lawyer, Paul Schaeffer, to make no deals or commitments that shareholders might view as detrimental to their interests, he paid—Guber-Peters Entertainment paid—for his own first-class airfare and for a room at the Hotel Seiyo Ginza, among the most expensive hotels in Tokyo. Ohga greeted him cordially at the hotel and mentioned that he came with strong recommendations from Yetnikoff,

Schulhof, and Mr. Morita. Perhaps, Ohga joked, he would be the dissenter. Good, Guber retorted, that would let him off the hook.

From the hotel, they rode in Ohga's chauffeured Jaguar to a tempura restaurant on the ninth floor of the Takanawa Prince Hotel, near Sony headquarters. The subject at dinner was filmmaking. Guber declared that the heart and soul of film was music, and recounted his experiences working with John Williams and Pavarotti. He likened his own sensitivity to film to Potter Stewart's definition of pornography: he couldn't define what it was that would make a film great, he explained, but he knew it when he saw it. "Here," he said, tapping his head, and "here." His hand moved to his heart. In his recounting to me of the interview, Guber continued to deny that virtuoso performances like the one he delivered that night were consciously designed to land him the job: "They had not made me an offer of any kind, so I was playing it carefully. Sure, if I had been completely uninterested, I wouldn't have gone. Remember, I had a big business of my own to run! But there was no strategy because I still didn't know what the target was. There was a tactic, and that was to be thoughtful, provocative, clear, and desirable, of course. I figured, who knows, someday this might be a resource for me, or I might want to come back. You get very few chances to sow seeds in such fertile soil, and that was the time to do it."

Whatever his conscious intention may have been, Guber delivered a knockout punch that night. Conjuring one of Hollywood's slickest jive-geniuses at the same table with Sony's icy and cultivated chief is a challenge to the imagination. But Ohga was never simply a businessman with a keen eye on the bottom line; he was also an artist with an artist's appreciation for creative passion, and, by his own account, what sold him on Peter Guber was what he perceived as his passion for filmmaking. As Guber talked about film, Ohga observed that "his eyes sparkled, and he forgot all about time. To create things you must be entirely engrossed, and I could feel that this man truly loved film. What I bought was his passion." Subsequently, Ohga accepted Guber's invitation to participate in a Q-and-A session in the class he taught at UCLA film school, and he recalls that his first impression was reinforced: "The classroom was jammed with young students, and you could see they truly respected him. As I watched them listening to him raptly, I felt that this was really Mr. Movie Man!"

The next morning, when Guber alighted from the limousine that had delivered him to the garage beneath the Sony headquarters building, Ohga was waiting for him. Guber perceived that he was being granted very special treatment, and he was correct: normally, even VIP visitors are shown to the formal guest room that adjoins Ohga's suite of offices and are served a cup of green tea to contemplate while waiting for him to make an entrance through his private door. When Ohga inquired what Guber would like to do that morning, he asked to observe the process of designing a Sony product, a masterly request. Ohga personally escorted Guber through the design center that he had created as a young man. Later, as they parted, Ohga shook Guber's hand warmly, and said, "This is interesting. We must consider this strongly."

The weekend following Guber's successful encounter with Ohga, Schulhof reached Herbert Allen to say that Sony was ready to enter a formal bid for Columbia Pictures Industries of $27 per share, a figure that represented a $10 to $12 increase from the last point in the negotiations. Allen must have been pleased: at this price, many of Columbia's shareholders, including himself, would earn millions of dollars. A question that remains is how Sony arrived at the number. Peter Peterson was, and is, appalled: "We had taken Ovitz's numbers; then one day Mickey Schulhof calls up and tells us he's gone up to twenty-seven. And Schwarzman says, 'Where did these numbers come from?' And Mickey says Ohga has approved. The idea of Schulhof making midnight calls to Norio Ohga that nobody knows about—this is just a level of informality that is not tolerable!"

What Peterson could not be expected to understand, because he had no knowledge of it, was that the inner circle of the organization had already resolved at an Executive Committee meeting to proceed with the acquisition in deference to Morita's wishes. Now that Peter Guber had passed muster and it appeared that he and his partner would be available to run the studio, Morita might have been expected to assert again his view that money was not the issue when a unique opportunity presented itself. With Morita in favor of acquiring, and

Schulhof recommending that $27 a share might be needed to ensure Herbert Allen's agreement, it is likely that Ohga would have approved even a price he considered high.

Meanwhile, a separate negotiation between Sony and Guber-Peters Entertainment, Inc., was hurtling forward. Following his return from Japan, Guber flew to New York with Jon Peters, Alan Levine, their personal lawyer and the man they would install as president of Columbia, and Terry Christensen, the entertainment lawyer who had made a name for himself in Hollywood at the time of his widely publicized negotiations on behalf of Kirk Kerkorian. Sony was being represented by a trio of lawyers from the firm of Skadden Arps, but consigliere Paul Burak was also involved, and so were Ted Masaki, Peter Peterson, and Steve Schwarzman. At the first meeting, Guber opened with a presentation of Guber-Peters Entertainment's assets: a projected $50 million in royalties over the next three years from *Batman* and *Rain Man* and net income of $15.2 million to be posted at the end of the current fiscal year in May 1990. Christensen introduced the notion that an evaluation should include royalty streams that would flow to the company from the *Batman* sequels already being planned. Though it was pointed out that Guber and Peters would not be producing *Batman* sequels if they left Warner, Christensen would continue to emphasize the phantom value of the unmade films.

Guber-Peters Entertainment Company was trading at $11; Schwarzman ventured an offer of $8. Christensen replied that Peter and Jon felt strongly that their shareholders deserved, and would expect, $22.50 per share. Moreover, inasmuch as Peter and Jon would be severing a valuable relationship with Warner, they expected Sony to demonstrate good faith by dealing with them first, before the Columbia acquisition. Sony objected that this was putting the cart before the horse, particularly in view of the necessity of obtaining a release from Warner.

Christensen stood his ground. Unless Sony agreed to conclude the GPEC negotiation first, "his people" would walk away. The Sony lawyers proposed simultaneous negotiations, and Christensen agreed. As a quid pro quo, he asked for a $100 million bonus pool to be accessed by Guber and Peters after five years. The figure was lowered to $50 million and accepted. At the end of the day, Sony had come up to $12, and Christensen remained in the low $20s.

Guber acknowledges that by this time he had resolved to seize the opportunity he was being offered: "This was the full-bore banquet, and if I passed this up I would never know how it feels to play in such an arena. If you've ever picked up a Sony-designed piece of equipment, it has a magic about it, and I liked the magic. The toys Ohga showed me in Tokyo: I thought, Wow! I'm part of the toy store! I never even thought about the movies and the television. It was never about the money. The *raison d'être* for the whole move was something intrapsychic as opposed to economic. Once they say, 'The winner is,' and you've won the Academy Award, and you've made thirty-five to forty pictures and done all the TV shows and the music—once you've done all that, it's just about more, and more isn't necessarily better. Different is better!"

Guber and Peters were now in the enviable position of being viewed in Tokyo as indispensable to the Columbia acquisition: as the negotiation proceeded, it was clear that the leverage was theirs. In the fiscal year that had ended on May 31, Guber-Peters Entertainment had lost a reported $19 million on revenue of $23.7 million; the stock, which had been as low as $5, was currently trading at $12. Projections from Blackstone indicated that $13 was a generous offer, but when Christensen threatened to walk away in the final moment, Sony agreed to pay $17.50 per share.* The total for the eleven million shares outstanding would be $193 million, and there was an additional $30 million in debt. Guber and Peters each owned 13.8 percent of the company; at the time of sale, their interest would bring them each $26.25 million. Sony also approved a compensation package that was, even by Hollywood standards, sumptuous. Guber and Peters would each be paid a salary of $2.75 million for the first thirty months of the five-year contract. In the second thirty months, the salaries would increase to $2.9 million, and they would be adjusted each year "in accordance with increases in the cost of living." Guber and Peters would also receive a

* For certain details of the Guber-Peters negotiation, I am indebted to *Hit and Run: How Jon Peters and Peter Guber Took Sony for a Ride in Hollywood*, Nancy Griffin and Kim Masters, Simon and Schuster, New York, 1996.

yearly percentage of combined Columbia and GPEC profits, calculated before taxes, on a sliding scale that began at 2.5 percent on profit in excess of $200 million and rose to 10 percent on $425 million or more. There was to be a $50 million lump sum payment in five years, to be divided among Guber, Peters, and "a group of executives not to exceed five" at Guber's discretion. Finally, Guber and Peters were to receive a payment equal to 8.08 percent of the studio's appreciation as measured in five years. The outlandish 8.08 percent figure was volunteered by Yetnikoff as an additional inducement when negotiations had stalled. When Guber asked, "Why 8.08 percent?" Yetnikoff replied, "So you'll remember me when this is over."

From the beginning, Guber had been very clear that he and Jon Peters were under contract to Warner Brothers, and had repeated his view that securing a release would not be a problem. He based his confidence on precedent: the English producer David Puttnam, under contract to Warner, had been granted a release so that Herbert Allen could bring him to Columbia as the head of the studio for what turned out to be three catastrophic years. Guber and Peters had a similar experience of their own. Two years previously, with money raised from private investors, they had maneuvered to acquire MGM from Kirk Kerkorian, and Warner Brothers had released them from their contract at the time: "I repeated it a hundred times as the depositions will show. 'I am under contract to Warner Brothers. I believe, because of my experience with MGM . . . ' I went through this whole thing with MGM, and Terry [Semel] and Bob [Daly] let me out of my deal to pursue MGM, and then we didn't do it because of the way the financing was set up. They said, 'If you ever get something like that again, we'll let you go do it.' And here it was, a year later! That was the reason that I honestly believed that I had these people's statement of fact and word that I would be released."

Sony's legal team was unwilling to accept Guber's word about a verbal promise: as a condition of finalizing the purchase of Guber-Peters Entertainment, they insisted on a deadline for securing a written release from Warner. They proposed three days after signing, but Christensen argued that this would provide Warner substantial leverage in the unlikely event that they chose to fight, and persuaded Sony to extend the deadline to October 25, just short of one month after the date of purchase: if Guber and Peters failed to deliver a release from Warner Brothers by that date "in form and substance acceptable to Sony," the Sony purchase of Guber-Peters Entertainment would become null and void.

At a meeting during the weekend of September 23 and 24, the Sony board approved the purchase of Columbia Pictures Entertainment for $3.4 billion and the assumption of Columbia's carried debt of $1.6 billion. The meeting was brief and, as always, uneventful.

The Japanese board of directors is a dramatically different animal from the American and even the European board on which it is modeled; and its distinctive makeup and function play importantly in the decision-making process that led to the Columbia acquisition. On an American board, outside directors, with no affiliation to the company, exercise decisive power. They are elected by the shareholders and charged explicitly with the responsibility of overseeing and protecting the shareholders' interests. In most major American and European corporations, the board is dominated by outside directors, who typically represent 75 to 90 percent of the total membership. To be sure, particularly in the United States, there is significant overlapping of membership among the boards of the Fortune 500 companies: and it is frequently charged that an elite club of business leaders with interlocking interests results in self-serving American boards. Nonetheless, the system of checks and balances built into corporate governance as practiced in the United States is focused on shareholders and does by and large function to protect their interests.

Under the Japanese Commercial Code of 1899, Japanese public corporations are also accountable to the shareholder: by a majority vote at their annual meeting, shareholders were even empowered to install or remove directors and officers of the company and to dictate business policy and practice (these executive powers were abrogated under the U.S. Occupation Revised Code). Until very recently, however, there have been few if any outside directors on any Japanese board to represent shareholders' interests. Japanese boards are also distinguished by two classes of directors, ordinary and representative.* While an ordinary director votes, a representative director, elected by the directors, is empowered to bind the corporation legally with his signature and could theoretically divest the organization of assets or sign it away entirely on his own authority (for this reason, the corporate seal is normally kept under lock and key). This hierarchy, unthinkable on an American board, where great care is exercised to ensure that everyone is on an equal footing, serves further to concentrate authority in the hands of top management: it is invariably a company's chief executive officers who become the representative directors.

In the early days of its history, the Sony board was composed of outsiders like Tajima, Maeda, and Mandai, who had the distance and the independence to cast a cold eye on some of Ibuka's wilder pipe dreams, and were acutely conscious of protecting shareholder interest, which happened to be identical with their own. The board that "approved" the Columbia purchase in 1989, on the other hand, was a textbook example of the insularity that has come to be associated with Japanese corporate boards. There were thirty directors, all employees of the Sony Corporation. Twenty-six were ordinary directors, executives on the upper rungs of the career ladder, including a number of engineers who have figured importantly in these pages so far, Toshitada Doi, Kozo Ohsone, and the wizard of Sony, Nobutoshi Kihara. At the head of the table sat four representative directors (Ibuka by this time had been designated honorary chairman and was rarely involved): Akio Morita, chairman and chief executive officer; Norio Ohga, president and chief operating officer; and two executive deputy presidents, Masahiko Morizono, the father of Sony's commercial video camera, and Morita's brother, Masaaki Morita. The Executive Committee,

---

* The notion of representative director—*daihyō torishimariyaku*—is first introduced in the Revised Commercial Code of 1938, and retained in the broader revision of the code promulgated by U.S. Occupation Headquarters under MacArthur in 1950. Based on the laws of the state of Illinois, this new code required for the first time the formal creation of a board of directors as the supreme governing body of the public corporation. Representative directors, empowered uniquely to represent the corporation, were to be elected by "mutual vote" of the board of directors. In actual practice, at Sony and other Japanese corporations, representative directors are appointed by the chairman.

which had already decided to go forward with the Columbia purchase, was controlled by the same people, and in particular by Morita and Ohga.

In characterizing the Japanese board, even Ted Masaki in his lawerly way acknowledges an inherent conflict: "A representative director is the highest-ranking executive officer under our commercial code. Now, that is the source of a lot of confusion and difficulties. Because Japanese corporate law places responsibility on the board members for overseeing the functions of the company's top management, its highest executive officers, who are the representative directors. . . . So from moment to moment, you're either functioning as senior management or you're functioning as an overseer of senior management and there's no way in the real world you can truly make that distinction. So there is a built-in ambiguity in the Japanese legal structure, which I think is continuing to cause difficulties in the Japanese company's ability to arrive at a corporate governance scheme that would be comparable to a U.S.-type situation where . . . the separation of power is so distinct."

In 1991, Sony invited two "outsiders" to join its board, both the chairmen of banks that had extended Sony loans to help finance the Columbia acquisition. The following year, Morita invited Peterson to become the first American outside director, and he agreed, resigning his seat on the board of 3M because it was a Sony competitor in the audiotape business. He continues to sit on the board as an ordinary director but is an outspoken critic of Japanese governance even as practiced by Sony: "They don't like me to say this, but I say it anyway. Corporate governance in Japan is a totally different phenomenon. In America, outside directors are very much concerned with anything that relates to the future, large expenditures of shareholders' money—that's the language that's used—and therefore that gives rise to very careful processes with regard to approvals. . . . In the Japanese companies the decisions are made—I don't know how in the hell they're made, but what I do know is that the outside directors are not involved. You more or less get informed, and the fact that you're a director, you can say what you want, but you're just not involved in making the basic decisions."

Relatively speaking, it is hard to avoid the impression that the Japanese board in general has functioned as a rubber stamp to ratify the decisions of top management, and Sony's board has been no exception. Ohga himself, while he normally advances the official position that the board finalizes decisions in service to the shareholder, recently told me, speaking of an earlier time, when Iwama was ill and he acted as company president, "The Japanese board of directors is a formality. Our Executive Committee was like our real board. That's where we made all our decisions about how much to invest and so on. . . . Consensus takes time! In actual practice, Mr. Morita and I would discuss things and then tell the others of the decisions we had reached. Many things were decided in that manner, between the two of us."

Late on Monday night, September 25, the Columbia board met to review the Sony offer, which was announced the following morning. At 8:30 A.M. on Wednesday, September 27, the Columbia board met again and approved the sale to Sony. An hour later, Mickey Schulhof signed papers committing Sony to make a cash tender offer beginning on October 2 and running for twenty days. Given the high price, it was considered unlikely that another bidder would appear.*

That evening, when late editions of the business press were reporting "rumors" that Guber and Peters were being considered to run Columbia, Schulhof went to Washington to attend a Kennedy Center honors dinner and sat at a table, as he recalls, "with the president of the United States, George Bush, Steve Ross, his wife, Courtney, Charlton Heston, myself, my wife, and there may have been one or two others." Schulhof continued, "I was sitting next to Courtney Ross, and she said to me, 'You know, my husband is really angry,' and I said 'Oh?' She said, 'Yes, you are going to be hiring people who are under contract to Warner,' and I said, 'It's not something that I am prepared to discuss.' And then Steve Ross made an oblique comment to the effect that in this country—unlike in Japan, I think he was implying—we observe the sanctity of contracts. So I knew there was going to be a problem."

---

* In the final days, Columbia did cast about for another bidder in hopes that an auction might drive up the price. Briefly, the oil magnate Marvin Davis seemed interested, creating a flurry of phone meetings, but Davis dropped out as abruptly as he had surfaced.

The following morning, September 28, Sony announced its intention to purchase Guber-Peters Entertainment for $200 million. In articles appearing in the press that evening and the following day, the Guber-Peters contract with Warner Brothers is described as "a potential obstacle."

On Monday, October 2, the Coca-Cola board approved the sale to Sony of its 49 percent interest in the studio; Allen and Company also gave Sony an option on its 3 percent interest. On October 3, Sony announced that it had commitment letters for loans totaling $3.5 billion from five Japanese banks, including the Mitsui Bank of New York, the Fuji Bank, and the Bank of Tokyo. Because businesses in Japan are often sold for multiples of thirty to forty times earnings, the Columbia price tag had not been an impediment to securing loans. "We consider it better to borrow funds locally [in the U.S.] as much as possible," Morita stated, "as current yen-to-dollar exchange rates are unstable. However, we will still use our own funds."

On October 4, a contingent from Warner Brothers, including Time-Warner president Nick Nicholas, Jr., and an attorney, Arthur Liman, met with Sony in New York to communicate the damages that Steve Ross now promised to seek if Guber and Peters breached their contract. Ross wanted Columbia to give up its fifty-year lease on 35 percent of the Burbank lot and the adjoining "ranch," a plot of land for filming exteriors, returning the space and facilities to Warner, and to sign over an office building just outside the main gate, the Studio Plaza, which Coca-Cola had purchased for TriStar. In return, Ross would transfer ownership to Columbia of a lot (the original home of MGM) in Culver City in West Los Angeles. (Ross calculated that the exchange would net Warner roughly $25 million in real estate value.) He wanted in addition a 50 percent interest in the record club business that Sony had inherited from CBS Records. Two years previously, while it still belonged to CBS, Yetnikoff had attempted to sell a one-half interest in the record club to Ross for $500 million. Ross also wanted cable-TV rights to Columbia's film library. As a final ornament on the settlement, Ross would ask Sony to endow a charity fund to be used at his discretion in the amount of $20 million.

Warner's unambiguously hostile intentions led to a meeting between the Guber-Peters and Sony teams which remains the most provocative and controversial episode in the Hollywood negotiation. It lasted an entire day, October 9, and deep into the night, and was punctuated by people hurling oaths across the table and storming in and out of the room. By the time it was over, Sony had acceded to Christensen's demand that the deadline for a release from Warner be waived, and had further agreed to indemnify Peter Guber and Jon Peters personally against the damages that Steve Ross seemed likely to seek. All players were present: Guber and Peters and their lawyers, Alan Levine and Terry Christensen; Schulhof; Yetnikoff; Peterson and Schwarzman from Blackstone; Paul Burak and Ted Masaki; and the lawyers from Skadden Arps. Christensen argued that the time had come for everyone in the room to stand united against Warner in order to achieve the desired outcome. Sony must waive the deadline and indemnify Guber and Peters, he insisted, in order to demonstrate solidarity to Steve Ross. If Sony continued to hedge its bet, Ross would perceive a disparity of interests inside the enemy camp and feel entitled to revenge.

Sony's lawyers were opposed to granting the waiver, not to mention the indemnification, and so were Peterson and Schwarzman. As he listened to them inveigh against the advisability of heeding Christensen, Yetnikoff moved to the other side of the table, as though he were joining Peters and Guber, and began screaming abuse at the Sony lawyers and at Peter Peterson, who recalled dryly: "Schwarzman and I said, 'You can't indemnify them.' We said, just intuitively, 'Why should Steve Ross, at a very sensitive time, when he's just settling down at Time-Warner, let these guys walk out from under his nose?' And it was 'What do you mother-fucking know about Steve Ross, you mother fucker!' And that sort of thing. Charming!"

Yetnikoff asserted thunderously that providing an indemnity made good strategic sense. His reasoning as he reconstituted it for me was persuasive: "I was the only real lawyer employed inside Sony in that room, I mean actually licensed to practice law in the state and city of New York"—Guber would certainly have laid claim to the same unique position— "and what I pointed out was that if you give a guy an indemnity, you're saying I believe in that guy, I'm acting on good faith. And that strengthens your position if and when they come after you. And needless to say, they were coming after us anyway. What point would there be in suing Guber and Peters personally for a billion dollars! Secondly, and I was certain of this, you could see it in their faces as they sat there listening, if we didn't indemnify them, they were going to leave and they were going to kiss Steve Ross's ass any way they could and beg him to take them back!" Guber confirms that Yetnikoff was reading him correctly: "We couldn't be out at both places—that would be taking a stupid pill. They were asking us to jump, and if we fell, we needed a net. Without it, I said to myself, we have to go home, and that would have been fine too."

Late that night, with the dialogue at an impasse, Guber and Peters returned to their hotel with Christensen and Levine to await word from Sony. Some time after 11:00 P.M., Yetnikoff and Schulhof both remember phoning Ohga, who was staying in his Metropolitan Tower apartment in New York but had declined to attend the meeting. "I told him," says Yetnikoff, 'You must do this! You must give them an indemnity, or they're out of here!' And I added that if he wouldn't indemnify them I wanted out of the movie business myself. I said I'd just go back to running the record business." Schulhof's account of his phone conversation with Ohga accords with Yetnikoff's recollection: "I said, 'We are at a point where we are going to have to make a very serious decision. The lawyers are advising us not to provide an indemnity. Walter is screaming that we have to indemnify them, and we are going to need to decide one way or the other, because,' I said, 'it is clear to me if we don't provide an indemnity, these guys are gone,' and I was 100 percent sure of that. And Ohga asked what the recommendations were around the table, and I told him, and he asked me how I felt—and I'd rather not go into that—and then we decided to indemnify them."

Those involved are still reluctant to talk about the indemnification decision in particular, and that is not surprising in light of the fact that Warner Brothers did sue for a billion dollars. Ohga himself has so far declined to address the issue, saying only that many key decisions were made in a brief space of time and emphasizing that he was ultimately responsible for all of them. Schulhof matter-of-factly positions the decision as a considered response to an alternative that "Sony" deemed unacceptable:

"Yetnikoff was pushing for it. Ohga viewed Yetnikoff as a key player in our software business. He had known Yetnikoff for twenty-five years. Walter was the one who made the first introductions out in Hollywood. I think he gave a lot of weight to how he felt. . . . He knows these people. He will lose face if we don't proceed because he brought them into the company. It [was] very possible we [were] going to have serious legal troubles if we [went] forward. But if we buy the company without management, we are going to be rudderless while we look. I've spent three months looking and so far these look like the most appropriate people for our particular needs at the beginning. Now it's a question of who holds weight. The lawyers or Walter."

Schulhof's recollection is interesting in two respects. His suggestion that the decisive vote was cast by Walter Yetnikoff and not himself seems out of character. Was he stepping back from the heat of full responsibility, as his disinclination to reveal his own comments to Ohga might also suggest, or is his an accurate assessment? There is no knowing, but it does seem possible, as Schulhof suggests, that Ohga would have listened to Yetnikoff on this occasion with particular attention, not only because he had more experience in the entertainment business than Schulhof, but also because Ohga seems to have felt about him, as he has often said, that he was an exceptionally cunning business thinker.

Schulhof in his account also neglects to comment on the process by which the decision was reached. It is as if he took for granted that critical decisions at Sony were formulated in the course of person-to-person phone calls between himself—and in this case Yetnikoff—and Norio Ohga. It is precisely this issue or, more properly, its suggestion, which is the touchiest area for Sony. In my conversations about the negotiation with Ted Masaki, he was always at pains to evoke a careful deliberation: "I will say, though, without getting into details of what transpired, it was not something that occurred accidentally. A situation arose which required us to make that decision. It wasn't done offhandedly, there was an extensive discussion, and it was a decision made ultimately because we had to at the time. I can say it was not an offhanded or negligent act. It was something that was thought through and debated. . . . In a large acquisition of that nature, no one makes a decision without the sufficient advice and input of counsel, and that's the way it was done."

Undoubtedly, as Masaki insists, "advice and input of counsel" were abundantly rendered. The question is whether such advice was heeded. Paul Burak, who was also at the meeting, in his own circumspect way implies otherwise:

"I can't talk about that room. Advice was given. I can't reveal as a lawyer what advice was given. A number of people in the room expressed an opinion on this question of waiver. The matter was considered. Why they did it, you're going to have to ask Ohga. The arguments were that they needed management; these people were the right people; they felt that they would be able to get released. But that was not a view that was shared by a number of people and a number of advisers. I expressed my view on that, but sometimes people don't listen to lawyers."

Speaking as the voice of Sony, Masaki makes reference more than once to a situation "which required [us] to make that decision." What can this have been? On October 2, the press reported that Congresswoman Helen Bentley, a Maryland Republican, had asked the Justice Department to look into possible antitrust violations in Sony's purchase. Was the company feeling the need to complete the entire transaction before the federal government interdicted it? Was Sony worried about another bidder showing up before its tender offer expired on October 30? Or was it simply Yetnikoff and Schulhof's apprehension that Guber and Peters were poised for flight which had required a hard decision? Clearly, Peter Peterson has no idea, though he was also in the room and close to the center of Sony power by virtue of his relationship with Morita. His reflection is colored by the bewilderment and dismay that characterize his view of the entire proceedings:

"This was another one of those telephone calls, I would guess. And in Norio's defense, this was complicated. Did Mickey say to him: 'Norio, Blackstone advises against this, the lawyers advise against this, here's why they advise against this, but we think we ought to do it anyway'? I don't know if he said that, but I doubt it! In a normal company, there would have been a fairness opinion, a presentation by the outsiders, the board would have heard their views, and so forth. But in Norio's defense, my guess is he wasn't told very much. Because I don't know why Norio would take a risk like that, you see. So I don't think Norio got the whole story . . . I prefer to believe that if Norio had been told that our financial advisers oppose this and here's why, Norio would have said, 'Wait a minute! Let's look into this! . . . I always found Ohga to be reasonable. I assume his personality structure didn't change overnight, and that there was something about the process that was highly superficial or highly something that was not typical of normal communication."

Ted Masaki is not the only Sony insider who implies a secret something that affected the acquisition process decisively and that he declines to usher forward from the shadows. Morita's brother Masaaki was in New York at the time of the acquisition, serving as chairman of the American subsidiary group. Though he was not directly involved in the negotiations, I asked him to explain why his brother had allowed the process to be dominated by a dialogue between Norio Ohga and Mickey Schulhof. "That's a good question," he said. "I'll tell the whole story after everyone is dead." On another occasion, when I was interviewing Ohga, he made a similarly provocative remark: "Ten years from now, I'll tell the whole story." I was sufficiently struck to tell him what Masaaki Morita had said and to inquire whether there was some deep, dark secret that everyone was at pains to conceal. "It's nothing like that," he declared. "There's no dirty story at all, certainly none of the kind of thing you sometimes hear about people taking money under the table. Nothing like that will ever emerge even a hundred years from now because nothing of that nature ever happened."

How, then, to account for Sony's behavior during the negotiation? According to Peter Guber, "There is an emotional momentum to a deal that drives it forward to happen. Once that momentum kicks in, nothing can stop the deal from happening. They had made a fundamental decision to buy the company, they believed they had someone in me that could manage it, the missing piece, and after that, any and all further obstacles were merely impediments. Nothing was strong enough to stop that momentum, which was emotional more than it was intellectual." Guber's intuitive grasp is the more remarkable in light of what we know about the Executive Commitee meeting in August.

On October 10, Sony announced its intention to proceed with the Guber-Peters purchase with or without a release from Warner. Three days later, Warner Brothers filed a lawsuit against Sony in California Superior Court which accused Sony and Guber-Peters of "secrecy, treachery, and fraud" and alleged, with familiar Hollywood bombast, that Schulhof and Yetnikoff had induced Guber and Peters to breach their contract by secretly offering them "the most lucrative package of financial inducements in the history of the film industry." Warner was seeking $1 billion in actual damages, punitive damages, and a permanent injunction prohibiting Guber and Peters from accepting any employment outside their contract with the Warner studio. The same day, Sony and Guber-Peters countersued for "sabotage" of their merger agreement. In their twenty-six-page brief, Guber and Peters asserted that their contract with Warner bound them as film producers but not as studio executives, and contended that Time-Warner cochairman Steve Ross and Warner Brothers chairman and CEO Robert A. Daly had "previously acknowledged several times" that they would be free to pursue opportunities elsewhere. Terry Semel retorted, in a court filing, "I do not make oral agreements to terminate written agreements, especially when they involve key talent and millions of dollars!"

I asked Mickey Schulhof whether he considered Peter Guber an honest man: "There are people who will tell you that Peter is manipulative and so bright that nothing happens by chance. There are others who will tell you that Peter slants the facts to go in a direction where he would like to go without ever lying. There is something probably in between. Peter is not a simple human being. I don't think anybody is who succeeds in the motion picture business." Peter Peterson was more pointed: "One theory holds that Schulhof was dying to be a big player in the business. This was his way of winning their support.

Maybe at some naïve level he may have believed them—he couldn't believe that these charming, wonderful fellows would lie to him. But we know that people from Hollywood do lie, a little."

In fairness, it seems likely that Peter Guber truly believed that Semel and Daly would let him go. I asked whether he was surprised that Steve Ross and company had brought suit. "It was more than a surprise. It was a deep shock. We had made them a lot of money, gigantic, gigantic sums of money, and had been very successful with them over a long period of time. And this was certainly the opportunity of a lifetime that someone else would have looked at and understood. The only loathsome thing I'll say to you, and it is loathsome, and I regret having had this feeling: whether it was legally correct, and no doubt they were legally justified, their behavior was uncharacteristic, it was discompassionate, it was avaricious, and it was political, because they knew the American press didn't like the Japanese."

In a public statement, Robert Daly maintained that Sony's initial requirement that the producers obtain a release from Warner was proof that the binding nature of their contract had been recognized. Steve Ross in his prepared statement said icily, "We regret that Sony has attempted to enter the U.S. motion picture industry by illegally and willfully raiding key talent under exclusive contract to a competitor. This conduct cannot, must not, and will not be condoned."

Such statements, and the anti-Japanese sentiment they barely concealed, were deeply disturbing to Sony. From the earliest days of Sony America, Morita had taken great care strategically to ensure that Sony would not be perceived as a Japanese intruder in the U.S. marketplace. Until now, he had succeeded wondrously in camouflaging the company's identity, to the extent that many consumers had no idea that Sony was Japanese. The illusion had been shattered at a time when uneasiness and resentment had intensified, which made the company more than ever vulnerable to the kind of adverse publicity that the lawsuit generated.

From the moment the Columbia purchase had been announced, in response to anxious speculation in the press, Schulhof, acting on orders from Ohga in Tokyo, had emphasized repeatedly that Sony's new film business, like CBS Records, would be handled by the American subsidiary as an entirely American company without interference from Tokyo. Morita dined with foreign correspondents in Tokyo on October 3 and was quoted in the *New York Times* the following morning: "I expected this adverse reaction and I am not surprised. These are days when Japan investment is much criticized, and I know the American people have a special affinity for the movie industry." He acknowledged that the only way to assure Americans that the Columbia purchase was not counter to their own interests would be to operate the business "purely as an American company," which he promised to do. Asked whether the studio would be allowed to make a film critical of Emperor Hirohito and his role in World War Two, Morita replied, "We would have no objections. Maybe the Japanese people and the Japanese government would not want to show it here. But if Columbia management believed the story was good, then we make the movie." Notwithstanding assurances of this kind, the national reaction to the purchase remained negative, and talk of the Warner lawsuit poured gasoline on the spreading fire.

The situation was made worse by the appearance in the United States of a book cowritten by Morita, *The Japan That Can Say No*, which was perceived as aggressively anti-American, "America-bashing of the highest order." Morita's coauthor was Shintaro Ishihara, a talented novelist who had given voice during the 1950s to a generation of bewildered postwar Japanese youth and had subsequently become a conservative politician, a member of the Diet since 1968, and an outspoken nationalist. During the late eighties, Morita more than once had accepted Ishihara's invitation to address his political support groups. *The Japan That Can Say No*, published in Japanese in January 1989 by Kobunsha—Yoshiko's family's publishing house—was a collection of alternating excerpts from Morita and Ishihara on the subject of Japanese-American economic relations. In late September 1989, an agency of the Pentagon circulated a partial, pirated translation, which was held up on Capitol Hill and in the press as evidence of Japanese perfidy. The most inflammatory section was Ishihara's analysis of U.S. trade policy vis-à-vis Japan as a function of American "racial prejudice."

Morita's portion of the book was by comparison bland, a rehashing of his standard set pieces on the inequity and bias of America's position on foreign trade. He was, nonetheless, extremely agitated and chagrined by the abrupt appearance of the unauthorized English version. It was as if a mask he had always been careful to wear in the presence of his Western friends had been momentarily torn away. He insisted that the incendiary remarks had come from Ishihara and not himself, and orchestrated a public relations campaign to disavow and dissociate himself from the book. When the Japanese edition, which had sold more than a million copies, was scheduled for reprinting, and again when Simon and Schuster published a complete English translation in 1991, he demanded deletion of his portion of the text.

A court hearing in the Warner suit was set for November 2, but there was never much likelihood that Sony would allow the dispute to go to court. Having paid $5 billion for a studio that was basically idling and an additional $200 million for Guber and Peters, the pressure to settle would have been crushing even without the added threat posed by mounting anger and bad publicity. Yetnikoff had been involved in the negotiation, but was told to stay away after he became embroiled in a dispute with Ross's wife, Courtney, about a Michael Jackson song she wanted to use in her documentary about Quincy Jones. Possibly, as Schulhof represented, Ross had declared an end to further discussion with Sony so long as Yetnikoff remained in the picture. Perhaps Schulhof took the opportunity to emerge from Yetnikoff's long shadow. In any event, Yetnikoff's "demotion" at this time was the beginning of a fall from grace which would end a year later when Ohga abruptly fired him.

Late in October, Ohga and Schulhof visited Ross in his new offices at Time-Warner in New York, sending a signal that Sony was ready to settle and move on. Ohga remembers the meeting as cordial: "But guys like Steve Ross are usually all smiles when you meet them face-to-face," he told me, "and the next day the lawyers deliver you a big book full of new demands." A settlement was signed on Tuesday, October 31. Sony would accept the exchange of Burbank for Culver City. Warner would obtain some cable broadcast rights to Columbia's film library, and a 50 percent share of Sony Music's record clubs. Ross's request for a $20 million charity fund was dropped. The settlement added about $400 million to the cost of the acquisition.

On November 16, 1989, Peter Guber and Jon Peters went to work as cochairmen of Columbia Pictures Entertainment, presently to be renamed Sony Pictures Entertainment. Chairman Akio's dream had come true.

9

HOLLYWOOD CONTINUED: TAILSPIN

The cochairmen's first job was to ready the Culver City studio for occupancy. "The place was a dump," Guber recalled, "horrible! You wouldn't want to work there. In a creative environment you want a campus that's exciting and fun to come to. This lot was technologically horse-and-buggy! We had to paint the building, so to speak, to announce, 'We're here, Sony is here!'" Guber's approach was characteristically grand. He restored the exterior and interior locations on the lot's Main Street and the commissary to their original state during Columbia's heyday under the Cohen brothers; he installed a high-definition TV facility and a state-of-the-art digital effects and animation studio built with Sony equipment; and he lavished particular care and money on the Irving Thalberg office building, an art-deco masterpiece and a Hollywood landmark. The production designer on *Batman*, Anton Furst, was retained to oversee the project. Original plans called for acquiring a mortuary adjacent to the lot and razing it to make room for a Japanese koi pond, but this and other Jon Peters extravaganzas, including a theme park to be called SonyLand, were abandoned early on. All told, Sony approved a capital allocation of $110 million, to be spent over three years, for transforming Culver City into the "West Side's premier facility," a goal that Guber achieved.

Columbia continued to operate on the Burbank lot until April 1990; between April and December, Guber and Peters and their staff were commuting to Culver City. The move was not fully accomplished until the spring of 1991. Meanwhile, the chairmen moved into the penthouse floor of the Studio Plaza building owned by TriStar, which they refurbished in a Japanese motif at a cost of $250,000. The redecorated offices included water pools, floor-to-ceiling bamboo blinds, and Japanese "mud walls" installed by a crew of craftsmen brought in from Japan. When the Louis B. Mayer suite in the Thalberg Building had been restored to its former grandeur, the temporary office in Burbank was abandoned. Ohga was mildly alarmed: "That was the first indication I had that they were not good businessmen. They built fantastic offices for themselves in Burbank, and then, before the work was completed, they told me they were moving to Culver City. I failed to understand that."

Having prepared a building full of designer offices, it was incumbent on the new studio heads to fill them with executives who could manage the business day to day, a challenge that neither Guber nor Peters seemed eager to accept. Within a month of taking over, Guber had hired their entertainment lawyer, Alan Levine, forty-two, to be president of the film entertainment group. Levine was a tightly wrapped Republican and Rotarian who dressed in conservative business suits and was a member of the Hillcrest Country Club. The third-generation entertainment lawyer in his family, he had represented filmmakers all over town, but had not done business with or managed them, and would shortly earn a reputation for ineptness when dealing with the creative community. It was hard to imagine anyone less like the voluble, flamboyant, ponytailed Peter Guber than Levine, who nonetheless would prove himself to be a cunning power player and, above all, a survivor: when the dust had settled after the upheaval at the studio in 1994, he alone would remain.

In February 1990, Guber persuaded Mike Medavoy, forty-nine, an executive at United Artists who had founded Orion Pictures in 1982, to accept the position of chairman of Columbia's TriStar Pictures unit. The following month, Frank Price, fifty-nine, who had previously headed both Universal and Columbia, and who had been briefly considered for Guber's job and had been rejected by Walter Yetnikoff as stodgy, was installed for the second time in his career as chairman of

Columbia's film unit. On April 15, 1990, the new triumvirate agreed to an interview with the *Los Angeles Times.* In hindsight, their exchange prefigures the infighting and confusion in the chain of command which would characterize Guber's administration:

Times: Alan, you have a couple of big guys to your left and right. Where are you going to fit?

Levine: Right here! Each of these fellows has the ability to run his own shop, and I've managed a lot of people . . . on the basis of giving them help and assistance, whatever they've needed to flourish. That's what I did with Jon and Peter, and that's what I'm going to do with these guys.

Times: Where are all three of you going to fit vis-à-vis Jon and Peter? Is there an agreement about how the process will work?

Levine: As far as day-to-day management of the company is concerned, that's my job. Jon and Peter are running the entire company. What I believe Jon and Peter really are looking to do is help set the map, or help set the plan for the future. Frank and Mike have autonomy when it comes to developing, producing, marketing, and distributing their pictures.

Times: Mike and Frank, what will happen when you decide you're going to make a film? Do you expect to make that decision jointly with Peter or Alan?

Medavoy: Both Frank and I have autonomy to go and make whatever film we decide to make.

Times: Any budgetary limits to that? Has anyone said, it's all yours if you're talking $25 million, but come talk to me if it costs $40 million?

Medavoy: We've never had that discussion because I don't think we've ever contemplated not having a discussion about movie projects, and I don't mean asking for approval. We both feel there's a resource in Peter, Jon, and Alan, and we want to counsel with them.

Over time, Guber continued to install expensive film executives between himself and ultimate responsibility for key decisions. In August 1990, Frank Price, over the heated objections of Alan Levine, who was theoretically his boss, and with secret approval from Guber, who was predisposed to choose nonconfrontation when it was an option, brought in Jon Dolgen, a former president of 20th Century–Fox known to be a skilled cost-saver and abrasive as a nail file, to be president of Columbia Pictures. Before long, Dolgen would be functioning, with considerable effectiveness (albeit inadequately given the seriousness of the problem) as a kind of financial controller to the studio overall. A few months later, Syd Ganis, a former studio chief at Paramount, was hired, at a salary reported to be $650,000, to be head of "strategic planning for global strategies." By this time, Columbia Pictures was being likened to a "graveyard for elephants" because of the number of senior executives it employed at top salaries. The same executives received exorbitant severance fees on their way out a revolving door, which soon was turning at vertiginous revolutions per year even by Hollywood standards. All in all, the new chairmen's approach to managing the studio resulted in an annual overhead estimated at twice that of the other studios.

As overlord of Sony software, Mickey Schulhof was Peter Guber's American boss. On Ohga's orders, production budgets over $40 million had to be approved by Schulhof, who also had approval of national marketing campaigns and executive hiring and firing when large sums of money were involved. It has been said that Guber held Schulhof in thrall and could manipulate him at will. Schulhof admits to being dazzled by Guber's charm and by the glamour of the film business to which he had been granted access with the preeminent status of "player." He denies that he was blinded by the relationship: "It's very difficult in any entertainment business to have the same kind of relationship you would in a financial business. There is a requirement for—collaboration is the wrong word, creative involvement maybe. I'm not a filmmaker. I refused to take a house in California. I went in, I had meetings, and I left. But we weren't exactly at arm's length. That just wasn't how it was with Peter Guber."

There is no measuring Schulhof's degree of objectivity where Guber was concerned. They became friends, hosting each other at parties on both coasts and taking family vacations together around the world. And it is a fact that Schulhof approved—some observers would say "rubber-stamped"—every project Guber submitted. At the very least, Guber had a way with Schulhof, just as Schulhof had a way with Ohga. Early on, Guber made it clear to his staff that communication with Schulhof was to go through him. Schulhof was equally zealous about protecting the exclusivity of his channel to Ohga. And so it went, from the lips of Peter Guber, via Mickey Schulhof, to the ears of Norio Ohga. Given Ohga's predisposition to accept Schulhof at his word, this closed-circuit communication seems to have delivered to Guber the equivalent of a license to spend.

The new administration's mandate, as articulated by Schulhof in consultation with Ohga, was to jump-start the studio and to recover market share, which stood at a lifeless 9 percent, last place, when Columbia was acquired. When Guber took over, he was dismayed to find the production pipeline nearly empty, with two projects in development at Columbia and two more at TriStar. He and Peters and the two studio heads reporting to them, Price and Medavoy, went shopping for hot properties to rush into development. Guber lured a number of A-list producers and directors to the studio with long-term production agreements, which were considered the richest deals in town. Danny Devito, Tim Burton, and James L. Brooks were all invited to move their production companies onto the lot and promised funding for their projects. Peters brought in Barbra Streisand to direct and star in a film she had been shopping, *The Prince of Tides.* As she would not be singing in the film, Guber had reservations, but Peters insisted and got his way. He also created a sensation by paying $700,000 for a script called *Radio Flyer*; the story of two brothers who escape into fantasy from their abusive stepfather. The script had put him in mind of his own childhood when he read it and had made him weep. Peters laid out an additional $300,000 as a first-time director's fee to the young author of the script, David Mickey Evans. Frank Price had come to Columbia with a pet project, *Return to the Blue Lagoon*, and was also working on *A League of Their Own*, starring Tom Hanks and Geena Davis; *Boyz 'n the Hood*, a first film with a modest budget of $6 million; and *Bram Stoker's Dracula*, to be directed by Francis Ford Coppola, already a friend of Sony's because of his interest in using its digital videotape as an alternative to film. At TriStar, Mike Medavoy quickly placed a number of expensive and highly visible projects in development: the director Paul Verhoeven's *Basic Instinct; Mr. Jones*, starring Richard Gere; and *Bugsy*, an expensive chronicle of the gangster Bugsy Siegel, produced by and starring Warren Beatty and directed by Barry Levinson. The feather in Medavoy's cap was Steven Spielberg, whom he had signed to direct *Hook*, a $62 million retelling of *Peter Pan*. Allowing for the eighteen months required on average to complete a film, the new slate of features would be ready for release in time for summer 1991. Guber planned to hit the jackpot during that year's Christmas season, with *Bugsy*, *Prince of Tides*, and Spielberg's family extravaganza. As the production was readied to begin principal photography, a life-size pirate ship was erected on the same soundstage where Dorothy had arrived in Munchkinland in 1939 when Culver City had been home to MGM.

On May 9, 1991, before any of the new films had been released, the business press reported that Jon Peters had "resigned" as cochairman of Columbia TriStar after fifteen months on the job. "Few expected Peters to remain in management," wrote the *New York Times*. "His flamboyance, volatility, and general loathing of bureaucracy reportedly rankled Sony executives." The *Los Angeles Times* quoted a producer on the Columbia lot: "We all chuckled as we saw Jon trying to adapt his producer's life style to a corporate mentality. A studio executive's job is basically to say 'no' and hold things down. A producer's job is always to ask for more, and this guy is a born producer."

Needless to say, Peters had not resigned: Mickey Schulhof had summoned him to his office in New York and fired him. The previous day, Alan Levine had visited Schulhof, and there was speculation that Levine was responsible for Peters's undoing. Schulhof says it was clear to him that Levine was merely fronting for Guber, who was at the end of his rope: "Jon had a growing reputation for disrupting meetings and browbeating people, and it was wearing Peter down. Peter is a manipulator, a very good one, and a very nice one, but he has never been able to fire people easily, particularly not a partner. I checked to be sure this was nothing personal, not just Peter trying to get rid of a former partner, then I called Jon to New York and told him he had to go. There was a bit of stammering at first, and then 'Why?' and then

'What's wrong?' But when I fire somebody I don't go into extensive explanations because it's just an invitation for trouble later on."

If Guber and Schulhof had come to view Peters as a liability, it wasn't simply because of his abrasive management style and fickle enthusiasm. From the moment he had taken office as cochairman, his headlong extravagance had been a constant source of delight to the press. There were stories of former wives and lovers with no experience in film being awarded huge production contracts, of platoons of beautiful blondes employed at outlandish salaries as secretaries and personal assistants, of chefs and chauffeurs, and, as if the new job at Sony allowed time to dabble in real estate development, of buying and selling estate properties in affluent Santa Barbara, north of Los Angeles.

The deliciousness of the subject—unimaginable luxury purchased for wicked sums of money—invited by its nature a degree of hyperbole and distortion. There was, for example, a widely reported rumor that Peters had dispatched a Sony jet to London empty of passengers but filled with flowers for a former girlfriend. Schulhof, whose authorization was required before an international flight could leave the ground, says it never happened.

To be sure, Guber's own appetite for luxury and his unabashed sense of entitlement were easily a match for his partner's, but Guber was savvier than Jon Peters, and went about enriching his professional and personal environment less conspicuously. Peters had flaunted his good fortune from the beginning. Perhaps his self-indulgence would not have hurt him so badly if he had not also taken every opportunity to boast to the press about his access to Sony's deep pockets and the wealth and power it was affording him. Journalists were happy to quote him at length. Word got back to Tokyo that Peters was forging a reputation as a braggart and, worse, an exploiter of Sony generosity and this, from Peter Guber's point of view, as much or more than his erstwhile partner's orneriness as a comanager, was a genuine liability.

From the outset, Walter Yetnikoff and Schulhof and later Guber himself had done what they could to isolate Peters from the Japanese, whom he was certain to offend with his repellent and, to them, inexplicable behavior. But Peters was not easily contained. He was seen more than once running his professional hairstylist's fingers through Morita's silken hair or rushing across the commissary to take Norio Ohga in his arms and kiss him. According to one close observer: "If you were Sony, you could not conceivably accept the madness of Jon Peters absent your previous successful experience with Walter Yetnikoff. With Yetnikoff in mind, having made money in your collaboration with him, you could shrug and tell yourself, "They're *shvartzers*."

The terms of Peters's severance deal remain undisclosed, but it is likely he received in the neighborhood of $20 million dollars, an amount roughly equal to money he had spent on development projects that had not materialized. He also received what Schulhof described as "a modest production deal" for two years and space on the Sony lot. In a prepared statement, Schulhof declared that Sony expected Peters to make "an even greater contribution to Columbia and the Sony family" in his new position. Two years later, one of many small ironies that dot the landscape of Sony's Hollywood adventure, Peters moved back to Warner Brothers.

The summer after Peters had departed, Guber proposed to Schulhof that the name of the studio be changed from Columbia to Sony Pictures Entertainment. As always, he was being strategic: "I thought to myself, These guys are not committed to this company! What can I do to get them to commit? Paint their name on the building! It was going to be much harder for them to dump something called Sony Pictures Entertainment than Columbia Pictures. I felt if ever we were going to get them to go public, if ever we were going to get them to go further and build the company, they'd have to own it, *de jure* ownership. Paint their name on the building." Schulhof agreed, and made the recommendation to Ohga. On August, 7, 1991, the board formally adopted the new name.

In September, Frank Price followed Peters out the door. He had never been at the top of Guber and Peters's list of candidates to run Columbia. They had preferred an executive they had known at Warner Brothers during the height of their success there, Mark Canton. For obvious reasons, a second raid on Warner was out of the question, but Guber had been negotiating secretly for Canton's release since June. Once Terry Semel and Bob Daly had agreed to release Canton from his job as executive vice president of worldwide production early in September, Price's days were numbered. He left in the final week of September, generating negative press about the tumult at Sony with bitter remarks that he shared with journalists.

Canton moved in at once. Guber had intended to make him executive vice president of Sony Pictures, but Canton insisted on and received the title of chairman of Columbia. With Canton's volatile presence on the scene, together with Medavoy, who would also be forced out in early 1994, and Syd Ganis and Alan Levine and Jon Dolgen all reporting in principle to an increasingly distracted Peter Guber, it became more difficult than ever to ascertain who was responsible for green-lighting a film at Sony Pictures Entertainment.

The summer of 1991, when Guber's new slate of films began appearing in theaters, delivered mixed results at the box office. *City Slickers* became the number-one comedy of the year, *Boyz 'n the Hood* was a profitable sleeper, and *Terminator 2*, a production belonging to Carolco, which Guber had undertaken to distribute, was hugely profitable. On the negative side of the ledger, *Return to the Blue Lagoon*, Frank Price's pet project, lost money, and *Hudson Hawk*, starring Bruce Willis, was a disaster.*

That Christmas, *Hook* was a major hit despite unfavorable reviews, and so was *The Prince of Tides*, which opened on Christmas Day, but *Bugsy*, a much touted and very expensive project, lost heavily.

The 1992 season opened in February with *Radio Flyer.* Six weeks into production, Mickey Evans, the youngster whom Peters had hired to direct his own script, had been yanked off the project and replaced by the veteran producer-director team of Lauren and Richard Donner, who had received an additional $6 million for their services. Two years

---

* For details of box-office performance, I have relied throughout on *Hit and Run: How Jon Peters and Peter Guber Took Sony for a Ride in Hollywood*, Nancy Griffin and Kim Masters, Simon and Schuster, New York, 1996.

in the making at a cost of $32 million, *Radio Flyer* lost a substantial $32 million. Frank Price's *Gladiator* lost another $20 million. The following summer, 1992, *A League of Their Own*, Guber's *Single White Female*, *Mr. Saturday Night*, *A River Runs Through It*, and *Bram Stoker's Dracula* were all profitable to differing degrees, but TriStar had a number of failures, including *Chaplin*, another very costly flop.

In public, Guber and Schulhof were continuously avowing "deep satisfaction" with the film unit's performance. They invariably pointed to Sony's gain in share of the domestic box office, which had climbed from 9 percent to nearly 20 percent by the middle of 1992. Echoing their satisfaction in Tokyo, Ohga proudly declared that a 20 percent share of market represented "an all-time high for a single movie company in the last ten years."

The problem was that, market share notwithstanding, Sony was overspending on production and losing money as a result. The studio's Christmas 1991 releases alone had cost an estimated $225 million, fully half the $450 million annual investment in twenty to thirty films that was standard at other studios. For the fiscal year ending March 1992, Sony Pictures Entertainment was projecting total costs of $700 million. While the negative numbers were concealed by consolidating the performance of Sony Pictures with the profitable music business, under the single heading of "Entertainment," it was clear by mid-1993 that the movie business was losing money at the alarming rate of $250 million annually. To critics who suggested that his costs were out of control, Guber responded defiantly. "This place was a mess when I arrived," he told the *New York Times*. "I have put Humpty-Dumpty together again. This is a creative process. It's not like making widgets. We brought talent into the studio. We got Spielberg! That's like being blessed by the Pope. You know what signal that sends? That we are back!"

At a meeting in New York in the summer of 1992, Ohga himself ordered a retrenchment. Jon Dolgen, the cost-cutter, now president of the Motion Picture Group just under, or alongside, Alan Levine, was charged with clamping down on production spending at TriStar and Columbia. While agents in Hollywood observed that Sony was buying fewer scripts for less money, Medavoy and Mark Canton both denied that they were feeling pinched.

All winter and into the spring of 1993, which was to be a calamitous year for Sony on many levels, the talk of the town was the June 18 opening of a Mark Canton project called *Last Action Hero*, starring Hollywood's reigning emperor of "marquee value," Arnold Schwarzenegger. Though it had been budgeted at $60 million, by the time it was released, at a total cost, including promotion and advertising, of over $120 million, *Last Action Hero* had become the most expensive film in Hollywood history to date. Schwarzenegger's fee of $15 million, plus a percentage of profits, established another record. Everything about the production, including "Arnie" himself, was larger than life. The marketing campaign reached an interplanetary scale: Sony paid $500,000 to have the film's title and Schwarzenegger's name painted on the tail of a NASA rocket to be launched in May (at the last minute, the launch was canceled).

When *Last Action Hero* opened in June, audiences walked out in the middle. Early reports projected a loss of $20 million, but the real figure was more than twice that amount. The laughter at Sony's expense was at least as damaging, and the untrammeled extravagance and hyperbole of the marketing campaign fanned the fire. Sony's chagrin was aggravated by the fact that Steven Spielberg's film for Universal, *Jurassic Park*, had opened just days earlier and was making history at the box office.

Later that summer, James Brooks's long-awaited musical, *I'll Do Anything*, and another Frank Price selection, *Geronimo*, both performed disastrously. TriStar offset the losses with a series of hits, *Cliffhanger* with Sylvester Stallone, *Sleepless in Seattle*, and Clint Eastwood's *In the Line of Fire*. But the Canton-Schwarzenegger debacle continued to hang over the studio like a pall. "*Last Action Hero*," Schulhof told me, "became the lightning rod and symbol of excess and overspending and zero results. The company was developing real problems, and we knew we had to change gears."

Observers during the Guber years were at a loss to account for what appeared to be Sony's reluctance to become actively involved. It is clear from the following exchange in the *Los Angeles Times* interview in April 1990 that Sony headquarters in Tokyo had vanished from the management picture of its new business within a month or two of the acquisition:

Times: Mike, when you were hired, did you talk to Norio Ohga first? Was he part of that process?

Medavoy: Nope, never.

Times: Was anyone in Japan part of that process?

Medavoy: Nope.

Times: Was Walter Yetnikoff part of that process?

Medavoy: Nope.

Times: Who was?

Medavoy: The conversations were with Jon, Peter, and Alan.

Times: Alan, on your side, did someone get approval from a Japanese executive to make these hiring decisions?

Levine: No, no approval. We had our mandate when we came into this company to build it and to run it.

Times: Has Sony set any mission for the next two years?

Levine: No, as a matter of fact, they told us to get our arms around the company and assemble the best management team at all levels that we possibly can and then come and tell them what we want to do with the company."

Eighteen months later, Frank Price would leave Sony without ever having met Norio Ohga. Peter Peterson, with perplexity and dismay, ruminates on this period in Sony's history: "It's a cultural, social, psychological issue here, as to why they would so distance themselves from hands-on. I mean, what was the fear? What was the insecurity? What did they think would happen if they got too involved? What was it about the cultural, psychological, personal, that would have permitted this degree of passivity?"

Part of the answer may have been strategic. It is well to remember that Sony had purchased Columbia at a time when the Japanese buying spree was being viewed with mounting hostility as a silent invasion, and that Sony had gone out if its way to offer assurances that the studio would be run as an American company. "If we act like the American occupying army that controlled postwar Japan," Ohga told the *New York Times*, "we will be bashed, but if we manage to keep it totally as an American company, everything will work out fine." As it turned out, Ohga's confidence was unfounded: the press remained relentlessly hostile, jeering at every Sony misstep and loss that it managed to uncover, and predicting continually that the Japanese company was on the verge of admitting defeat and selling out. Sony was and remains acutely sensitive to public opinion, and the attitude of the press continued to function as a deterrent to interfering with the American team.

But Akio Morita's insistence on the long view may have been more decisive than any strategic considerations. From the beginning, it was Morita who had sponsored the move to Hollywood, and so long as he felt that Sony Pictures was being established as a presence in the American marketplace, the staggering cost is not likely to have concerned him. Besides, when he found the time to pay attention to his expensive new toy, which he managed to do infrequently, it afforded him great pleasure. A born showman himself, one of the world's great grandstanders, Morita had a weakness for the special variety of glamour that only Hollywood could provide, and Guber was very good at staging the kind of event that he could be counted on to relish. During the filming of *Hook*, for example, he arranged an intimate Saturday brunch for Morita and Yoshiko on the deck of the great pirate ship and invited Spielberg and the stars of the film, Dustin Hoffman, Robin Williams, and Julia Roberts, to join them. Morita snapped digital pictures and videotaped with the latest Sony gear and had the time of his life. From his personal vantage, at that moment, his film studio could hardly be doing better.

In the past, particularly if he was in any way concerned, Morita had kept his hands on the business as though he were a potter turning clay. But beginning in early 1990, he was virtually out of the picture: from this time he began to focus his vast energy on international business relations and, in particular, on his consuming campaign for the chairmanship of the Keidanren, the Federation of Economic Organizations. There is no exaggerating the critical importance of his disengagement: from 1990, the burden of responsibility for overseeing Sony business fell principally to Ohga. And while Norio Ohga can hardly be said to have been "hands-off," he was inclined in times of difficulty to hold himself at a remove. Nor was Hollywood by any means the only problem demanding his attention. The American electronics business was foundering, and Mickey Schulhof on the East Coast was creating unrest inside the organization as he augmented his "software" empire. Nonetheless, Schulhof remained very much the apple of Norio Ohga's eye. In fact, a major reason for Ohga's disinclination to intervene in Hollywood, even as the business spiraled downward, may have been his heartfelt reliance on the man he referred to as his kid brother, and, remarkably, on Peter Guber himself, who had secured his confidence at their first meeting.

Nevertheless, by the summer of 1993, Ohga was expressing his concern. "I spoke to Alan Levine because he was president, and I told him in no uncertain terms that budget control had to be applied even in the movie business . . . Levine may have been a first-rate Hollywood lawyer, but he was no businessman." It is worth noting that Ohga chose to address himself to Levine, "because he was president," rather than to Guber. To Ohga's way of thinking, the president of the studio was responsible for its financial health. As a filmmaker and an artist, Guber was not to be held accountable in the same way.

From around the time of the *Last Action Hero* debacle, however, Guber had been evidencing a disaste, not only for the problems of running a studio, but for his métier, producing films. He began talking obsessively about diversifying into sports franchises and real estate development. And he increasingly deferred to a gang of Sony executives—Ganis, Dolgen, Canton, Levine—who were themselves locked in imponderable power struggles. "It was black-holes-ville over there," I was told by a producer with a project at Columbia. "When you finally got hold of Peter, he'd tell you he didn't get into this and that, he didn't do movies, he didn't do scripts, he didn't get involved with production problems—that was for the *vantzerei.*"

Schulhof observed Guber falling away with growing concern: "Peter was burned out. He was personally troubled by all the negative press, and with Peters gone he couldn't deflect criticism anymore. He began delegating more and more to Canton, things he was much better at doing himself. He was great at reading a book or a script and seeing the film in it. And during the first three years, he designed his marketing personally, came up with tag lines and concepts for trailers. When he rolled up his sleeves and put his hand in, things were always visibly better. I told him his performance was skidding, and for a while he seemed better, but it didn't last."

Ohga himself was dumbfounded by what he perceived as a sudden change in Guber's attitude, and remains perplexed: "We had proof of his superior ability in the fact that we climbed to number one at the box office after he took over and remained there for three years. And then one day, suddenly, he *lost enthusiasm*"—Ohga used the English words—"for making movies. I didn't understand at the time, and I still have no idea how someone like him, who had lived for movies—movies was what he had talked and breathed, and now all of a sudden it was all sports teams. He wanted to acquire—football was it? volleyball? Maybe it was a basketball team or a baseball stadium. I told him we weren't in the business just to make money at the box office, we needed films as video software, and he should keep his eye on the goal and stop wasting his time on other things. But he couldn't stop talking about it."

On learning of Ohga's judgment that he had "lost enthusiasm," Guber exclaimed: "Yes! That's precisely what happened!" His explanation, that Sony's refusal to see beyond the box office had worn him down, seemed genuine and was persuasive: "They understood the magic of electronics products, but they didn't get that show business has to be *show* business and not just business! Ohga kept whispering to me, "Make hits! Make hits!" He didn't understand that the movie business alone is a crappy business: the movie god's sun rises and sets. I kept telling them we had to take the business public to get some leverage or acquire more assets. That's why I talked about sports, because sports is a franchise entertainment that allows you to leverage yourself into cable! We had no way into media play in the U.S. because we were a Japanese company and we had no stock for the same reason. But we could have bought the Lakers, the Kings, and the Forum in the biggest media market in the world. It would have made us a major player. It's what Murdoch did and Ted Turner did, and Disney has just done! But they wouldn't hear of it. They had bought their film company and they just checked it off their list: make hits! I began to realize I wasn't going to get through."*

At the end of 1993, Guber told Schulhof he wanted to renegotiate his contract to include a production agreement for himself in the event that he was terminated for cause. The terms he proposed were bountiful. He would continue to receive his $2.9 million annual salary for a period of three years. He would be installed in a production building of his own on the Sony lot, and Sony would extend to him a revolving line of credit in the amount of $275 million. Furthermore, he would have the option to produce inside his own company, at his discretion, designated projects in development at Columbia and TriStar which he could lay claim to having initiated. At the end of a conversation with Ohga during which Schulhof emphasized that granting Guber's demands involved a calculated risk, they agreed to accept his proposal. According to Schulhof, "Peter was the kind of person, because he's talented, who needs to feel comfortable. Besides, we didn't want to lose a creative connection to him."

In November 1993, word that Guber had purchased a $6 million co-op apartment in New York gave rise to rumors that he was planning to leave Sony. Schulhof hastily called a press conference and declared that he had just signed a new "long-term contract" and was "totally committed." The other executives on the team, he asserted, had also "re-upped" for extended terms. The picture he painted was of stability and shared commitment.

Earlier that year, Mike Medavoy had lost an arcane battle to Mark Canton, or Alan Levine, or both, and had departed from TriStar. Throughout the spring, others followed: Tim Burton and Danny De-

* Many people at Sony Pictures today would agree with Guber that Sony must take the film business public if it hopes to optimize its potential value. "It's utterly essential," I was told recently, "because just making twenty to thirty films a year in an ever more costly universe is a fool's game." The point Guber and his successors are making is that Sony Pictures would have to pay cash to acquire new assets so long as it remains a wholly owned subsidiary of a foreign company. A public offering would result in stock that could be traded for acquisitions. As of December 1998, Sony's current president, Nobuyuki Idei, continued to maintain that he had no interest in a "partial sale" of Sony's entertainment businesses.

vito took their business elsewhere, explaining angrily that it had become impossible to get anyone's attention, let alone approval, for a project. Jon Dolgen, who was said to have been jockeying for power with Levine, announced that he was leaving to run the entertainment division at Viacom. Morale declined to a new low and with it went financial performance. In the first six months of 1994, Sony Pictures released only two films, *City Slickers II* and *Wolf*, an expensive project directed by Mike Nichols and starring Jack Nicholson as a werewolf. Both were severe disappointments.

In June, Schulhof visited Guber at the studio and delivered a gentle ultimatum: "I said, 'Peter, you're not running the company and you're not happy. If you want to stay you can stay, but you must do what you're good at doing because I can't let the company continue to slide. If you're burned out, I'll understand it, but you have to be honest."

In fact, with the studio back in last place at 9 percent of market share and hemorrhaging money, Schulhof by this time had resolved that his friend Peter Guber should go, but Ohga counseled patience: "Mickey was impatient to lose Guber. For a time they were like brothers, traveling together around the world. Suddenly Mickey began saying bad things about Guber, but I said it would be hard to replace someone with his talent. He was a great producer with a great nose for opportunities. *Men in Black* was his idea. And six months after we bought the studio he asked me to get him the rights to *Godzilla* and I personally went to TOHO Studios and negotiated for him. He saw opportunities fast. But I had to agree with Mickey that he was not suited to run a big business, that his talent as a filmmaker would take priority with him. And I had come to feel that if we let him go on this way, Sony Pictures would be in big trouble. It was all the spending. Until now we have never acknowledged this. We've continued to maintain that both Peters and Guber were very talented. And they were. Hiring them was not a mistake. Projects they initiated are still making money for us."

As in this equivocal reflection, Ohga declines to criticize Peter Guber without qualification. Granted, he is a proud man, disinclined to acknowledge his own shortcomings and mistakes. At the same time, he conveys in repeated conversations what seems to be his genuine perception that responsibility was not Guber's alone: "What happened between us and Peter Guber," he told me recently, "wasn't just his fault. In our own way, we must also accept a measure of the responsibility. But I'd rather not talk about that now."

Early in September, Schulhof rendezvoused with Ohga in Europe to state his case about Guber. "I said, 'I've reached a point in my mind where I think, even though you want it to take longer, I don't think we can wait,' and Ohga asked me why, and I told him the reason. I said, 'It's not personal, I still like Peter, but he's not a manager and he's too hands-off and the company needs a different kind of management style right now.'" This time Ohga authorized Schulhof to take action. He flew to California and in Guber's office in the Louis B. Mayer suite, conveyed "Sony's" decision that he should leave his job. On September 29, 1994, the business press was full of the news that Peter Guber had "resigned." Guber insists that he was not fired. He recalls going to Schulhof several times that summer to tell him he was burned out and wanted release: "I said in so many words, 'I'm done. Done! Done! Done! Done! Done! Done!'" Schulhof recalls a variant reality: "Some people scream by the way they act. Peter got his head handed to him just as I did. But he was screaming to be let out."

Characteristically, Guber's statements at the time had to do with staying attuned to the delicate balance inside himself: "Are my interests, my passions and concerns, with this kind of job? Clearly, right now, they are no longer," he volunteered in an interview. In a subsequent reflection, he remained sharply focused on himself: "I wanted passion and curiosity to be at the center of my experience. I wanted to enter the third act of my life with passion and commitment. It was time to follow my dreams."

Guber's dreams led him out of the Louis B. Mayer suite in the Thalberg Building to production offices at the end of Main Street which he remodeled in rattan and teak and lazy overhead fans for his new company, Mandalay Entertainment (named after his ranch in Aspen). With a nearly $3 million salary guaranteed for another three years, a $275 million credit line, and a slate of projects he had preempted, Guber was ready to go back into business for himself, and promptly did so. No one will disclose what happened to the $50 million bonus pool to be allocated in the fifth year of his contract, at his discretion, to a group of not more than five executives. At the time of his dismissal, the press reported that he was "apparently in line to receive an undisclosed portion" of a bonus pool. In view of negative financial results during his term, it is likely that the size of the bonus had diminished, but it seems a reasonable assumption that he did receive a portion of whatever money had been set aside.

Two weeks after Guber's departure, on November 17, 1994, Sony shocked the international financial community by announcing in its quarterly financial report a $2.7 billion write-off of its investment in Columbia Pictures. The official explanation was poor box office results, a series of top-executive resignations, and rising costs of running the studio. The company also disclosed an additional $510 million operating loss at the studio for the period June through September 1994, which it blamed on "a combination of unusual items, such as abandoning a large number of projects in development and providing for settlement of outstanding lawsuits and contract claims." While the total quarterly loss of $3.2 billion was much smaller than the $8 billion announced by IBM for the second quarter of 1993, or the record loss of $21 billion absorbed by General Motors in the first quarter of 1992, it was among the largest write-offs ever announced by a Japanese organization, and its impact on Sony's financial performance was cataclysmic: for the six-month period ending on September 30, the company reported a net loss across all its businesses, consolidated, of $3.1 billion, compared to a profit of $578 million for the previous year. The price of Sony stock fell sharply on the Tokyo market, in New York, and in London. The company asserted an expectation of reporting full-year profits by the end of the fiscal year on March 31, but later revised its projections to indicate a continuing loss.*

---

* Certain Wall Street analysts were angry at Sony's abrupt disclosure of the huge loss, and charged that the company had masked the failing performance of the film business by consolidating film and the highly profitable music business under the heading of entertainment. In August 1998, the SEC fined Sony $1 million for effectively misleading investors for several months prior to the write-off, and ordered the company to allow independent audi-

Interpretations of the write-off were polarized. Some analysts agreed with Sony's announced position, that the "cleansing of the books" at this time was a wise provision for the future if the company intended to restore the business with additional investment and new management. Ohga himself represents unhesitatingly that the decision to write off, ultimately his own, was sound: "Basically we wrote off goodwill value and start-up costs, the huge fees to lawyers at the time of purchase, fees to Michael Ovitz [he does not mention Blackstone or Guber and Peters]. To carry costs like that on our books and depreciate them over forty years would have been harmful to us from the point of view of the future. . . . I had already announced inside the company that I would be choosing my replacement as president in 1995, and I wanted to make some healthy room for the man who followed me by cleaning up the books. I don't consider that we really lost anything, not when you think about what Mitsubishi was about to really lose on Rockefeller Center . . . or what Matsushita lost when they sold Universal with the yen so high!" Others viewed the write-off as a last-ditch, desperate move to ready a failing business for a sale. Sony denied this. In fact, Schulhof had quietly retained J. P. Morgan and Blackstone to consider merger and public offering possibilities more than a year earlier. The notion of taking a write-off had first surfaced in conversations with the investment bankers about necessary preparations for a sale of stock.

Beneath the rhetoric about clearing away rubble to make room for new growth and investment, Sony was acknowledging publicly that it had overvalued Columbia at the time of purchase. At the very least, the write-off declared that the Columbia asset was now worth $2.7 billion less than it had cost. In the harsher words of a Japanese analyst who was widely quoted at the time: "What this move says basically is that Sony's president, Norio Ohga, wasted $2.7 billion of the shareholders' money." There was speculation that Ohga would be obliged to accept responsibility for the loss by resigning as president, CEO, and chair-

---

tors to examine the section of its annual report where management provides details of business results for investors.

man. To be sure, there is an expectation in Japanese society that leaders will bear the brunt of responsibility for actions taken by the group. In another organization, the president might have resigned at this moment. Not inside the Sony family. Hollywood was the cherished dream of its progenitor, Akio Morita, who had been felled by a stroke a year earlier. Morita's dream had been fulfilled by his chosen heir, Norio Ohga: no matter how disastrously the misadventure may have proceeded, and whatever thoughts may have been furtively entertained, no one in authority inside the organization was emotionally prepared to take the family to task.

10

OHGA AND SCHULHOF: A TALE OF LOVE AND HUBRIS

In the last, luxuriant years of his career, between 1990 and 1995, when Sony's new president prevailed on Ohga to dismiss him, Mickey Schulhof was Sony's lord high commissioner of synergy. For a man whose passion was entrepreneurship, it was a heady time: Ohga insulated him against all opposition even as he extended his authority to encompass the entire American operation. By mid-1993, Schulhof had direct control of twenty thousand people in the United States and another ten thousand around the world, and was overseeing a portion of the business which accounted for just under half of Sony's $34 billion in worldwide sales.

Schulhof's consolidation of power was accelerated by the removal, late in 1990, of the only other American with a claim on Ohga's reliance and affection, Walter Yetnikoff, who had dreamed of emulating Steve Ross by becoming head of Sony's music and film businesses. For a brief moment, it appeared that his dream would come true. In the weeks before the completion of the Columbia Pictures acquisition, there was talk of a "steering committee" to oversee the new businesses. The committee was to be manned by Schulhof, Guber, and Yetnikoff, who claims that he was given to understand that he would be in charge. "We agreed that I'd be running the show," he told me, "and Morita and I shook on it!" But shortly after the new cochairmen had taken office, Yetnikoff was informed, by Peter Guber in his version of the story, that each of the three committee members would have one equal vote. Yetnikoff contacted Ohga and complained angrily that business could not be run by committee. Did Norio Ohga run his business at Sony by committee? he claims to have asked rhetorically, or did he reserve final say for himself? While acknowledging that he had the last word, Ohga dryly informed Yetnikoff that he would not be enjoying similar, unilateral authority. To this day, Yetnikoff maintains that the committee had been proposed by Guber as a means of ensuring that he would not have to account to anyone. Schulhof suggests that Ohga doubted Yetnikoff's ability to run the film business and had devised the committee as a stratagem for neutralizing his authority. Until April 1990, the press continued to refer to Yetnikoff as "chairman of the new software committee," possibly because that is how he represented himself to reporters, but according to both Schulhof and Guber, the steering committee was abandoned before it had ever convened.

While he was troubled by Yetnikoff's drug and alcohol dependency, Ohga had regained a measure of confidence in his friend when he installed him as worldwide CEO and president of the newly acquired record business in February 1988. After all, Yetnikoff had made good on his promise to deliver CBS Records to Sony intact. Subsequently, he had played a decisive role in the Columbia Pictures acquisition, and Ohga by his own account had relied heavily on his business acumen as a check against Schulhof's inexperience. On the other hand, Yetnikoff had also done himself considerable damage during the same period. In the first year, the record company had performed poorly under his leadership; by late 1989, Columbia Records had slipped to second place in domestic market share, far below Warner Records, its principal competitor. Worse, as a result of erratic behavior and aggressive nastiness, which had not been cured by drug and alcohol rehab, Yetnikoff also had managed to alienate a number of his own recording artists, including superstars Bruce Springsteen and Michael Jackson. The trouble with Springsteen began when he agreed to participate in a benefit tour for Amnesty International. Yetnikoff considered Amnesty International anti-Israel and demanded that Springsteen decline. When the singer refused, Yetnikoff treated him and his agent, Jon Landau, a personal friend for years, to some of his corrosive abuse.

On August 24, *Billboard* published what was to be Yetnikoff's last interview as CEO and president of CBS Records, along with a sidebar written and signed by Landau: "Walter Yetnikoff was a good friend to Bruce Springsteen and me for many, many years. We enjoyed a superb professional relationship and a pleasant social one. For reasons that remain obscure to us, that relationship ended not long after CBS was purchased by Sony. Neither Bruce nor I have had a significant conversation with him in nearly two years." Yetnikoff dismisses the *Billboard* blurb contemptuously as a setup designed behind the scenes by his archrival, David Geffen. How is it, he demands to know, if Bruce Springsteen had been so offended, that he continues to release hit albums on Sony's Columbia label?

Yetnikoff charges that his relationship with Michael Jackson was also misrepresented as part of the same campaign to discredit him, engineered, at least in part, by Geffen. Jackson and Yetnikoff had been associated for years and had been particularly close since 1982, when Jackson's *Thriller* had become the largest-selling record in music history, earning $60 million. In the summer of 1990, Geffen was seeking permission to use in the sound track of a movie he was producing a Beatles song that Jackson owned and had recorded. According to Yetnikoff, Jackson phoned him at his home to say he preferred not to release the song to Geffen and wanted Yetnikoff to tell him he couldn't have it as if it were his decision. Yetnikoff obliged, making of Geffen a sworn enemy. Doubtless, Yetnikoff's public insinuations about his rival's homosexuality inflamed Geffen's anger.

News of Yetnikoff's feuding, however exaggerated or distorted, reached Ohga in Tokyo and was a source of consternation. But there was nothing to indicate in advance that he was preparing to terminate his friend. On the contrary, throughout the summer, Ohga went out of his way to assure Yetnikoff of his support. Late in June there was a bizarre episode involving the use of one of Sony's corporate jets. Sony Aviation in New York asked Yetnikoff to authorize a flight from Los Angeles to Singapore. Under instructions from Ohga not to permit frivolous use of Sony's planes, and assuming that the request was coming from Schulhof, Yetnikoff declined to authorize the flight. The following day, he received an angry fax from Ohga informing him that he had withheld a plane requested by Akio Morita. Yetnikoff replied with three faxes of his own, increasingly heated, in which he characterized the confusion in New York which had created the embarrassment as typical of Mickey Schulhof's environment and voiced his suspicion that Schulhof had set him up for a fall. In closing, he demanded to know whether Ohga thought him "so stupid" that he would knowingly have denied Mr. Morita one of his own planes! Ohga responded with a fax, dated June 29, 1991:

TO: *Walter R. Yetnikoff, CBS Records, Inc.*
FROM: *Norio Ohga, Sony Tokyo*

Dear Walter:

Walter, of course I know you are not that stupid. I have not yet talked about this matter to Mr. A. Morita, so please don't worry.

However, I am quite concerned about the troubles going on about airplanes recently. I would like to talk about it with you when I go to New York and explore the formalities to solve the problem.*

In any case, I feel very relieved to read your yesterday's fax messages. You are my old friend, and we will stay as friends forever.

Best regards, signed "Norio Ohga"

There is nothing in this conciliatory note to suggest that Ohga was thinking about getting rid of Yetnikoff. In July, he extended the term of his lucrative contract another three years. And in August, following an article in the *Wall Street Journal* which reported Yetnikoff's break with Springsteen and Jackson, Ohga went on record with a statement of support: "We could not be more pleased with CBS Records' performance or with Walter Yetnikoff."

One week later, Ohga showed up in Yetnikoff's office in New York when he was supposed to be on vacation off the coast of Alaska and informed him that he was finished at Sony. According to Mickey Schulhof, "There was a palace revolt and Walter lost." Assuming that Schulhof is correct, the victor would have been Yetnikoff's protégé, Thomas "Tommy" Mottola, a former pop singer and recording artists' agent. In

* Presumably, Ohga refers to reported abuses of the corporate jet at the film studio.

1988, shortly after Sony had acquired the company, Yetnikoff had hired Mottola to be president of the U.S. division, and during his two years on the job, Tommy had done well. He had inherited a band of working-class rockers from Boston called New Kids on the Block and the British singer Michael Bolton, and both acts were flourishing under his management. He had also signed a bright new star, Mariah Carey, who was already moving to the top of the charts (and whom he would shortly marry). And he had been at pains to replace the staff of the U.S. company with executives and A and R managers who were loyal to him but owed no allegiance to Walter Yetnikoff. After Yetnikoff was forced out, Mottola stayed on as president of the U.S. company until, in 1993, with Ohga's approval, Schulhof promoted him to president and COO of Sony Music Entertainment worldwide. He is currently president and CEO (Norio Ohga has retained the chairman's position). In his plush office on the twentieth floor of the Sony building in New York, Mottola, who dresses for the part of pop-music czar in a black leather jacket and ostrich boots, his jet-black hair slicked back, offered an explanation of what happened to his mentor: "It's really simple. Walter brought me in, was incredibly supportive of me in every area of the company and the business and with Japan, and had some unfortunate problems that were happening to him personally, which have been written about for years, and you can read them to your heart's content.

"And, truthfully, those circumstances, and I think his problems with Mickey and with Tokyo, were the things that led Tokyo to make decisions about what they were going to do. And at that point in time, it was so far gone to a point where there was nothing that anyone could do about anything. Even my close friendship with him, and at the same time Ohga himself had a twenty-five-year relationship—so, you know, we all felt really badly, but we had a business to run, and a company to run, and that takes precedence and priority over any individuals. And Ohga and the board made a decision to make a change, and it was their decision, and they let me know the day of, what was going to happen, and that was that."

Mottola's portrayal of himself as a concerned bystander may be disingenuous. Schulhof suggests that Tommy had been pushing his boss hard for more authority, and that Yetnikoff had attempted to fire him in a last-ditch spasm of self-defense. Yetnikoff adds details of an episode that occurred just days before his dismissal and seems likely to have been relevant. Mottola was preparing to create a record label for Michael Jackson inside Sony Music and was negotiating the deal with Allen Grubman, a legendary manipulator in the music industry who represented both Jackson and Mottola. On the Friday before Labor Day 1990, Yetnikoff exploded at Mottola, accusing him of giving away the company store to Grubman in a negotiation that involved a gross conflict of interest. Maybe he would teach Grubman a lesson by releasing an album of Michael Jackson's greatest hits, which was within his contractual right to do. Yetnikoff's secretary since 1981, Eileen Nawrocki, remembers Mottola bursting from Yetnikoff's office in high agitation and exclaiming to no one in particular, "He's gone crazy! He wants to get rid of Grubman!" It was at that point that Mottola went to Schulhof and threatened to walk with his staff and a number of top recording artists, including Mariah Carey, if Yetnikoff were not stopped. Between Schulhof and Yetnikoff there was no love lost: they were natural born rivals for Ohga's favor. Moreover, during the negotiations with Steve Ross, before and after he had been removed from the process, Yetnikoff had openly derided what he judged to be Schulhof's limitations as a negotiator, rolling his eyes and contemptuously referring to him as "Na-Mickey." Accordingly, when Schulhof relayed Mottola's message to Ohga, it is likely that he added a recommendation of his own. Given Ohga's concern about Yetnikoff's instability, Schulhof may have persuaded him that it was time to take action.

Yetnikoff remains angry, bitter, and apparently perplexed about what really happened. The day after Labor Day, he received a phone call from Paul Burak requesting his presence at 9 West 57th Street for an important meeting. His suspicions aroused, Yetnikoff took his lawyer, Stanley Schlessinger, with him. As he stepped out of the elevator, he saw Ohga and Mottola in conversation through the glass wall of a conference room, but Burak ushered him past the room to an office used by Ohga when he was in the city. Ohga entered and closed the door behind him: "He says to me, 'I say this with tears in my eyes,' and there were, but that didn't stop him from reading me his verdict. He tells me 'the board' had recommended that I should take a sabbatical leave! What board? I was on the board of Sony Music Entertainment, and there hadn't been time for a board meeting in Tokyo, that had to be nonsense, he'd been on vacation in Alaska. I knew Mottola and Namickey had been working on him, but I could never understand how he could buy it just after renewing my contract. I still have no idea. Maybe I wasn't fired, maybe I resigned."

According to Ohga, firing Yetnikoff was one of the most difficult moments in his Sony career: "Not only were we close friends, we had worked together since 1968. I had been working with Walter since his days as in-house counsel at CBS. I thought about it a long time. It was very painful when the time came." The bitter experience of continuous litigation for twenty years in its American colonies had left Sony hypercautious, more legalistic than many conservative American organizations: at Ohga's request, Paul Burak had written out for him a brief speech in English to read aloud to Yetnikoff, indicating words which could be safely used and words to be avoided.* "We must be careful at times like this to align ourselves with American legal conventions," Ohga explained. "Otherwise there are lawsuits and all kinds of trouble." Sony had traveled a long way from the days when Morita would agree to make or break complex deals in informal conversations.

Ohga was a sentimental man, but it is important to remember his magical switches. Recalling his subsequent dismissal of Mickey Schulhof, perhaps even more difficult and painful than firing Yetnikoff, he described a capacity for detachment: "It's not a question of wet and dry [a Japanese figure of speech for "sentimental" and "rational"]. To manage a business you must do what is necessary. If it was best for Sony, in Sony's interest, then it had to be done no matter what."

Schulhof's interpretation seems to echo Ohga from his own perspective: "You should view Ohga's turnaround toward Walter Yetnikoff in the same light as you view his behavior toward me. Ohga is a friend

* Burak was in the room, and it was an awkward moment for him as well: Yetnikoff had been one year ahead of him at Columbia Law School, his editor at the *Columbia Law Review*, and someone he has continued to admire. Hearing this, Yetnikoff has responded "That's heartwarming—tell him to send money!"

and loyal, but above all, he is a creature of his own company. And what happened with Walter was that he lost totally the support of everybody underneath him. And Ohga's view of a company is that one man can make a difference, but no one man can do it all by himself. At a certain point, the pressure builds on him, and he decides, He's my friend, but this is no longer personal."

Days after Yetnikoff's ouster, announced as his resignation on October 4, the name of the music company was formally changed, at Ohga's recommendation, from CBS Records to Sony Music Entertainment, Inc., SME. Yetnikoff had opposed the change heatedly, arguing that the business should leverage what was already among the best-known labels in the industry by calling itself Columbia Records, but Ohga rejected the idea when he learned that the name could not be used in Japan or Spain, where it was already owned, respectively, by Sony's rival in music, Nippon Columbia, and the German media group Bertelsmann.

Ohga took the title of chairman for himself, leaving Tommy Mottola for the time being in his job as president of the U.S. record company, and asked Schulhof to run the overall business on his behalf. "He didn't ask my opinion," Schulhof recalls. "He tells me, 'I want you to oversee the record company day-to-day, but I'm chairman so call me for approval on everything,' and I say 'Fine.' It doesn't trouble me because that's what I'm used to doing." Apart from his work on the CD-mastering operation in Terre Haute, Schulhof had little knowledge or experience of the record business, but he was a famously quick study. After only three months, in January 1991, Ohga informed him that he would take the titles of chairman and CEO. "Ohga never gave me a choice. He said, 'Here's what we will do with you,' and I said 'OK.'"

The understanding between Ohga and Schulhof, that responsibility would be conferred suddenly and without explanation and unquestioningly accepted, recalls the mode of operation employed by Morita and Ohga. Schulhof says that he had no contract with Sony at this time, and that a contract was never mentioned until he asked for one after 1993. Given the nature of his privileged relationship to Ohga, that is not surprising. The other Americans at the top of Sony's management ladder, including Walter Yetnikoff, had contracts, but none of them, not even Yetnikoff, had been admitted to the inner circle of Sony family. In that circle, loyalty and commitment were assumed: Ohga himself has never had a contract.

Early in 1991, Schulhof established Sony Software, Inc., an umbrella subsidiary from which he planned to manage, in addition to the record and film companies, all his software projects. His explanation of the concept reveals his ambition and the sweep of his authority: "It was a construct that allowed me to send a message to everyone about where we were going to take the company. I wanted to get into interactive entertainment. And I wanted a vehicle that would let me subsidize projects which moved across the businesses from the top of the organization, but without miring headquarters in operations. If Sony Music or Sony Pictures said, 'That's a great idea but who will pay for it?' I could say, 'Not a problem, it's a Sony Software project and Sony Software will bear the expense.'"

He moved first into the CD-ROM business. He formed a new division inside Sony Software, Sony Electronic Publishing, and he hired a twenty-seven-year-old physicist and electronics wizard named Olaf Olafsson to run it. Schulhof's plan was to establish a leading-edge position in content creation for "everything distributed electronically." Olafsson's first assignment was acquiring intellectual property rights to software "books" for use with Data Discman, a portable CD-ROM reader with a flip-up screen and a miniature keypad, which Sony was preparing to launch in the United States. Discman had sold well in Japan; Schulhof predicted that its success in the United States would be determined by the software packaged with it. Olafsson prepared a first offering of twenty-five titles on disc, including the King James Bible, a Webster's dictionary, *Roger Ebert's Movie Home Companion*, and an encyclopedia (each disc could store up to a hundred thousand pages of information). Discman was scheduled for release in time for the Christmas season, 1991, and would be priced at $500. Invited to present to a class at the Harvard Business School in October, Schulhof used the product as an example of creative synergy between Sony's hardware and software businesses, and predicted confidently that a hundred thousand units would be sold in the first year.

The timing of the launch proved unfortunate: Matsushita, Sanyo, and others had their own electronic readers on the market for Christmas that year at prices lower than Sony's; and an article in the *Wall Street Journal* had just reported concerns that CD-ROM discs might not have nearly the life span of the papers and books they were replacing. But the real problem was simply a lack of interest in the product. The Japanese were using Data Discman to satisfy their curiosity about the world in almanacs and encyclopedias as they rode back and forth to work in crowded trains and subways an average of 2.4 hours daily. The American consumer was less excited about accessing data on a tiny screen. In the words of John Briesch, president of Sony's Consumer Products Group, "Discman died a horrible death."

Schulhof wrote off Discman as a lesson in how to acquire intellectual property rights, and moved on to packaging CD-ROM titles for other publishers at the CD-mastering plant. Mobilizing the Terre Haute sales force, he built a $20-to-$30-million publishing business. He also sent Olafsson in search of a software development company to acquire; Sega and Nintendo were moving into electronic games and publication, and there was talk inside Sony of video games. In 1992, Schulhof purchased a small company in Liverpool, England, called Psygnosis for $15 million and put Olafsson in charge. As always, Ohga sheltered him from opposition: "There was zero understanding of this in Tokyo," Schulhof recalled, "but Ohga said, 'Leave Mickey alone, he's going to do an acquisition. Let him manage it and let's see what happens.'" Psygnosis began by creating software for the Sega and Nintendo platforms and later provided some of the best-selling titles for Sony's own PlayStation when it came to market in 1994.

Early in 1991, as Jon Peters and Peter Guber supervised what was known in Hollywood as the "face-lift" at Culver City, Schulhof prepared to relocate his own headquarters from 9 West 57th Street to a much grander location, the "Chippendale" building, so called for its resemblance to an English highboy. At 550 Madison Avenue, it stood between 55th and 56th streets, a thirty-six-story monument erected by AT&T to its own power and prosperity. The time had come for a physical consolidation. General headquarters and Schulhof's burgeoning operations were housed on several floors of the Revlon building at

9 West 57th Street and had spilled over into other buildings in the vicinity; the record business was crowded into twenty floors at "Black Rock" (the building is named for its black granite exterior), CBS headquarters on 51st Street. A five-year lease had been included in the acquisition agreement, but staffers were already clamoring for more space and complaining about the inadequate heating system. Schulhof retained an agency to find 600,000 square feet in an appropriate space, and the AT&T building had turned up.

In the buyers' market created by the recession, he negotiated a lease option with a fixed price of $200 million, generally considered a bargain. He then invested upward of $80 million in renovations. The original architect, Philip Johnson, had agreed with the city, in exchange for higher floors, to design the ground floor as an open space, and it was a cold and drafty place used by no one but the homeless; it was also the only block on Madison Avenue which had no retail space. Schulhof extracted permission from New York to enclose it with a curtain wall, and hired Charles Gwathmey to design an internal plaza with retail space for rent. From the mezzanine upward, each floor was gutted and enlarged: AT&T had configured the building for a thousand employees, and Sony moved in with fifteen hundred and currently has eighteen hundred people there. AT&T had already spent lavishly on the top two floors, thirty-four and thirty-five, which Johnson had designed as an executive penthouse on a grand scale. The offices of the chairman, CEO, and chief financial officer were on the thirty-fourth floor, which Schulhof left untouched. He moved into the CEO's office in the southeast corner, with windows looking south toward Wall Street. AT&T chairman Robert Allen's office, in the northwest corner of the building, with windows fifteen feet high overlooking the Hudson River to the west and Central Park to the north, was, and remains, reserved for use by Norio Ohga on his occasional, increasingly infrequent visits to New York. A sweeping marble staircase, which Schulhof left intact, connected corporate offices and a corporate dining room. On the thirty-fifth floor, he preserved the main corporate dining room and several others, but added the Sony Club, an elegant penthouse restaurant for the exclusive use of Sony executives from anywhere in the world when they were in town. In addition to its own main dining room overlooking the city, the club featured a bistro with a pizza grill and a beautifully appointed sushi bar, which could accommodate up to eight people and was decorated with miniature barrels of Morita sake and, on the walls, framed calligraphy by Norio Ohga, ink-brush variations on his favorite Chinese character, *mu*, meaning dream or dreams.

To some observers, Schulhof's renovations were seen as copycat duplication, and possibly even emulation, of his friend Peter Guber's lavish studio remodeling. Predictably, Peter Peterson represents the critical view with an outspoken mordancy: "It was one of the most elegant buildings around. Next thing you know, they hire this guy Gwathmey to redo the thing, millions of dollars in a totally different venue, style. This was good enough for AT&T. And incidentally, we're doing very poorly, it isn't as though we're doing well. And buying a restaurant to get the services of an individual. I was speechless." Peterson refers to Barry Wine, a chef and restaurateur who owned the Quilted Giraffe, located on the ground floor of the AT&T building before Sony moved in. But Peterson's dismay at Schulhof's purchase of the restaurant to acquire Wine's services may be founded on a misunderstanding. Schulhof maintains that he had to buy out Wine's lease when he built Sony Plaza. He structured the payout over three years, and when Wine asked what he was going to do with himself, Schulhof proposed that he help design the club he was planning for the top floor. Wine came along, in other words, as part of a package designed to move his restaurant out of Sony's way.

Understandably, the heaviest fire was drawn by the Sony Club and some of the other refinements that appeared to have no bearing on doing business, including a floral designer on permanent call to care for flower arrangements in executive offices. Peterson continued: "I was constantly being asked by my friends to account for what was going on over there. 'Pete, you're a conservative fella,' they'd say. 'I had dinner over there the other night, and I was served wines that you only read about in books! Then there was that sushi bar—this guy stood there with all this immensely high priced fresh fish that was going to be thrown away, and nobody came in. Couldn't someone have taken reservations!' What could I say!"

Certainly Mickey Schulhof, like his new friends in Hollywood, had a propensity for indulging his expensive tastes. On the other hand, as was the case with Peter Guber if not Jon Peters, some of the charges leveled against him appear to have been exaggerated and unfair. Schulhof himself particularly resents criticism about his handling of the new headquarters. "That building is one of the best investments Sony has made. . . . If Sony chose to rent the stores and the basement and the plaza, which it doesn't do, it would produce a healthy return on investment of the $50 million we spent on building them. But even without renting, just add it all up and look at the market value of the building today. If Sony doesn't like it, Idei should sell it. My guess is they could make $100 million profit!"

By 1992, with backing from Ohga, Schulhof was creating new businesses prolifically in every corner of the Sony map. With Warner Records, a Sony Music partner since Steve Ross had extracted a 50 percent interest in the Columbia Record Club, he launched VIVA, a TV-music channel in Germany, and SW Networks (Sony and Warner), a syndication business that sold dedicated music channels to radio stations across the country. He took Sony into the concert amphitheater business, purchasing interests in twenty amphitheaters. He created Sony Signatures, a retail business for merchandising posters, caps, T-shirts, and whatever else could be personalized by Sony recording artists and movie stars. He put Sony's name on Loews Theatres, which had come with the Columbia TriStar acquisition, and pushed through a Peter Guber idea to build a Sony multiplex across the street from Lincoln Center in New York City. Sony Retail Entertainment, which he formed to manage the movie theater business, also designed a cinema complex for San Francisco in Yerba Buena and the huge entertainment center that will be a part of Sony's new European headquarters currently under construction at Pottsdamer Platz in Berlin.

There were numerous other, smaller investments. General Magic, for example, was a Silicon Valley start-up, which Schulhof helped Apple and Motorola fund with $5 million. John Sculley spun off the company from Apple to allow the software wizards on the original Macintosh team to develop a "palm-top" connector to the Internet, a variation on the U.S. Robotics Palm Pilot. The project failed when Sculley developed a competing product inside Apple. "It seemed like a very good bet," says Schulhof, "and Ohga backed me on it and we lost. I always felt, and he agreed, that failures were inevitably going to be a part of our game."

To many people in and outside the organization, Schulhof's headlong activity appeared not only reckless but random. Predictably, Peter Peterson took a dim view: "He got into all these businesses, just a whole array of these software and electronics this and the Internet that and the retail this and so forth. Partly I think because he'd never really been an operating person. One of the things you learn in business after a while, it isn't just concept that matters. It's execution, and it takes a lot of time to execute even a good vision. So even assuming his vision was right, the thought that Sony could get into ten or twelve businesses over here with this management infrastructure—I mean a seasoned manager would have said, 'Please stop!' "

Schulhof may well have filled his plate greedily, particularly in view of his distaste for managing operations. At the same time, in fairness, every step he took appears to have been consistent with his mission as he construed it: the quest for synergy across businesses. To this end, he was a constant meddler, forcing, or attempting to force, Sony companies to reach beyond themselves and connect productively. To Sony Pictures, he insisted that sound tracks must feature recording artists at Sony Music. Conversely, he required the record company to use actual clips from Sony films whenever possible. People resented his interference, but Schulhof declared that cooperation was the law of the land and forced the issue. An example he recalls with pride is Will Smith's first big film, *Bad Boys*, a $16 million action-adventure produced by Don Simpson and Jerry Bruckheimer which was among Columbia's biggest hits in 1995: Schulhof strong-armed Sony Music into using action sequences from the movie in the music video to promote the sound track, and the resulting video was a hit in its own right.

Schulhof's efforts to synergize on a larger scale often failed to deliver on their promise. An example was a many-spoked deal with Michael Jackson. In a press release dated March 20, 1991, Sony Software trumpeted a "historic unprecedented multimedia partnership among

Michael Jackson and Sony Software's Sony Music Entertainment and Columbia Pictures Entertainment." The joint venture was to include six new Jackson albums, his own Nation Records label, motion pictures, television, and short films, and was projected to generate $1 billion in revenue. It was Jon Peters who had wooed Jackson into signing, but after Peters had departed in May, the superstar would deal only with Mickey Schulhof in person: the staff at Sony Music began referring to him as "Michael's project manager." That summer, a new Jackson album, *Dangerous*, was released on Sony's Epic label and, by previous Jackson standards, was only a modest success with worldwide sales of nine million copies. Jackson's private record label, MJJ Music, found a home at Sony; and one short film was produced, based on *Dangerous*, and ran briefly in theaters. The first full-length feature film for Columbia Pictures, "a musical action-adventure based on an original Jackson idea," was abandoned midway through a development process that involved a dozen rewrites and cost several million dollars.

With software, Schulhof achieved a measure of success. But the combination of software and hardware was tougher to manage. An early experiment involved Spielberg's film *Hook*, which was scheduled for release in time for Christmas 1991. Schulhof's idea was to create a series of promotions and premiums designed to augment the film's appeal and ultimate profitability. At Sony Electronic Publishing, Olafsson was developing a video game, which would be ready in advance of the film's release; and the electronics company would provide Walkman models to be awarded to theater audiences in lotteries. For the first time, Schulhof declared expansively, the release of a Hollywood film was being designed to deliver a cornucopia of synergistic possibilities. In the end, Spielberg was concerned that the ballyhoo would compromise the film's chances of recognition as a *succès d'estime*, and the campaign was scaled back.

If the record and film companies resented Schulhof's interference, executives at the electronics business hated him for it. Their hostility was sharpened by the fact that, on paper, until June 1993, though he was on the electronics company board, the business was organizationally outside his control. But this never stopped him from barging in when he saw a possible connection to exploit. On the occasion of a new

TV commercial campaign for CD players, for example, he "recommended" that Sony Music artists be approached to provide the music. At first he would try to persuade, and, by his own account, when persuasion failed he would "bluff" his way into making management believe that his suggestion was a corporate decision. But always, he insists, he viewed himself as working on behalf of the entire organization.

On March 30, 1988, the electronics business relocated to its own headquarters in a new facility across the Hudson River in Park Ridge, New Jersey. The building, Kenji Tamiya's last project before he returned to Japan, was entirely paid for by the gross profits from Walkman sales in the United States for a single year. The business was managed by Neil Vander Dussen, an electrical engineer who had been recruited from RCA in 1981 to run Sony's broadcast products division. In 1983, Vander Dussen had moved to the larger consumer products division, and in 1985, on the recommendation of Tamiya, had been named president and COO of the electronics organization.* Vander Dussen belonged to the genus of American manager characterized by Norio Ohga as "boy scouts." In fact, he was a trustee on the national board of the Boy Scouts and the YMCA. He was thoroughly competent and hardworking, putting in long hours and expecting no less from his subordinates, whom he inspired with confidence and loyalty. Someone described him admiringly as hailing from the George Patton school of management, and certainly he seems to have been as American as flapjacks, a very different animal from the ironic and contentious Jews who drove Sony's history decisively. One of his goals, which he achieved, was elevating sales of electronic products to $1 billion. To keep his team pumped up and on course, he passed out pocketknives bearing the inscription AWHTDIDI: "All we have to do is do it!"

Vander Dussen bitterly resented Schulhof's aggressive interference,

* When the business was relocated to New Jersey, it was reincorporated in Delaware for tax reasons and assumed the name Sony Corporation of America. Simultaneously, the parent subsidiary in New York, the base of Schulhof's software operation, was renamed Sony USA (SUSA). In June 1993, when his authority was extended to encompass hardware, Schulhof renamed the hardware business Sony Electronics, Inc. (SEL).

and remains in touch with his resentment: "After the acquisition, Mickey was always pushing us to help make the pictures operation more successful but without taking any credit and at actual cost to ourselves. There has to be an honest and fair distribution of money on both sides, but that wasn't what was happening. For example, we might want to include a disc with a tape machine as a premium. Warner would cooperate with us and sell us the discs at a sizable discount. But Mickey would insist that we pay list price to Columbia, and do all the work besides. I said, 'Fine, but let's recognize up front so everyone knows it that this will impact our profitability in a negative way.' In other words, we didn't want this to happen without recognition that we were bending over backwards on behalf of the company as a whole. The next thing I know, Mickey turns around and complains to Ohga that we were not cooperating. There was constant ongoing controversy rather than a spirit of cooperation."

Schulhof's own recollection of this period seems to confirm, if obliquely, that there was substance to Vander Dussen's complaints: "There were a lot of times when it would be really helpful to leverage the fact that we were a big electronics company to enhance our image as a growing player in the software business. For example, Sony has a terrific division for broadcast equipment, and some of the companies I was doing business with were in that field. So where possible, I tried to get Neil's help in making an accommodation in the availability of certain multitrack tape recorders, for a recording studio. I don't recall anything about accommodations on a major scale, but he may be totally right. But the overall effort was always to try to leverage assets for the benefit of both groups."

By mid-1991, the rift inside the American subsidiary between Schulhof's software operation and the electronics business had become visible to outsiders and drew attention from the press. Sony hardware executives were said to be openly critical of Sony Software for alienating outside software suppliers critical to the success of Data Discman by driving greedy business deals. Journalists reported surprise at the intensity of criticism from the hardware side. An outside publisher, for instance, ventured that Park Ridge executives "relish the prospect of a conspicuous failure by Sony Software." Judging by his current view, which is philosophical, Schulhof himself took the hostility in stride as inevitable: "These were guys who spent their whole career, twenty years at Sony, working on electronics, working very hard, trying to build up the Sony name, and all of a sudden here comes Mickey and Ohga acquiring these other companies, and the spotlight of attention suddenly shifts to us. Nobody writes an article in *Business Week* about the wonderful job Sony is doing with its picture tube factory, but the front page has an article about Sony's new entertainment colossus, and so does *Forbes*, and there are jealousies that run rampant through the company. And they were mysterious and they were troubling and I discussed it many times with Ohga. I never invited it, but they were there."

The conflict intensified, as did the turbulence it caused, when Ron Sommer was brought from Sony Germany to replace Vander Dussen. An Israeli educated in Vienna, Sommer had earned a doctorate in mathematics at the age of nineteen and had gone to work in the German computer industry. In 1980, he had been hired by Schulhof's counterpart in Europe, Jack Schmuckli, then head of Sony Germany, to run the nonconsumer division in the German business. When Schmuckli became chairman and CEO of Sony Europe in 1986, Sommer succeeded him as president of Sony Germany. From there, at the end of 1990, he was transferred to New Jersey as president and COO of the electronics company.

The decision to bring Sommer occurred in the context of a severe, global downturn in Sony's earnings. The steep rise of the yen against the dollar since the Plaza Agreement of 1985 had driven up the cost of exports, placing Japanese manufacturers at a competitive disadvantage. At the same time, catastrophically, the U.S. market, which had been generating 30 percent of the company's worldwide revenues, was in serious recession: by the end of 1991, growth in the electronics business had slowed from its historical 20 percent annually to 4 percent for the year. Sales of CD players and Sony's 8-millimeter-video Handycam recorders had dropped sharply. Matsushita had reached the marketplace first with outsized, flat TV screens. And Sony's huge investment in high-definition TV had been reduced in value when the U.S. broadcast industry agreed to adopt domestic HD-TV format standards. The impact of the downturn was manifest in the company's disheartening financial report for fiscal 1992 (ending on March 31, 1993): operating income had declined 45 percent over the preceding year, and Sony announced a pretax loss of $207 million, its first operating loss since Ibuka had founded the company in 1946. The seriousness of the loss was magnified by a debt burden of $13 billion, which required interest payments of $728 million annually and limited the company's ability to acquire additional loans.* Plans called for capital spending to be reduced by $500 million in the following year; the investment in research and development, Sony's lifeline, was to be curtailed, and a halt was called to factory building. An expensive high-definition television plant was to be retrofitted for conventional TV.

Clearly, it was felt that a more aggressive and strategic leader was needed to revitalize the American operation. To make room for Sommer, Vander Dussen was superannuated to the office of vice chairman, where he remained for a year until he retired in December 1991.

From the moment they met, Doctor of Physics Michael Schulhof and Doctor of Mathematics Ron Sommer saw each other as dangerous rivals. For as long as Sommer remained in the United States, nearly three years, they were adversaries, circling each other warily and abruptly engaging in heated territorial battles. In Ohga's view, they got along so badly because they were so much alike. Ron Sommer, he maintains, was the only man he ever met who was a match for Mickey Schulhof in intelligence and drive. According to someone else who knew them well: "Ron hated Mickey. But they should have loved each other. They were both smart, very smart, both Jewish. If they had ever managed to work together instead of always at odds, they would have flown circles around Tokyo, I assure you. But they were too alike." Sommer, currently chairman of the board of management of Deutsche Telekom, declines to comment on his years at Sony. Schulhof's account is summary: "Ron was very tough and very opinionated. . . . He tried to

* Since the acquisitions, Sony had been twice to financial markets in Japan. In February 1990, the company had raised nearly $3 billion in convertible bonds and warrants, and an additional $1.7 billion in stock. In November 1991, 29 percent of the Japanese music subsidiary—the original CBS/Sony Records—was sold publicly for $915 million.

lay claim to things which I disagreed with and sometimes fought for. He was like a bull in a china shop."

Almost at once, Schulhof and Sommer were at loggerheads over the marketing strategy for Sony's first digital recorder, the DAT Walkman. DAT—digital audio technology—enabled digital playback of CD quality from a specially designed cassette with a capacity of 120 minutes. The idea was to release the DAT Walkman together with prerecorded digital cassette tapes from Sony Music and other record companies. The problem was that the digital technology also allowed perfect, hissless recording from other formats. The music industry, in a replay of their initial resistance to CDs, declined to release hits in DAT format because they were afraid of widespread copying. The National Music Publishers Association went so far as to file suit against Sony for providing consumers with means of infringing copyright.

Ironically, in view of the fact that he was Sony's designated synergy advocate, Mickey Schulhof, wearing his hat as chairman of Sony Music, chose not to break rank with the record industry and refused to authorize Sony Music tapes in the DAT format. Ron Sommer, responsible for marketing the new product successfully, was furious. "Sony is a company that does not want consensus on everything," he told the *New York Times*. "If Mickey does not want to support DAT, fine. I'm not going to kill him for that." The locution seems curiously charged, even in the negative. Schulhof commented, somewhat abstrusely, "Synergy does not mean coercion." There were moments before and after DAT when he would not hesitate to force the issue as he saw it, but in this case, as he explained to me, "Neither Ohga nor I ever expected DAT to be such a large, mass-market product, and we didn't feel it was worth arm-twisting our record executives. CDs were worth forcing down their throats. DAT was not." The following year, Sony, Matsushita, and others agreed to pay record companies a 2 percent royalty on each digital recorder sold, and Sony Music Entertainment (Japan) did issue a number of classical titles in the DAT format. But the industry's overall refusal to cooperate had the effect of limiting sales of the hardware to audiophiles and professionals willing to pay the steep price of over $800. Some saw the failure of DAT as a textbook example of the tantalizing but ultimately empty promise of synergy.

In July 1991, Schulhof told a reporter that Sony had decided to manufacture a digital tape recorder based on new technology developed by Philips of the Netherlands, the company's old collaborator and sometime rival. Sommer went public at once with an angry rebuttal that was widely reported: such product decisions were the province of the hardware group, he declared, and Sony had not yet decided on this introduction: "Maybe we both would have preferred that he hadn't made that comment," he told a reporter. The story of the row appeared in the Sunday, August 11, edition of the *New York Times*, under the headline: "Will Intramural Squabbling Derail Debt-ridden Sony?"

Now it was Morita's turn to be angry. Schulhof remembers: "I got a phone call at home at midnight our time, and it was Akio Morita telling me he was not happy that disputes between two of his key executives were spilling into the press and that it was not to happen again. He said his next phone call was to Ron Sommer. He was eminently fair. I said I understood, and I agreed with him, and I was sorry."

Ohga seems to have viewed the infighting in New York, and in Hollywood, where it was occurring simultaneously, with a degree of philosophical detachment: "Everyone felt upstaged by Mickey, and that did become a problem. But when they stated their case against him, our executives never spoke about their own inadequacy. And the fact is that superstars and more ordinary people just aren't going to get along.

"Ron Sommer was different: he was Mickey's equal, as smart or smarter than Mickey, and that's why they collided. It was really all about who was the true boss of the American business. I didn't take sides. And Morita and I agreed that allowing them to compete wasn't necessarily a bad thing for the business. When two extraordinary people are vying with each other there are definite benefits. You will never have peace among everyone. You have to consider what will be the fastest route to getting where you want to go and then handle the talent skillfully."

Early in 1993, Ron Sommer was transferred back to Europe. Sony Germany had been losing money, and, in Jack Schmuckli's judgment, management was at fault and should be replaced. It was Ohga who proposed recalling Ron Sommer. Schulhof suggests that Ohga jumped at the chance to remove Sommer because he was a thorn in Schulhof's side. Ohga dismisses this notion as typical of Schulhof's arrogance, but Jack Schmuckli agrees: "I know that Ohga's interest was to get Ron out of Mickey's way." If so, it was not the first, nor would it be the last time that Ohga acted to give Schulhof adequate room in which to exercise his expanding authority.*

Notwithstanding the freedom Schulhof enjoyed during these years to act on his own authority, he was in fact, on paper, reporting to Morita's youngest brother, Masaaki Morita. Morita had dispatched his brother to the United States to head the American organization as chairman in July 1987, just as Yetnikoff and Schulhof were negotiating for CBS Records. Sixty years old at the time, Masaaki was senior general manager of consumer products worldwide, sales and manufacturing, a huge and important job; he was also a representative director, one of two executive deputy presidents of the corporation, and a regular member of the Executive Committee. On the organization chart, after Ibuka, his brother Akio, and Ohga, Masaaki Morita—"M.M.," as he was known—was the fourth highest ranking officer in the company. To many observers outside Sony, his appointment as governor of the American colonies signified their importance to the company in general and to his elder brother in particular. It may also have been true that the American mission was a convenient and, given its appropriate importance, seemly opportunity to extract Masaaki from the hothouse environment of headquarters where, five years before his mandatory retirement, his career had reached its ceiling.

---

* Sommer, whose ambition was easily a match for Schulhof's, had his eyes on a bigger job than president of Sony Germany: he informed Ohga that he wanted, in addition to running the German company, to be president of Sony Europe. Ohga prized Sommer's intelligence and strategic acumen and agreed that he should be president of Sony Germany and president and COO of Sony Europe just under Jack Schmuckli, who was chairman and CEO. Schmuckli knew Sommer well enough to be apprehensive, but had to accommodate Ohga's request. A power struggle ensued; Sommer moved to consolidate his own power at Schmuckli's expense, stepping on toes and causing considerable unrest. Schmuckli prevailed on Ohga to intercede, charging that Sommer was going to destroy his operation. When Sommer was in Tokyo, Ohga told him that he was "breaking people's hearts" and should reflect on his behavior. Within a year, Sommer left Sony for the chairman's office at Deutsche Telekom.

When Masaaki was interviewed for a job at Sony—Tokyo Telecommunications Engineering at the time—he had been admonished by Ibuka's young cousin by marriage, Shozaburo Tachikawa, to forget that he happened to be Morita's kid brother, and had given his "word of honor as a gentleman" that he would never expect favors. He had kept his promise: by prearrangement, he reported to Ibuka throughout his career and consciously avoided direct contact with his brother. Morita for his part was careful to keep him at arm's length. Ohga maintains, no doubt correctly, that he himself was closer to Morita than Masaaki and knew more of what was in and on his mind.

Even so, Morita must sometimes have felt conflicted about Masaaki's role. After Kazuo Iwama died in 1982, for example, and before Morita had chosen Ohga to succeed him as president, the possibility of appointing Masaaki instead must have occurred to him. His youngest brother had proven himself to be an able manager and a gracious and appealing if not particularly charismatic leader. There, perhaps, was the rub: at the far end of the family table, Masaaki had been taught deference, reticence, self-effacement, and now perhaps he lacked the forcefulness Morita wanted in his successor. Furthermore, Konosuke Matsushita was drawing criticism at the time for having chosen his son-in-law as his successor, and Morita wanted badly to avoid any semblance of nepotism. Nevertheless, Sony was a family business, his family business, and Masaaki was his brother. But Morita had chosen Ohga, and the moment when Masaaki might have been moved into the president's office had come and gone forever. In that context, Masaaki's assignment to the United States may have represented, in addition to its other obvious merits, a move to help ensure that his long and selfless career at Sony would end on a properly distinguished note.

As the first head of the American business with a background in engineering rather than sales, M.M. had a mandate to strengthen Sony's R and D and manufacturing operations inside the U.S. market. He had not been involved in the music and film business acquisitions and continued to have little to do with software. He was, nevertheless, nominally Schulhof's boss. And while Schulhof seems to have felt free to ignore Masaaki, this proved to be a lapse in judgment. When Schulhof explained in a *New York Times* interview that he reported to no one but Norio Ohga, Morita phoned him angrily from Tokyo to let him know in no uncertain terms that he was expected to demonstrate appropriate respect for the executive chain of command, particularly where M.M. was concerned.

In June 1992, M.M. turned sixty-five, retirement age at Sony, and returned to Japan to assume the honorary title of Sony Corporation adviser.* He was replaced in New York as titular head of the U.S. subsidiary by Ken Iwaki, a man who stood with him on the top rung of the executive ladder and was widely rumored at the time (as it turned out, incorrectly) to be Ohga's chosen successor. Once again, Schulhof blithely ignored the man who was nominally his new boss: "He was M. Morita's successor, so he was technically in charge of the electronics side. At that point I had very little to do with electronics. I was focusing on the entertainment company, on film and music and the interactive groups, all of the software businesses. So I viewed myself as side by side. He probably viewed himself as my superior, and that was OK. . . . The necessity to interact was minimal enough. Probably one of the things that grated everybody was that I didn't go to ask permission all the time because I never viewed my authority as having flowed from them."

Ohga represents that Schulhof's insubordination was beginning to trouble him by this time, and he smilingly brushes aside the suggestion that Ken Iwaki's authority was limited to electronics. In any event, Iwaki had been in the United States for only a year when Ohga reassigned him to Singapore in June 1993.

On June 1, Sony USA took back its original name, Sony Corporation of America, and was reconstituted as the U.S. corporate parent. The same day, Schulhof was named president and chief executive officer, and the electronics business in Park Ridge, renamed Sony Electronics, Inc. (SEL), was included in his purview. In Tokyo, among the powerful senior general managers—*jigyō-honbuchō*—of Sony's major

---

* In July 1992, he was appointed chairman, president, and chief executive officer of Sony Life Insurance Company, Ltd.

divisions such as consumer products and semiconductors, there was strong resistance to handing Schulhof control of electronics in addition to entertainment. Ohga overrode the opposition: Schulhof, despite his arrogance and his hubris, was still the only man he knew and trusted who could bring him ten when he asked for one. In this heady atmosphere and at the pinnacle of his career, Schulhof, for his part, must have felt, and certainly appeared, unstoppable. In fact, his nemesis was waiting in the wings in the person of Nobuyuki Idei, the man Ohga would choose to succeed himself as president at the end of 1994.

The early nineties were years of darkness and tumult in Sony history, but the company sustained its most painful blow on November 30, 1993, when Akio Morita suffered a stroke. When he was in Tokyo on Tuesdays, at 7:30 A.M., Morita gathered with other Sony tennis players at the Shinagawa Prince Hotel for an early-morning game of doubles. Regulars included his brother M.M.; the Sony engineer Toshitada Doi; and Minoru Morio, currently chief technology officer. That Tuesday morning, Morita arrived with a new racket he was eager to try, a gift from his friend Yotaro "Tony" Kobayashi, chairman of Fuji-Xerox, Ltd., a company started by his father (among Japan's power elite, bestowing gifts on friends seems to be a defining behavior). In the first match, he was paired with his brother. Masaaki remembers waiting for the ball to be served behind him as he stood at the net, and, when nothing happened, turning to discover Akio on his knees: in the midst of his serve, Morita had been overcome by dizziness. Masaaki wanted to take him to a doctor, but Morita assured him it was only a severe cold and insisted on being driven home. The dizziness persisted and Morita felt worse. That afternoon, he was taken in an ambulance to Tokyo Medical and Dental University Hospital—the teaching hospital where Ohga's schoolmate, the eminent heart surgeon Akio Suzuki, had performed his coronary bypass—and was found to have suffered a cerebral hemorrhage. In an operation that lasted four hours, a blood clot the size of a golf ball was removed from his brain.

When he regained consciousness after two days in intensive care, the left side of Morita's face and body was paralyzed and he had lost the power of speech. He was conscious, and responded to his name with a squeeze of his hand. At a press conference on Friday, December 3, having suppressed the news for two days, Sony announced that Morita had undergone an emergency operation and was "recovering satisfactorily." Perhaps that was accurate given the seriousness of his condition: it was far from clear how complete his recovery would be or if he would ever return to the chairman's office.

Cruelly enough, the announcement that Morita would become the next chairman of the Keidanren was scheduled for the day he fell. He had campaigned hard for the position for three years, depleting even his extraordinary reserve of energy. His motive is likely to have been complex. Morita was a nationalist, and there was no better position from which to advocate Japan's business interests. It was also bound to be useful in promoting Sony, and Sony was never far from Morita's thoughts. Whatever the balance may have been between his patriotism and his desire to benefit Sony, there was in addition Morita's Olympian ambition. Founded in 1946 to represent Japan's corporate interests to the rest of the world, the Keidanren, the Federation of Economic Organizations, was the private sector's most powerful entity, with more than a thousand corporate members. Morita longed unreasoningly for the prestige, not to mention the influence, conferred by the office of chairman.

When he began his campaign, his chances of success were slender. To be sure, in 1988 he had been appointed one of several vice chairmen and chaired the Keidanren Council for Better Corporate Citizenship in the United States. The top position, however, had been reserved historically for business leaders from the establishment: steel, finance, public utilities, and heavy industry. Sony the postwar upstart did not qualify. Moreover, some viewed Morita as an opportunist who was overly attentive to his influential friends in the West. In 1986, during the semiconductor negotiations between Japan and the United States, for example, Morita in his role as mediator had cultivated a friendship with the leader of the U.S. negotiation team, Robert Strauss, hitching rides with him in his plane to Washington and inviting him to his home for dinners with business leaders in the industry. He was also something of a loose cannon: one minute, as in *The Japan That Can Say No*, he was lambasting the United States for pressuring the Japanese, always too compliant in his view, to limit their automobile exports; in the next breath, he turned a disapproving eye on Japan's own protectionism, decrying what he called "the Japanese fortress."

On the other hand, it was well known that the ruling Liberal-Democratic party regularly sought Morita's advice on foreign affairs. Between 1969 and 1987, he had a direct phone line in his home to three prime ministers, Sato, Ohira, and Nakasone.* In the early nineties, with globalism increasingly in the air after the fall of the Berlin Wall, the importance of Morita's visibility and his unique viability as an international figure had to be acknowledged.

In 1990, Morita created an office to handle his external activities in a penthouse apartment in a newly built residential high-rise called the Ark (the developer was a friend and had the apartment next-door). He organized a task force to help him network the financial community and handle the press. He developed his views on a free global market in a series of articles written for Japan's leading monthlies, and became a familiar presence on Japanese and international television. Most critically to his goal, he placed himself at the disposal of the Keidanren head at the time, Gaishi Hiraiwa, chairman of Tokyo Electric Power Company. For three years, at Hiraiwa's bidding, Morita hurled himself around the world on emissary missions for the Keidanren: a typical itinerary took him from Tokyo to New York with Hiraiwa for meetings with James Robinson of American Express and George Fisher of Motorola, friends of Morita's for years; on to Uruguay to attend the GATT talks; from there to Europe for meetings with political and business leaders in the EC; back to Washington; and, the same day, home to Tokyo. By mid-1993, it was clear to Morita's staff and friends that he was pushing himself beyond his limits. His schedule was determined a year in advance. In the two months before his stroke, he traveled from Tokyo to New Jersey, Washington, California, Texas, Britain, Lisbon, Barcelona, Düsseldorf, and Paris, met with Queen Elizabeth, Jack Welch, Jacques Chirac, and Isaac Stern, among many

---

* Yoshiko Morita says that Prime Minister Sato sent his wife to her in 1971 to inquire whether Morita might accept an appointment as foreign minister, and that he was later offered the post of ambassador to the United States.

others, traveled all over Japan, played nine rounds of golf, and spent seventeen full days in his office at Sony headquarters.

Ohga was deeply concerned, and very unhappy that Morita was increasingly distracted from Sony business. On the other hand, by June 1993, *The Atlantic* was describing him as "the best-known Japanese citizen in the world," and Hiraiwa was deciding that Morita should be his successor. The weekend before he fell, Morita invited Tadashi Yamamoto and his wife to join him and Yoshiko at their home in the Hakone Mountains. Yamamoto was a longtime friend and, as founder and president of the Japan Center for International Exchange, had functioned since 1970 as an impresario of Japan-U.S. business relations. Hiraiwa had confided his intention that Morita should become the next chairman, and he wanted Yamamoto to organize a series of study sessions for him with younger executives across the country. Yoshiko served a beautiful Japanese dinner. The following morning, Yamamoto and Morita went for a walk in the woods, and Yamamoto recalls that Morita seemed to have trouble keeping pace.

Following his stroke, Morita spent months in physical rehabilitation at the hospital. As it happened—one of the small, unaccountable convergences in history—Masaru Ibuka lay in the adjoining room, recovering from a heart attack he had suffered three weeks earlier. During the months they spent in rehabilitation, Ibuka and Morita were wheeled back and forth to each other's rooms and sat together in silence, their hands clasped.

When Morita returned home in the spring, he was confined to a wheelchair. Gradually, he regained some mobility, though he was not able to stand or walk, and for a time recovered a partial ability to speak. His New York secretary since the early seventies, Hiroko Onoyama, recalls him eating his lunch with his own chopsticks and speaking with difficulty but comprehensibly when she visited him for the first time at his home in Aobadai, Tokyo, in November 1994, a year after his stroke.

On November 25, 1994, the Sony board of directors in Tokyo formally accepted Morita's tender of resignation as chairman, submitted November 16. Coming as it did, just one week after the Columbia Pictures write-off, there was speculation in the foreign press, and even among Japanese observers, that the timing was intended as a signal to shareholders that Morita had accepted responsibility for Sony's Hollywood misadventure. This interpretation seems unlikely for manifold reasons, foremost among them Morita's unquestioned and unassailable preeminence as master of the Sony house. An additional strike against this reading was the new title awarded Morita by the board, founder and honorary chairman, not a likely appellation for a man who had resigned from office under a cloud. The chairman's seat was, for the time being, left empty. In an equally improbable reading, some construed this void to represent Morita's dissatisfaction with Ohga for failing to achieve profitable hardware-and-software synergy following the Columbia acquisition.

When he was able to travel, Morita began spending time convalescing at his resort home on the beach in Kahala outside Honolulu. Gradually, despite his improvement during the first year, his condition reversed. By April 1995, his speech had deteriorated to a point where only Yoshiko, and a few other intimates, including Ms. Onoyama, could understand him and had to translate for the benefit of visitors. Prohibited by his doctors from flying in an airplane since 1997, Morita remains in Hawaii, confined to a wheelchair, unable to speak or feed himself but conscious and aware.

In December 1997, Masaru Ibuka died at the age of eighty-nine. On January 21, 1998, following family services, a bereaved Sony Corporation held a "company funeral" for fifteen hundred executives and family friends at the New Takanawa Prince Hotel. Ohga officiated. Ohga's wife, the pianist Midori Matsubara, played the funeral march from Chopin's Piano Sonata no. 2 as Ibuka's son entered the hall with his father's ashes and installed the urn on the Christian altar. Following a performance of "Eulogy for Two Trumpets," composed by Ohga in Ibuka's memory, Yoshiko Morita delivered a message from her husband. Behind the scenes, this seemingly simple gesture had caused an upheaval at the top of the organization. When Ohga saw Yoshiko's text, he had taken exception. He felt that it was inappropriate to touch on personal memories and feelings at a Sony Corporation funeral. And because he was officiating, he did not want Yoshiko's remarks to precede his eulogy, the first of three. According to Yoshiko, Ohga telephoned her to communicate his concerns, and she agreed to delete details of personal memories of Ibuka. Ohga pressed her to limit her remarks to the final page of her text, a short "Message from Morita," which she had composed on his behalf. Yoshiko was angry but agreed to comply with Ohga's wishes. On the day of the funeral, she refused to attend the timing rehearsal in the morning. When she took her place at the front of the hall, in first position, she read the entire text, taking seven minutes instead of the one minute she had been allotted.

Her eldest son, Hideo, had urged her to take as much time as she wished. Masao was also on his mother's side: "I'm often opposed to what my mother does," he told me, "but this time I think Sony was wrong. There is a complex relationship between Sony and 'Mrs.' [inside Sony, the English word refers exclusively to Yoshiko]. "She is a powerful presence, and it's been hard for them to handle her since my father is gone. But she was building Sony with Ibuka and my father before Ohga or Idei were even around." The contretemps is at the surface of a lifetime of history between Yoshiko and Sony in which, in her role as Morita's wife, she still inspires fear. With this history in mind, and because her "eulogy" invokes so vividly the bond between Ibuka and Morita, I translate at length. In accordance with the Japanese custom, Yoshiko addresses her words to Ibuka directly, facing the smiling, larger-than-life photograph installed above and behind the garlanded altar:

> Ibuka-san:
>
> I first met you forty-seven years ago, before I was married to Morita. We were young: you in your forties, Morita in his thirties, and I in my twenties. I remember vividly my impression when we met at the Morita house. A somewhat frail and easily embarrassed young man, you seemed pure, earnest, and very gentle. As time passed, you and my husband and I shared many experiences known only to us, precious memories.*
>
> Four years ago, your condition suddenly worsened, and Morita and I

* In her original draft, Yoshiko had included details, such as Ibuka's wedding present to Morita, the first H-Type tape recorder to come off the production line, which he proudly lugged into the reception.

had many sleepless nights. Before long, Morita was also stricken. When he was moved to the hospital room next to yours after his close call with death, you encouraged each other. When you left the hospital, you visited back and forth between your house in Mita and ours in Aobadai. Though you could barely speak, you called aloud to my husband to cheer him, "Akio! Akio!" Morita's head was lowered, his gaze on the floor. But you gripped each other's hands. Three years ago, Morita moved to Hawaii to convalesce, and you telephoned him there. Your voice at the other end of the phone was strong: "Akio, hold on!" Recently, your voice had grown weaker, and then you ceased to call.

On December 18, I returned to Tokyo to get ready for the New Year. For three years, I had always gone straight to your house from the airport, and when I phoned from the car, you were already hovering on the brink of death. I suppose you were waiting for me to return. I spent a sleepless night at home, and just as I dozed off, the telephone rang. It was 3:40 A.M.; you had taken your last breath two minutes earlier, at 3:38. I rushed to your house and gripped your hand. I rested my cheek against your face. It was still warm. Your face was peaceful.

I knew what I had to do but was unable to do it. In Hawaii, the sun was already up. But how could I bring this news to my husband on the phone? It was all I could do to instruct the nurse who answered to hide the newspapers and keep Morita away from television and the radio. The night of the wake, I was asked to write something to read for my husband at the family funeral. I worked all night until dawn, in tears, struggling to compose a brief message that I felt my husband might have written if he were able. The day of the family service, I asked the priest to read it.

I returned to Honolulu immediately. Morita met me at the door, but I was unable to tell him the news. The following morning, I took a deep breath in our garden and said as casually as I was able, "Ibuka-san has finally gone to his rest." He peered into my face. Then he gasped—"Aah!"—and burst into tears. He looked down and continued to gasp and sob. I read him the message I had composed. "Was this acceptable?" I asked him. "If so, please squeeze my hand." Morita gripped my hand tightly, looked up at the white clouds in the blue sky, and quietly wept. As I was leaving again to be here today, I explained that I intended to read the message myself this time and asked for his approval. He nodded his head vigorously, as if he were entreating me. Dear Ibuka-san, please listen once again to my husband's words to you.

"Ibuka-san: you've finally set out on your journey to a new world. I met you first during the war, more than fifty years ago. Fifty-one years have passed since we made our company. In good times and in bad, we were always together. Now we are apart, but let us watch together over the next generation as they make their way through these difficult times.

I will not say good-bye. I am certain that we shall meet again one day. It is only a matter of waiting.

Ibuka-san, please accept my heartfelt gratitude for the wonderful life you have given to me. I thank you from the bottom of my heart."

On Morita's birthday in January, and at least once in the summer, Ohga and others from Sony, including Nobuyuki Idei, pay their respects with a brief visit, and Morita's friends from around the world—Peterson and Henry Kissinger and others—drop in on him when they can. In 1998, the family built a guest house to accommodate visitors adjoining the grounds of the nearby Waialae Country Club. Beginning in January 1999, Sony took over from United Airlines as host of the first tournament of the year on the pro golf tour, played at Waialae. The company intends the Sony Open as an annual tribute to its stricken founder. According to Ohga, visits to Morita have become increasingly painful as his decline advances. Many of his Japanese friends in private life, the journalist and editor Hiroshi Yamaguchi, Tony Kobayashi, the social critic Chieko Akiyama, and others, fearing Morita might be discountenanced to be seen in his infirmity, have refrained from visiting him. "A lot of his friends," says his son Masao, "want to remember him as he was when he was in his prime."

Like other men with heroic appetites, Morita's fear of death seems to have been obsessive. Before his mother died, following an illness that hospitalized her for several years, he asked his son Hideo to prepare a tax strategy and returns designed to minimize the huge inheritance taxes levied in Japan. Hideo labored for months, exploiting every loophole he could find. When he took the documents to his father, Morita refused to look at them: "He just didn't want to see it. He didn't want to think that his mother was going to die." Later, when the documents prepared by Hideo and his CPA were filed, they held up to scrutiny by the Tokyo Tax Bureau's director. "That's the only time ever in my life my father called me on the phone and said you did a great job, I appreciate it. That's the only time. I cried because he never said that ever in my life."

When Hideo urged his father to formulate a strategy in anticipation of his own estate taxes and volunteered to do the work himself, Morita reluctantly consented and then refused to cooperate: "People all over the world think my father is the wisest man in the world. But he says, 'I'm never gonna die.' He hated when I would say when the time comes for you to die you have to think of us. He hated it." Hideo's frustrated attempts to enlist Morita's help led to bitter confrontations over a period of three years until just days before Morita was stricken. Speaking in English, Hideo remembered: "He wanted me to do it behind the scenes, but I couldn't do it, it was so complicated, without his help. So I had to fight with him for the whole three years. It was a hundred times more complicated than my grandmother's job was. And the whole nation knew about it. This would go past the Tax Bureau straight to the Ministry of Finance. So it had to be done very carefully. So I had to deal with my dad, and he never wanted to talk about it. Anything related to his death he hated, and so I had to fight. 'Don't talk about it, Hideo. I'm never gonna die.' That's it. Actually, three days before he had his stroke, I had a huge fight with him, and he was saying, 'I'm never gonna die.' That's why everyone, including my brother, thinks he had his stroke because of me."

Morita's crippling stroke was devastating to his family and to his friends and associates around the world. It was difficult to imagine, and horrifying, that this luminous man had lost his light. Hideo's younger brother, Masao, expressed his loss in terms that echo Hideo's picture of Morita as a demanding father who withheld the approval his children yearned for: "It's not fair: now I'll never hear him tell me I'm doing a good job no matter how hard I work or how I succeed. It's not fair!"

Inside the Sony family, no one was more deeply shaken than Norio Ohga, who had spent his career luxuriating in the affirmation that Morita's sons had lived without. For forty years, Morita had been

Ohga's mentor and unwavering champion. He had chosen Ohga as his heir and successor even before Ohga entered the company formally in 1959, and he had never even momentarily withdrawn his support. Ohga admired a number of men who had figured importantly in his life: his boyhood mentor, Ichiro Iwai; his voice teacher, the German baritone Gerhard Hüsch; Herbert von Karajan; Ibuka; certainly Kazuo Iwama. But Morita seems to have occupied a unique place in his heart and mind. Perhaps he loved him; certainly he felt connected by what he perceived as a mystical bond: "I never understood what Mr. Morita saw in me that made him pursue me with such determination, but one thing I know is that we were very much alike." As evidence, Ohga recalled a personality test that he and Morita had once taken (his selective memory of the contents has a certain interest of its own): "There were one hundred questions about how you make judgments and decisions. Do you lie? No. Can you deny you know something even when you do if it's important in business? Questions like that. When we were finished, my graph and Mr. Morita's were identical! And I remember saying to him, 'We'd better be careful: it's probably not such a good thing to have the same type of human being as chairman and president of the same company.' We discussed it, and he agreed we should be careful."

Ohga says he knew within weeks that Morita would not be returning to the business. He felt bereft, and he felt isolated; he had advisers, men he trusted and respected, Hashimoto, Iba, Ohsone, and others. But now the big decisions about what to do at Sony were his to make alone, without the support from Morita which had always somehow validated the process. One looming decision was the write-off. An even larger question was who should succeed Ohga himself as president. Sony's corporate bylaws called for the incumbent to leave office on reaching the age of sixty-five; Ohga had a little over a year, until January 29, 1995, to reach a decision. He had no succession plan. And, by his own account, he had had no indication about what Morita might have been thinking: "I would bring it up from time to time, but he had never once expressed any clear opinion about who he thought it should be. Now the responsibility for this decision would be mine alone to bear. I thought about it day and night from the day Mr. Morita fell ill."

Ohga agonized over the choice of a successor for a full year, a time he characterizes as the darkest and the most lonely in his Sony career. Unthinkably in the context of an American corporation, where the selection process would have been largely in the hands of outside directors, he kept his own counsel. If he revealed his deliberations to anyone at all, it was to Tsunao Hashimoto, Sony's chief of personnel, who was at the time one of four executive deputy presidents, the company's highest-ranking corporate officers below the president. A graduate of the Law Department of Tokyo University, Hashimoto had been in human resources since he entered the company in 1958, and had a reputation as a skilled mediator who held himself above the political fray. While Ohga has often expressed his respect for what he describes as Hashimoto's "fairness and objectivity," he did not consider him a candidate; he was already sixty-two in 1994, and was not known to be a risk taker or even an initiator so much as a respected keeper of the peace.

Another of the four executive deputy presidents was Ken Iwaki, who had been widely rumored from about 1989, in and outside the company, to be Ohga's likely choice. Iwaki was a Tokyo University graduate in economics who had begun his career at Toshiba. His interest was international business, and when he decided after six years that Toshiba would not afford him the opportunities he wanted, he wrote to Morita requesting an audience and was invited to join Sony in 1966. After four years in management at Sony's Shibaura video plant, he announced his desire to study in the United States and persuaded Morita to send him, first to the three-month advanced management course at Harvard Business School, and then on to Stanford for a one-year master's in international business in the Sloan Program. Clearly, Morita was impressed with Iwaki's intelligence and ambition, and pleased to have a young lion whose passion was mastering global business strategy.

Between 1971 and 1976, Iwaki was in charge of budgeting and product planning strategy at Sony America in New York. His first year coincided with Kazuo Iwama's last year as president of the American subsidiary, and he became Iwama's protégé, traveling to meetings of the Executive Committee as his representative. After 1972, when

Harvey Schein took over as president, Iwaki became an even more familiar presence at Executive Committee meetings, where he went repeatedly to argue on Schein's behalf for lower FOB prices in the United States and gained an understanding of the committee's personal dynamic which would serve him well when he began consolidating his authority in the eighties. From New York, Iwaki filed a report on what he called "true internationality" and the importance of installing an American-style corporate planning function at Tokyo headquarters.

Morita was sufficiently impressed to recall Iwaki to Japan in 1976 to design and manage a new Corporate Planning Division. Subsequently, with patronage from Iwama and Morita and, after 1982, as a member of Ohga's corporate staff, he built the division into a personal fief with immense power. By 1985, he had consolidated under his direct control corporate R and D planning, information systems, the office of the corporate controller and corporate strategic planning across all of Sony's businesses worldwide. As the company's top planner, he was also recording secretary to the Executive Committee even before he became a regular member himself, in a position to interpret the committee's deliberations and decisions to sector chiefs, revealing or concealing details as he saw fit, or shaping discussions that had been left open or resolved ambiguously, with emphases to suit him. In Iwaki's hands, the secretarial function became a powerful tool for leveraging personal power: "I wouldn't show people the minutes," he told me, "but I'd say, 'There was general support for such and such' or 'There was talk of investing here and cutting back there,' and so on. Often they'd leave out details like figures and I'd fill them in myself—I understood their main concern was a decent return on investment.

Some time in the course of 1990 or 1991, Ken Iwaki seems to have stumbled. Precisely what happened is impossible to uncover; probably he was guilty of the same variety of hubris which would contribute to Mickey Schulhof's eventual downfall, with the significant difference that in his case it was Ohga and Morita themselves, the source of all legitimate authority, who took offense at an exercise of power which they perceived as insubordinate. Iwaki began manipulating across sector lines, calling conferences among chiefs and coordinating them without bothering to consult with Ohga or Morita, and using the privileged information he possessed about top-level priorities for R and D investment to achieve the kinds of consensus he wanted. In August and September 1991, following coronary bypass surgery, Ohga spent forty days in the hospital and another month convalescing at home. In his absence, Iwaki seems to have become increasingly aggressive in his manipulations, and rumors of lines being crossed and toes stepped on reached Ohga and troubled him. Morita himself may have been even more offended; Masao Morita was working for Iwaki at the time, and may have reported Iwaki's activities to his father. By this time, Morita was no longer involved in details of the daily business, but ignoring his authority however slightly would always be a fatal mistake. A Sony executive who was close to the scene told me privately, "Iwaki was overconfident, as if he could run the company without Ohga or A. Morita, and he touched upon the nerve of A. Morita with his excessive taking of power."

There is no knowing whether Ohga had ever considered Iwaki seriously, but he was probably out of the running by the time Ohga dispatched him to New York in June 1992 as chairman, Sony Corporation of America, the electronics company, and executive vice chairman and COO, Sony USA, Schulhof's entertainment and software divisions. Iwaki's assignment was to check on the feasibility of establishing a second worldwide Sony headquarters in New York. Ohga had been playing with the notion for several years: "Three-fourths of our business was coming from abroad, and I thought it was *nonsense* to try to control it all from here. That's why I sent Iwaki to the U.S., to explore the possibilities. But he didn't take the project seriously. I don't know why." The likely explanation is that Iwaki sensed the New York assignment was a demotion. Friends recall that he was unhappy in New York, caught between Schulhof and Ron Sommer, both of whom theoretically reported to him, and often expressed anger at the man he referred to as "our company's crazy president." One year later, he was recalled to Japan and demoted from representative director to director. In June 1994, he was appointed to the position he currently holds as president and CEO of Sony Life Insurance Company, Ltd., an important post outside the mainstream.

The fourth-ranking executive on the roster, an executive deputy president and director just below Ken Iwaki, was Minoru Morio, a brilliant engineer from Tokyo University who had played key roles in developing Trinitron and subsequent color TV, 8-millimeter video, and DVD technology. Morio was a relatively youthful fifty-five, and was considered an extraordinary technical team manager. He had led Sony's consumer video and personal video systems groups and, in April 1994, was appointed president of the newly formed Consumer A&V Products Company. Ohga can hardly have failed to include him in his deliberations, especially given that Morita had more than once declared that the next company head should be from the technology side of the organization. Ohga maintains, however, that Morio was never a serious candidate any more than was Iwaki. In any event, in the course of 1994, Morio created personal situations, once in Hawaii during a company retreat, and again while traveling in China, which resulted in public embarrassment both to himself and to Sony, and effectively disqualified him for the president's job. It appears that his career was not otherwise damaged: still an executive deputy president and representative director, he is among the group of seven which runs the company and, as chief technology officer, oversees all of Sony's product R and D.

In the spring of 1994, when Ohga was spending days at a time poring over his executive roster, the list of senior managing directors (*semmu torishimariyaku*), the next rung down on the ladder, was nine names long. It included Kozo Ohsone, sixty-one, a man with a reputation for outspoken business savvy who was the fiery team leader on many of Sony's most successful audio products, beginning with the original Walkman; and, as of April that year, president of the Recording Media and Energy Company; Tamotsu Iba, fifty-nine, a charter member of the legal and international relations group created by Morita in 1960 and the principal architect of long-range financial planning; Junichi "Steve" Kodera, first manager of the San Diego Trinitron plant and, since April, president of the newly formed Systems Business Company; and Kenji Tamiya, Harvey Schein's protégé, who had served as president and then chairman of Sony America and would soon be appointed to head Aiwa, Sony's subsidiary and second consumer electronics brand. Ohga had trained and nurtured these men himself during the more than twenty years of his dominion at Sony, but in the end he decided none of them was successor caliber.

One of the very few people other than Hashimoto with whom Ohga shared some of his thinking about candidates was Mickey Schulhof, who recalls a conversation late in the summer of 1994: "We were having dinner together at Spago in Los Angeles. He was being criticized for not having chosen a successor. He and I were talking often at the time and we did discuss candidates. And I looked at him at that dinner, and I said: 'Do you want me to think about moving to Japan?' He said, 'Yes, if you think you could learn Japanese.' I said, 'That would be very difficult.' It was the closest I came to telling him that if he wanted me involved at more than just the board level or the subsidiary level, that I would consider it. I didn't want to say to him I wanted to run the company because I thought that no matter how much he liked me, the rest of the company could never accept it."

Ohga merely smiled when he heard the story, declining to confirm or deny it, but it is unlikely that Schulhof was entirely delusional. By Ohga's own account, he had considered dividing the company into two world headquarters, one in Japan and the other in New York, and Schulhof was aware of this idea: "Originally I think he envisioned a pure Japanese company run by a Japanese that would control electronics, research, and manufacturing, and a Sony International, possibly in New York, possibly run by me, that would handle worldwide marketing and worldwide software. The Sony of the future as he saw it, of the twenty-first century, would have two heads. But that game plan required Morita's approval of his choice, so that no one would dare to question it. After Morita's cerebral hemorrhage, all of his plans unraveled: it became clear that he would have to pick a Japanese."

11

IDEI THE HERETIC: EMPIRE'S END

At the end of 1994, weeks before his sixty-fifth birthday, amid furious speculation in every corner of the company, Ohga privately resolved his dilemma. Despite the importance of seniority even in a company as eccentric as Sony, he reached past sixteen senior executives, past the rumored front-runners and dark horses, to Nobuyuki Idei, a member of his corporate staff in charge of advertising and marketing worldwide. Late on a dreary afternoon in January 1995, Idei received a message that Ohga was waiting to see him in his office. In the elevator on his way to the seventh floor, he reviewed the events of the day, searching for a cause for the summons and finding none. The prospect of a private meeting with Ohga was not pleasing to Idei. With his boss, communication tended to be one-way and exhausting.

Ohga was peremptory. "I told him," he says, offering an English equivalent, "I have decided that you must be my successor." Idei was dumbstruck: he had had no warning, not an inkling. He remembers feeling principally an urgent need to remove himself from Ohga's presence. Neither man recalls what more was said, but each confirms the other's memory that the meeting was over before fifteen minutes had passed.

Idei was formally introduced to Sony and the world at a press conference held in March. The previous week, Ohga and Tsunao Hashimoto had traveled quietly to Honolulu to visit Morita at his beachfront estate in Kahala, where he has been confined since 1994. The corporation represents that Morita was advised of the decision and conferred his approval. In answer to a journalist's question about Morita's response to learning that Idei was Ohga's choice, Ohga replied: "He said something like 'Ah yes, I see. I suppose that would be the best solution.'" In fact, Morita had been aphasic since his stroke and, in view of his increasingly veiled consciousness, the pilgrimage was a ritual formality. For Ohga himself, it was more deeply personal, a gesture of his respect.

Before the public announcement, Idei visited Ibuka at his apartment in Mita to pay his respects. Since his stroke in 1992, Ibuka's condition had gradually declined; he was by this time largely bedridden and unable to communicate without help from someone in the immediate family or his secretary, Hiroko Kurata. When he learned that Idei was coming to see him, he insisted on getting out of bed and dressing in his suit and tie so that he could receive him properly in his wheelchair. "This is the president of Sony," he told the family. "He must be treated with respect."

Idei's appointment stunned the company. Many people across the empire had never heard of Nobuyuki Idei. Others were uneasy: he was known to be a naysayer and had a reputation for keen, analytical intelligence, disdainfulness, and a biting tongue. "Out of a hundred people at Sony," Ohga says, "ninety-nine opposed my choice. 'Not him,' they said. 'You can't be serious about him!' People felt he couldn't be trusted."

Those familiar with Idei's career were among the most baffled: one of the remarkable things about Ohga's choice is that a survey of Idei's thirty-five years with the company appears on the surface to usher forth no striking evidence that he possessed either the business sense or the leadership skills he would need to lead the Sony leviathan. Ohga maintains that Morita himself had never championed Idei, and needles Idei from time to time with assurances that he could not have been chosen as president had Morita been in good health. This may not be entirely accurate. In the early seventies, Idei and his family, stationed in Paris, had looked after Morita's daughter, Naoko, who was at school in France, and had provided a home away from home to

Morita's sons when they visited the continent from their boarding schools in Wales and Colchester. According to Yoshiko Morita, the families remained close. To the Moritas, Idei and his wife, Teruyo, were "Shin-chan" and "O-Teru-san," affectionate nicknames normally used by close family and childhood friends. The Ideis were regular guests at parties at Morita's house, and often accompanied them to concerts. Teruyo became a member of Yoshiko's support group, among the first to show up at the house when Yoshiko needed help packing for a move. Idei chose books for Morita to read, bringing them to him with important passages highlighted.

When Morita made good his promise to Ohga that he would be Sony's president one day, he had charged him with choosing an engineer for his successor, and Ohga had begun his deliberation with this constraint in mind. Apparently, he confided to Hideo Morita that he was having a terrible time: in the summer of 1994, Hideo ufged his mother to help Uncle Ohga by creating an opportunity for his father to convey his permission to widen the field of candidates. Yoshiko did so at the Moritas' summer home in Hakone. When she suggested, with Ohga in the room, that the search should not be limited to engineers, Morita, who had temporarily regained his speech in a limited way, responded, "Good idea!" Yoshiko took the opportunity to add a thought of her own: "I said I thought whoever it was should be married to someone with some social skills, entertaining and foreign languages, which the wife of an engineer rarely has a chance to learn. I didn't mention any names, I was careful about that, but I was certainly thinking of an international couple like Shin-chan and O-Teru." While there is no evidence to suggest that Morita ever advocated choosing Idei specifically, it is equally unclear that he would have vetoed his nomination as Ohga continually suggests.

Ohga and Morita before him were the sons of affluent merchants. Nobuyuki Idei grew up in a scholar's house filled with books. His father was a professor in the Department of Government and Economics at Waseda University, one of Japan's most prestigious private schools. His field was global economics. Before the war, he had worked as the first Japanese staffer at the International Labor Organization in

Geneva, an agency of the League of Nations; between 1941 and 1943, he had lived in China and worked as director of the Japanese Institute of Commerce and Industry in Harbin. Idei's own internationalism may have been in part a legacy from his father.

There was wealth in the family: Idei's forebears had traded in rice and lumber (Ohga's father was also a lumberman) in northern Tochigi Prefecture for generations until his father left the business and came to Tokyo. Idei grew up the youngest of five children in Seijo, an affluent, tree-lined residential district on the outskirts of Tokyo which is an enclave of industrialists and celebrated artists like Toshiro Mifune the movie star and the Nobel laureate Kenzaburo Oe. As the baby of the family and the only boy following the early death of his elder brother, Idei was pampered and adored by his mother and father and his three elder sisters in a manner reminiscent of Ohga's childhood circumstances. The family summered in Karuizawa, a wooded plateau atop a mountain pass three hours by car north of Tokyo. For generations it has been a vacation home to Japan's dynastic families, the Japanese royals, high government officials, and a community of writers and scholars to which Idei's father belonged, one of many benefits of his status as a university professor.

During his summers at Karuizawa, in the company of wealthy youth whose families belonged to the Karuizawa Golf Club, Idei developed the style and demeanor that led people to characterize him, somewhat inaccurately, as "high society." His life at home resembled Ohga's boyhood in other respects. His mother, a doctor's daughter and a graduate of the elite Ochanomizu Women's College, was a cultured and accomplished woman; there was art in the house and informed conversation about world events. Nobuyuki, known to the family and friends as Shin-chan, was a quiet, sensitive child, an avid reader—a habit he retains—and a violinist. As a student at Seijo Middle School, he played in a quartet with another Seijo resident, Seiji Ozawa, now the music director of the Boston Symphony. Idei went to Waseda High School and from there entered Waseda University and majored in economics. His father hoped he would become a scholar and a teacher.

Idei's initial interest in Sony had nothing to do with electronics; he had no technical background and had never been a gadgeteer. His graduation thesis, which was published in the *Waseda Journal*, was an analysis of the dynamics of a European common market, and he looked for a company that was expanding into Europe. When he applied to Sony in the spring of 1960, fresh out of college, he was also considering the seasoning company Aji-no-Moto. At his Sony interview with the redoubtable Akira Higuchi, he asked, cockily enough, to see someone higher up, and met briefly with Ibuka and Morita in the same room. Idei explained that he planned to go to graduate school in Europe but hoped to secure a job before he left. Morita agreed to employ him with the understanding that he would be allowed time off for school: it was just at this time that he was laying plans to launch Sony into the West, and a graduate in economics from one of Japan's top universities with a stated interest in Europe rather than America was a find.

After one year in the newly formed department of export sales, soon to be named Sony Overseas, Idei enrolled in the graduate program at the Institute for International Affairs in Geneva and studied there for eighteen months with the German economist Theodore Roepke. From Geneva, he filed a long report, an extension of his thesis, on the process of European coalition, which was read and valued by Morita and others at Sony. Idei's father was also impressed, and urged his son to enter the doctoral program. By this time, Idei was more interested in the real world of business than in economics as an academic discipline. He was beginning to envision himself as an economist working inside a corporation—today he laments the absence of such a person on his own staff—but not necessarily at Sony. In fact, as he prepared to return to Tokyo in 1963, he had decided to leave Sony and look elsewhere. When he arrived at Haneda Airport, someone from Higuchi's personnel office was there to meet him, and chided, "There's talk that you'll be leaving us." Idei maintains that his embarrassment at being caught out compelled him to deny the rumor and to resolve on the spot to remain at the company. Perhaps, as Molière might have said, Sony's new president was, in the early years, a Sony man "in spite of himself."

Idei worked at Sony Overseas in Tokyo until he was reassigned to Europe in 1967 as a sales manager based in Switzerland, with responsibility for France, Belgium, Luxembourg, and Holland. In 1968, Morita moved him to Paris with instructions to prepare the ground for a sales company, a Sony subsidiary, in France. Until then, as in the United States, where Morita's friend Adolph Gross was acting as a distributor for Sony products, Sony's only access to the European marketplace was through local distributors. The French distributor, a man named Georges Trenchant, was both a businessman and a politician with a seat in Parlement. Trenchant was unhappy at the prospect of a Sony subsidiary in France to compete with his own business, and he used his government connections to apply pressure to the Ministry of Industry and Commerce to block Sony's investment. Idei remained in France for five years, until 1972, attempting to establish a company in the face of government opposition. In the current version of the story, he is credited with the creation of Sony France, originally a joint venture between Sony and the Group de Suez Banks, but Morita was also actively involved in the negotiations and used his relationships with the finance minister and with powerful business leaders like Antoine Riboud of the Danone Group to achieve the Sony presence he wanted.

Idei's six years in Europe were formative. He developed serviceable French and English and a love of good wine. And he acquired familiarity if not ease in dealing with Western businessmen. His character as a Sony man was also shaped by the experience of "growing up" outside the capital on the farthest border of the empire. During these earliest years of expansion, there were fewer than ten Sony employees in Europe at any given time, no guidelines to speak of, and little supervision from Tokyo. Idei grew accustomed to acting on his own initiative and to feeling unidentified with headquarters or the corporate entity. In effect, he was being given room to develop as an eccentric.

Returning to Japan, he worked in the TV export section of Sony Overseas until, in 1979, he was given his first career break by Kazuo Iwama, then president of Sony, who promoted him to general manager in charge of audio products. One of his responsibilities in this important job was overseeing the launch of the CD player, which the engineers in his group had labored frantically to develop in time to meet Ohga's deadline of November 1982. On the day of the launch, Idei was home in bed with a fever. Ohga, recalling the moment with a smile, suggested he had succumbed to stress.

Between 1984 and 1986, as general manager of the Micro-Computer Office Automation Group (MCOA), Idei managed Sony's first attempt to enter the Japanese computer market. The plan was to create a multimedia home computer, and Idei organized a collaboration that was to include Sony, Philips, Apple, and Microsoft. The project fell apart when its champion at Apple, Wayne Rosing, was fired by Steve Jobs and moved to Sun Microsystems. Subsequently, Idei's group designed the MSX, a home computer that used MSX Basic and was driven by a Motorola chip. Idei was in charge of launching the machine, but was unable to win market share. In 1991, Sony withdrew its MSX products from the Japanese market.

In 1988, as senior general manager of Home Video, Idei championed the videodisc recording of von Karajan's complete operas which Walter Yetnikoff had declared nonsense and which had ended by costing Sony a sizable amount of money. The following year, 1989, he was appointed a member of the board. In 1990, Ohga placed him in charge of advertising and marketing worldwide; notwithstanding the disdain he professes for the advertising business, Idei is a natural born adman and has a masterly touch with the media. In 1994, he took over Corporate Communications and "creative communication." It was not until July 1994, less than a year before he was appointed president, that he was promoted, at the age of fifty-seven, to *jōmu torishimariyaku*, managing director, the lowest rung in the hierarchy of top management. Ohga himself, by way of comparison, albeit Ohga's ascendancy under Morita's patronage was unique, had attained the same elevation at age forty-two.

Ohga has said repeatedly, in public and private, that he arrived at Idei by a process of elimination, and there is no reason to doubt him. Over the course of the lonely year he devoted to his deliberation, he doubtless made his way down the list of his top managers, crossing off names as he went: Ken Iwaki was a keen strategist with a broad grasp of the global business, but lacked the fortitude and the charisma to lead. Minoru Morio, rumored to be the front-runner, was a brilliant engineer, but had no feel for the software business. And so on down the list until he paused at Idei. Ohga is the first to acknowledge that past performance was not what stayed his hand: "Logic would have demanded that he be rejected. Certainly his career didn't qualify him. He was not very successful as audio manager, and he fell on his face with our first attempt to enter the home computer business. I bought his future, not his past."

Ohga knew that Sony was at a crossroads. To lead the company into a new age, he wanted a man with strong, unorthodox views and the certainty and courage to act against the grain. He had known Idei as a fighter with a brashness to match his own since the early days of Europe, when Idei had had the temerity to engage him in heated arguments about marketing strategy in France. Ohga was also deeply concerned that Sony's image in the world should be preserved. It is well to remember that he had single-handedly refashioned not only Sony's advertising, but the look and feel of Sony products, and that he counts the perceived value of the Sony brand as perhaps the signal achievement of his career. Since 1990, Idei had managed Sony's advertising worldwide, reporting directly to Ohga as a member of his staff: Idei was an image-maker, a skillful marketer of Sony dreams, and Ohga must have felt that he could be trusted with the care of what he likes to call "the four letters, S-O-N-Y."

In explaining his choice of Idei, he told me as much. It was a Sunday night, and we were seated in the empty penthouse apartment in the Ark building from which Morita had coordinated his Keidanren campaign. His logic was impressionistic, roundabout, and in its way self-serving, but there was no question that he was speaking from the heart: "Sony must always be an extraordinary company. Mr. Ibuka and Mr. Morita were extraordinary people. They were a pair of geniuses. I did my part to grow Sony in their footsteps, and I always asked myself what was essential to the company. Recently, I find myself thinking about the Chinese character 'san,' which means to shine dazzlingly like the sun—*san-san to kagayaku.* It's not simply a matter of brightness, 'san' means an extraordinary radiance. Ever since I became involved with the company, I've worked hard to ensure that Sony would shine with that blinding light. It's been my main focus and my theme. And when I began to think about my successor, it was clear to me that he would have to be an extraordinary person too, someone who would make sure that Sony continued to radiate, to shine like the sun. I agonized over who might be capable of achieving that. When I heard that Iwaki had his eye on the job I thought, unfortunately, it couldn't be him. When everyone was talking about Morio, I knew in the same way that he wasn't right. Whoever it was had to have the taste and the discrimination to understand the nature of Sony's blinding light and something extraordinary that would permit him to sustain it. I pored over the list, and when I got to Idei, I thought, Maybe, just maybe, he might have what it takes."

Ohga's vision of Idei as a radiant being tests the imagination: his energy, which is palpable, is in the nature of an impending storm, glowering rather than sunlit. Nonetheless, his achievements during his first four years as president shine with a brilliancy all their own. Unquestionably, he has taken command: he has set the company on a strategic course into the future, winning confidence along the way; he has established peerage with business leaders, particularly the American digerati, whom he views as strategically important to his vision; and he has de-sentimentalized Sony's relationship to its American business. In that process, he has demonstrated something about himself which Ohga may have understood intuitively, in the manner of an artist: that his capacity for irreverence will allow him to take whatever action he deems necessary, even deconstruction of Sony's traditional organization and modes of operation, in order to equip the company to meet the challenges of a new age.

At Sony's annual management conference in May 1995, less than two months after his appointment, Idei unveiled before two thousand of Sony's top executives from around the world his first guidelines for what he would soon be calling "the new Sony." He had chosen not to discuss his speech with Ohga in advance—this was to be his debut performance as the new president, and he wanted it all to himself. Privately, he had consulted with a group of personal advisers outside the organization, and he acknowledges that they provided him with much of the material he used. Sitting in a Tokyo cafe with his American friend John O'Donnell, a venture capitalist from Harvard whose wife had worked for him, Idei was searching for a shibboleth for the new era, a distillation of the challenge he faced as Sony's first "salary-man president," when O'Donnell proposed "regeneration," now a key term in the evolving Idei glossary.

Another longtime friend, Takahiro Yamaguchi, president of Overseas Debt Service Company, promised him a gift of words in celebration of his promotion and offered, in English, "digital wave," "digital wave front," and finally, after a conversation about Masaru Ibuka's innocence and curiosity, "digital kids." When Idei carried the phrase back to his advertising department, one of his managers suggested adding the word "dream," and Idei's best-known catchphrase to date was born: "digital dream kids." The other members of his unorthodox team were Bernard Allien, a French philosopher and sociologist; an American business consultant and author of *Intuition at Work*, Gigi Van Deckter; and Kozo Hiramatsu, who had worked for Idei on the MSX project before he left Sony to head Japan IDG, a publisher of computer books and magazines.

As a first step in the direction of a new business strategy that would soon lead him to value chains, Idei planned to emphasize the integration of audiovisual products and home computers, but Hiramatsu persuaded him to focus more broadly on IT, information technology. With characteristic candor, Idei credits Hiramatsu: "If I hadn't introduced the notion of AV plus IT, we might be on an altogether different track today. That was Hiramatsu's contribution. If you want to be global, you need an eclectic, international team like mine."

"Regeneration," as construed by Idei in his keynote speech, was at once conservative and revolutionary: it involved resetting to the same zero point from which the founders had launched the business from the rubble of postwar Japan, stepping back into their innocence and passion, and discovering from there the tools and attitudes that would be needed to drive the company through a new and entirely different cycle of success. In "digital dream kids," Idei uncovered a double meaning: to compete in the digital age, Sony must appeal to the young, or the young at heart. At the same time, Sony's technical innovators and marketers must rediscover in themselves the excitement of creative children: "Ibuka-san was a transistor kid," he explained. "Morita-san was a Walkman kid. Ohga-san was a CD kid. And we must now be digital dream kids." Idei has continued to sound his thematic emphasis on Sony as dream-makers: the company's corporate image campaign worldwide in 1998 was built around the question: "Do you dream in Sony?"

Peering into the future has become increasingly Idei's obsession, but unlike Ibuka and Morita, his visionary forebears, his formulations are the work of a business theorist. Idei articulates the difference bluntly: "There are two people in this organization who have no power of abstraction, Akio Morita and Norio Ohga. For Mr. Morita, what mattered was the position and proper function of a switch, a start button, on a Sony product. If he took one home and tried it and the switch was hard to reach, that was an "insult," the engineer responsible had "insulted" him. He never generalized, and neither does Ohga. Mr. Ohga is an artist: he has an artist's pride and an artist's jealousy, like Kathleen Battle. If he is moved by something, it moves him purely and simply and he feels no need to explore its theoretical or even general significance. Our tradition of product manufacturing resists abstraction—and that allergy we have for generalizing has been a strength in our electronics business. But the top management of tomorrow has to develop the capacity to think abstractly."

Since coming to office, Idei has accelerated through iterations of a theoretical blueprint for Sony's business strategy in the next century. In his quest for vision, he has developed relationships with relevant leaders across the business spectrum: Jack Welch of GE, Percy Barnevik of Europe's ABB, Lou Gerstner of IBM; Microsoft's Bill Gates; Intel's Andy Grove; and Disney's Michael Eisner. He furthermore has found time in an unthinkable schedule for feverish reading, skimming his way through valises full of books on endless flights back and forth across the Pacific and sleepless nights in hotels around the world. He studies eclectically, and his reading at any given moment tends to figure centrally in his kaleidoscopic progress reports to the company. Beginning in 1996, at successive management meetings and off-site strategic retreats, he has assailed top management with a bewildering array of theoretical notions: a world of atoms being displaced by a world of bits (Nicholas Negroponte's *Being Digital*) complex systems (Santa Fe Computer Institute), emergent evolution (natural science), semantic borders (Claude Lévi-Strauss), value chains (Michael Porter et al.), and the algorithms of portfolio and value-chain management.

Skeptics in and outside Sony have suggested that Idei allocates more time to pondering the future than to conducting business in the present, but that criticism is unfair, and Idei has the data to prove it. In the summer of 1998, he commissioned the Boston Consulting Group to analyze his use of time in the office for the first three years; the project input was the daily logs kept by his secretaries to graph his time allocation by category. He was gratified by the results. He has spent an average 9.6 hours a day at the office, a number that he interprets as proof that he is well organized. Time spent in meetings, 95 percent in the first year, has decreased by 30 percent, and so has time allocated to the electronics businesses. This shift has enabled Idei to devote increasing time to promoting Sony outside the company, particularly abroad, and to working on new business. As of October 1998, he was spending 22 percent of his time on corporate communications, including interviews and investor relations, and 22 percent on the future. Over half his business day is still allocated to what he calls "building a defense," improving Sony's current business operations. That said, it is clear that he is proudest of the fact that he is increasingly free to "plot an offense," for the new era.

At the annual management conference in May 1997, his third, Idei unfurled Version 1.0 of his blueprint for the future. He began by proclaiming the advent of what he calls the "Age of Networks." As he now conceived it, Sony's value chain had three major components: electronics, still the core business; contents (music, pictures, computer games); and a critical third link that would assume increasing importance, network services, which would encompass broadcasting, network distribution and the Internet, and electronic commerce such as home banking and shopping. Integrating and expanding these links in the value chain would enable Sony's "Big Bang," the emergent evolution of new business domains. Driven by a parallel big bang in IT and network technology R and D, these new domains would be the basis of "the new Sony's" growth strategy. Managing the "complex system" represented by this value chain would require a new corporate architecture: a different kind of board, multiple compensation systems to provide incentive to different sets of talents, new business unit management, and possibly a global headquarters radically different in structure and function from the current configuration.

All of Idei's major business decisions to date have accorded with the logic of this evolving vision. His first move was taking Sony back into the home computer market. In November 1995, Sony and Intel announced a joint venture dubbed the GI Project, "G" for Andy Grove and "I" for Nobuyuki Idei, in which Intel would manufacture chips for a Microsoft Windows–based personal computer. At the time, although Sony possessed CD-ROM storage technology and manufactured monitor displays, there was no computer project in development. Idei was certain of the PC's importance in the value chain, and was forcing the company's hand with the Intel tie-in. In April 1996, he increased the pressure by adding to the eight companies Ohga had created in 1994 an information technology company. For the first year, the new company, headed by Kunitake Ando, marketed a machine built with Intel, a standard mini-tower desktop, while its engineers scrambled to design a thoroughgoing Sony product. In July 1997, the VAIO line of mini-tower desktops and a laptop was released in Japan. The brand name, VAIO, was an acronym for video audio integration operation and reflected Idei's emphasis on the importance of personal computers as a terminal linking audiovisual products and distribution networks like the Internet. The machines are packaged with Sony software that enables video and still photos and optimizes connectivity with other Sony products, such as digital Handycams and a digital still camera. Despite a computer market that has slumped in 1997 and 1998, VAIO has been a huge success in Japan; the laptop model, the Superslim 505, currently commands a 60 percent share of the domestic market: a slender, platinum, flyweight beauty with impressive power and connectivity which sells for $2,000, the Superslim is attracting younger men and increasing numbers of young women who are drawn by its stylishness.

The U.S. market has been more resistant. Surrounded by Compaq and Dell, the desktop models have continued to founder despite national advertising coalesced around the "Size does matter!" campaign developed at Sony Pictures for *Godzilla.* As 1998 drew to a close, the

VAIO laptop had begun to attract attention. The new president of Sony Electronics, Teruo Aoki, says the Superslim 505 is currently building what he calls "mind-share."

In May 1997, simultaneously with his unveiling of "Sony's Future, Version 1.0," Idei invested $38 million in a 25 percent interest in Japan Sky Broadcasting Company, Ltd. ("JSkyB"), a subscription satellite broadcasting service. The business had been formed in December 1996, as a joint venture between the Australian media mogul Rupert Murdoch and Masayoshi Son, a young entrepreneur who had built a multibillion-dollar software business called Softbank, and was already being described as the Bill Gates of Japan. Idei became the third partner, and persuaded the Fuji Television Network to acquire another 25 percent. This was his first experience in negotiating an agreement with a world-class foreign player. Murdoch is well known for his aggression and cunning, but Idei seemed to thrive on the challenge: "Remember, we were dealing on my turf. Outside Japan, I wouldn't have stood a chance to get a fair shake with him, but since it was here, I had the upper hand and we came out fine."

Plans called for digital broadcasting over 150 channels, some pay-per-view and others by monthly subscription, to begin in April 1998. In March, before it had gone on air, JSkyB merged with PerfecTV Corporation, a competitor that had grown its subscriber base at roughly 50 percent per year over three years to a total of 912,000 subscribers. In the new company, SKY PerfecTV!, capitalized at $320,000, Sony and the original partners, major shareholders with 11 percent each, are joined by Sumitomo, Nissho Iwai, Mitsui, Itochu, and the Tokyo Broadcasting System. In May 1998, SKY PerfecTV! began digital broadcasting on 162 television and 106 radio channels. Programming for six of the channels is provided by Sony Pictures, Sony Music, and the Sony Group. Subscribers pay an average monthly rate of ¥4,000 ($33 at current exchange rates). At the present rate of enrollment, the business projects breaking even in two years, at 1.8 million subscribers.

There are those at Sony who remain skeptical about the broadcast business, but Idei dismisses concern impatiently. He sees SKY PerfecTV! as an opportunity to establish Sony's presence in the digital services business, and as a tangible first step toward using digital network distribution as the adhesive in Sony's augmented value chain. Short-term profitability is not the issue.

Idei has accepted a job that he considered, in his own words, "mission impossible": "I knew that running Sony was impossible because it was a company driven by the founders' vision: the founders' personal vision was the basis and the standard for every decision. . . . Normally, there is a clear line between the owners of a company and professional managers, but in our organization that separation never occurred. Ohga is not an owner, but he behaves as if he were. His logic is theirs. And Sony logic until now has been based on personal relationships, on friendship—and that produces phenomena like Mickey, a second tier of ownership. Sony is an extraordinary company in many respects, but one of them is certainly that we have grown to half a hundred billion dollars on the founders' logic! Now we need professional management."

Idei has attacked the challenge of normalizing what he perceives as Sony's sentimental logic on two fronts, at home in the capital and in the American subsidiaries. In Japan, where the founders' sensibility is deeply rooted and resistance to change is high, he has moved cautiously. One of his challenges has been to bring into his sphere of influence the circle of advisers he inherited from Ohga. In the weeks following his January 1995 meeting with Idei, Ohga called in his senior managers one by one, Iba, Morio, Ohsone, and others, men whose careers had elevated them more rapidly than Idei to higher places in the organization, and informed them of his decision. These were delicate meetings: everyone was astonished. Some, certainly Morio, who had been considered and considered himself the front-runner, were disappointed and angry. "These were people who had been passed over by a man who had worked for them and whom they may not have admired. They needed to understand my decision and buy into it." Ohga was gentle and very skilled. But he was not inviting dissent or even discussion—this was emphatically not an example of the time-honored technique of consensus-building known as *nemawashi.*

No one voiced resentment, but it was there nonetheless, and potentially a problem. Ohga might have made it easier for Idei by assigning members of his senior team to Sony subsidiaries like Aiwa and Sony Life Insurance—in fact, Ken Iwaki, at one time a front-runner, had already been moved out from the center into the president's office at Sony Life—but he chose not to clear the way. By his own account, it was his judgment that Idei would need help from these men in running the daily businesses. It is also possible that he wasn't ready to give Idei a free hand.

Superannuating the old guard was an option that was also available to Idei, and his advisers suggested that he should do exactly that. He chose to keep the senior group in place. "They were expecting me to get rid of them," he says, "but I assured each one of them that wasn't going to happen. You need senior people to run the business from day to day. Iba was probably thinking it would be nice to become the head of Sony Computer Entertainment, but I told him I needed him to stay as CFO and that he would actually function as my chief operating officer.* In a sense I did 'fire' Ohsone—he was the president of our recording media company, and I pulled him out of that position and put him on the new board to think globally about production. Morio was president of the consumer AV company, and he stepped up, too, after I fired him, to the board and chief technical officer."

In the beginning, certainly during his first year, Idei and his senior team related to one another gingerly, with what he describes as "mutual deference." At some point, Iba confided that Morio was feeling unappreciated and truculent. Idei, who can be disarming and considerate when he chooses, went out drinking with Morio twice and reassured him that he was needed and appreciated, despite the fact that the new technology Idei favored was not his specialty.

Over time, as he has imprinted the organization with a vision uniquely his own, Idei has grown less deferential and more commanding. And the old guard has learned to respect his clarity and decisiveness. Yoshiyuki Kaneda, Sony's chief production officer, suggests that

* Effective April 1, 1999, Tamotsu Iba was installed as (acting) chairman of Sony Computer Entertainment.

adapting to Idei's leadership has been made easier for himself and his senior colleagues by the example already set by Ohga: "Remember, when Mr. Morita put Ohga in charge he was younger than anyone around, but he was full of confidence, like a superman, and we accepted him. That precedent has helped Idei, and he makes it easier for us with his consideration and respect. With his upper-class background, he knows how to behave, and he's been very careful to make us all feel comfortable."

The old guard is nearing the end of its tenure. Tsunao Hashimoto, the seniormost vizier and Ohga's closest adviser, reached retirement age in November 1997 and replaced Morita's brother M.M. as chairman of Sony Life Insurance. Hashimoto resigned from the Sony board at the time but continues to sit on the boards of the American subsidiaries and has been granted a title that was created specially for him, corporate adviser.* Kaneda and Kozo Ohsone, the outspoken audio man who played a central role in the development of the Walkman and enjoys a reputation for business shrewdness, have one year to go; Tamotsu Iba, who has led Sony's strategic planning office since Ken Iwaki lost his footing, is two years from retirement. Minoru Morio, Sony's chief technology officer and the youngest of the seniors at fifty-nine, will have two years beyond 2003 when Idei himself retires or, more likely, becomes chairman. As the inner circle leaves, Idei will replace them with younger men whose loyalty to him is unobstructed. Early in 1998, he asked top management to submit succession plans but reserved the right to disagree with their choices.

Idei had been in office for fully two years before he felt secure enough in his authority to institute a sweeping change in the structure of the parent company. In May 1997, he moved thirty company directors of junior and senior ranks off the Sony board and into a newly created management position called corporate executive officer. His purpose was a distinct separation between what he considered the proper function of the board, company policy and oversight, and op-

---

* Two other Sony elders have been similarly honored: Akira Higuchi, adviser, and Masahiko Morizono, chief adviser for technology.

erational management. The new class of corporate executive officers would be operations managers with profit-and-loss responsibility for their business units. The function of the restructured board would be to determine "management policy" across the entire organization and to oversee operational management.

Idei's rationalization of the board gave rise to anxiety in a number of the executives who had been relocated. Some worried about their benefits as directors, including legal immunity from being fired by officers of the company. Letters signed by Ohga were sent to their families assuring them that preferential treatment would be maintained. Others were unhappy with the reform because they sensed they had been dislocated from the center of power. Among them Jack Schmuckli in particular, then chairman and chief executive officer of Sony Europe, let it be known that he was "not amused," and it turned out that he was right to be concerned: in December 1998, his "retirement" from Sony was announced. A Swiss, Schmuckli had come to know Morita in the early seventies, when he had been sent to Tokyo as general manager of Polaroid Japan with orders from Edwin Land to demonstrate every new Polaroid product to Ibuka and Morita. In 1975, knowing that Schmuckli wanted to return to Europe with his family, Morita had sent him to Germany to build Sony's business there, making him his first foreign manager in Europe. During his twenty years as the company's top European executive, the business had grown substantially—from 70 million to over 700 million deutsche marks. But Idei felt that Schmuckli lacked the dynamism and charisma to lead the revitalization of Sony Europe, which will be symbolized by the company's relocation to a new Sony Center in Berlin in 2001. As a Morita protégé, Schmuckli had been viewed as family; if Morita had been present, it is unlikely that Schmuckli would have been forced out at age fifty-eight, seven years before his mandatory retirement. Clearly, Idei is not subject to sentimental constraints.

The new corporate executive class included more than a few of Idei's contemporaries, men who sustained complex feelings about his new authority. Under the banner of rationalization, Idei had cleaned house and consolidated his own power.

The board and the committee were now identical, comprised of the same seven representative directors: Ohga, the chairman; Idei; Hashimoto, who had been named vice chairman when Idei was appointed; and the four executive deputy presidents, Morio, Ohsone, Kaneda, and Iba. There were in addition three outside directors: Peter Peterson; Kenichi Suematsu, the former chairman of the Sakura Bank, one of the company's earliest investors through its connection to the Morita sake business; and Hideo Ishihara, the chairman of Goldman Sachs, Japan. Following Ishihara's death in 1998, Idei announced his intention to add "five or six" outside directors; in March 1999, he invited a professor on the faculty of commerce at Hitotsubashi University, Iwao Nakatani, to join the board, subject to approval at the general meeting of shareholders in June. Appointing an academic to a corporate board is the kind of unusual move that observers have learned to expect from Idei, but it should come as no surprise, given his own intellectual background. He says he wants one or two additional Japanese directors who understand both Japanese and international business, and several leaders of global companies, people like Jack Welch and Percy Barnevik. Before that can happen, more radical reforms will be necessary: "If we want a Jack Welch, we must create a board he could accept and feel comfortable with. Our current board is an extension of the past. Jack Welch would find it bizarre!"

In the United States, where he has had a freer hand to exercise his authority, Idei has made radical changes in personnel and corporate structure. Although he had not been involved in the record or film business acquisitions, he was angry at the excess spending, and dismayed at what appeared to him to be a "complicity" among the principals in the American operation. His bluntness, even concerning sensitive matters, is startling: "They were all in cahoots—Burak, Pete Peterson, Mickey—each of them was doing his job responsibly as an individual, but together as a group it seemed to me they were very skillful at working the *generous entity* called Sony. The problem was, we had never truly dealt with foreigners. When they recommended something we would generally accept the recommendation. I had done lots of business in the U.S. and Europe, and I could see these people taking advantage of relationships to move the company to expend funds. They would all act together in this. Japanese of the generation before mine had an inferiority complex about foreigners. Akio Morita himself was a living inferiority complex."

Idei moved swiftly and, under the complex and delicate circumstances, with fortitude and courage, to achieve what is properly seen as the de-sentimentalization of Sony's relationships in the United States. Peter Peterson is a case in point. Early in his term as president, Idei visited Peterson to request advice and help in normalizing the U.S. business. Impressed with Idei and flattered, Peterson introduced him to Jack Welch and Lou Gerstner, American CEOs whom he had admired from a distance, and agreed to chair an American-style compensation committee to include Tsunao Hashimoto and Paul Burak. Currently, he is a director of the parent company in Tokyo, sits on three of the four U.S. subsidiary boards, and has been involved in drafting new charters. It was also Peterson who led Idei to both John Calley and Howard Stringer, the key executive supports in his new corporate architecture for the American businesses.

Historically, Peterson's role at Sony had been a function of his friendship with Akio Morita. In view of Idei's determination to unravel the skein of personal relationships at the center of Sony's traditional sensibility, Peterson's augmented importance is unexpected and confusing, until Idei reduces it to a simple matter of mutual interest: "Pete is always dealing, business, business, business, trying to get the jump on Goldman Sachs. He's an intelligent street fighter. In a way I see him as very similar to Mickey. Between us, it's give and take: I do business with him, and I'll continue to do business with him, because he's useful. Morita needed Pete; I don't need him, but he gives me good advice. I take advantage of his sentimental attachment to Mr. Morita, and he takes advantage of it, too. It's a fifty-fifty arrangement."

During Idei's early conversations with Peterson, the subject of Mickey Schulhof was raised. Peterson surely conveyed his sour view of Schulhof's activities to Idei, but it is unlikely that Idei needed prompting: from the time he became president on April 1, 1995, it took him just eight months to accomplish Schulhof's removal.

Idei remembers Schulhof in what he calls "Phase 1" as a gifted young professional; in 1983 and 1984, when Schulhof came to Japan, they met often to discuss strategy and marketing, conversations that

Idei enjoyed and considered "mutually stimulating." But as Schulhof's power increased, so, in Idei's view, did his self-indulgence, particularly after Morita's stroke, as though a brake had ceased to function. "Morita and Ohga themselves tended to enjoy flamboyance and showiness and that fed into it: the new building, the lavish spending. He became a *party boy.*" Over time, Idei also found himself increasingly at odds with Schulhof's business policies. Schulhof was opposed to Sony's entrance into the computer market. Even earlier, they had clashed over the issue of branding in the entertainment business. Carl Yankowski, then president of Sony Electronics, recalled a meeting in the boardroom at 555 Madison Avenue in December 1993, when Idei challenged Schulhof aggressively. Ken Iwaki was chairing an international operations review attended by Ohga himself and several senior managers from Tokyo, including Morio and Idei. The subject was the damage to brand equity caused by improper use of the Sony name. Ohga had performed one of his best-known arias, on the inestimable yet perishable value of "the four letters, S-O-N-Y," and several examples of pirating had been discussed, including "Sony underwear." The subject shifted to the Loews Theatres chain. Schulhof had already pushed through the Sony multiplex across the street from Lincoln Center in Manhattan—a Peter Guber idea that would soon become the highest grossing single theater complex in America—and was proposing to change all the Loews Theatres to Sonys, arguing the importance of reminding the marketplace continually that Sony had entered the entertainment business. Idei was opposed, on grounds that many of the theaters were run-down or in poor neighborhoods and that renaming them would result in a dilution of Sony's brand equity.* He spoke his mind with a force and directness that surprised Yankowski: "I remember thinking to myself, This is a very interesting dynamic here. We have a gentle-

* Schulhof's proposal was adopted. Subsequently purchased or built theaters were named, variously, Loews Theatres, Sony Theatres, Magic Johnson Theatres, and Star Theatres. In 1998, Idei arranged a merger between Sony Retail Entertainment and Cineplex Odeon. The new company, Loews Cineplex Entertainment Corporation, has more than 2,700 screens at 425 locations in North America.

man in charge of advertising and corporate communications being very aggressive with the leader of the U.S. business."

According to Idei, Ohga asked him many times how he was feeling about Schulhof, and at least twice suggested he should feel free to replace him if he felt they couldn't get along: "I told Ohga, 'Maybe it can work out, *but he must manage the American company as if it is an American public company.*' What I was thinking, and I said it to Ohga, was that Mickey was *enjoying* both the privileges of an American executive, the power and the salary, and the ambiguity of Japanese corporate governance. He was skimming the cream off the top of both worlds!"

Idei maintains that he was willing to find some way to keep Schulhof involved until he angered him beyond enduring in late October 1995 by suggesting to the *Wall Street Journal* that Sony was preparing to sell off its entertainment companies in a public offering. According to Idei, Schulhof "*intentionally overinterpreted*" his admonition about behaving as though Sony America were a public company. "There was nothing I could use as evidence in a court of law. He could say that he simply misunderstood me. But I don't think so." In response to Peterson's recommendation that Sony America should eventually go public, Idei had written a letter in which he agreed, in the long term and in principle, that it might be appropriate to take Sony Music and Sony Pictures public as a means of raising capital. Idei suspects that Peterson gave a copy of the letter to his partner at Blackstone, Steve Schwarzman, Schulhof's close friend, who passed it on to Schulhof, who used it as a basis for his revelation to the press. Idei was furious, not only at Schulhof but at Peter Peterson: "I told Pete I didn't think he should sit on our board if he couldn't manage his files. As for Mickey, I was completely fed up."

Schulhof recalls the details differently, and professes surprise that Idei should have taken umbrage in any event: "Steve Schwarzman may have given a quote to Laura Landro of the *Wall Street Journal*, and I don't remember if the quote was drawn from a memo that Idei sent me, but possibly it was, and possibly I showed it to Peterson—Pete was on my board and on the board of the parent company, so there was nothing irregular about that. Besides, Idei knew, or should have known, that we had already looked several times at scenarios about selling a piece of Sony Pictures to the public, doing a strategic alliance with one of the Baby Bells, combining music and pictures and taking Sony Software public. None of that was new. We had hired J. P. Morgan the previous year to look at possibilities—Dennis Weatherstone, the chairman of Morgan, and Akio Morita had met personally to discuss the idea of taking parts of the U.S. company public. So if Idei had mentioned it, and I think he did, it would simply have made me think, Oh, an old idea is finally coming back."

It is unclear in the offending article, dated October 30, 1995, where Landro got her information, but Blackstone is mentioned prominently and so is Steve Schwarzman, suggesting at least the possibility that the source may in fact have been Blackstone rather than Schulhof. In any event, Idei's emphasis notwithstanding, a newspaper article was by no means the only source of his dissatisfaction and anger. He was furious about Sony's production deal with Peter Guber, despite the fact that Ohga had approved it, and he took the view that the exorbitant terms of the arrangement had been a gift from Schulhof to his friend. It was also a fact that more than one of the new businesses spawned by Schulhof's feverish entrepreneurship were losing money. Whatever the triggering event, what seems clear is that Idei made up his mind sometime in November that he could no longer abide Mickey Schulhof in his organization.

It is worth noting that before going to Ohga, he still felt obliged to meet with Hashimoto and Iba to confirm support for his position. That afternoon, with Hashimoto and Iba present, Idei communicated to Ohga his strong feeling that Schulhof had to go. Idei remembers little about the meeting except that persuading Ohga was "not a simple matter." In the end, Ohga concurred, and said that he would break the news himself, on his next trip to New York early in December: "Everybody is always saying that Idei fired Mickey, but it was me. I called him to my office, and I said, 'Mickey, we've come a long way together, but our time has come and gone.' From that point on, as I got into the important details, I read what Paul Burak, our lawyer, had drafted for me."

What Ohga may have been feeling, the degree of his reluctance, is difficult to measure. He maintains that he observed Schulhof disregarding his nominal superiors in the American organization and in Japan with growing concern. He uses the Japanese term *gekokujō*, which means a subordinate wresting authority from a superior and connotes the social chaos of Japan's late feudal period: "Everyone felt they were being upstaged by Mickey, and I had decided even before Idei that he wasn't fitting in and that it might be better for the company if he left. . . . He must have sensed that I was displeased. We used to be on the phone constantly, but my phone calls to him were becoming less frequent. . . . Gradually, over time, people tend to grow arrogant. How do you say it in English: to *overestimate* yourself? To decide about yourself that you're special and irreplaceable to a degree that's over and above your actual capacity." By the time Idei showed up in his office seeking permission to dismiss Schulhof, Ohga says his own mind was already made up. As evidence, he points to his decision not to discuss with Schulhof in advance or even to mention his choice of Idei as successor. Ohga implies that he was sending a signal at the time, January 1995, and adds that Schulhof knew his time was up when he read the announcement of Idei's appointment in the press and realized Ohga had kept him in the dark.*

Mickey Schulhof confirms that he learned of Idei's appointment indirectly, and maintains that he responded to Ohga's signal by asking if he should think about leaving the company: "I told him that Idei was going to want his own team, his own group of people, and that I thought I was too closely identified with him and Morita, and Ohga replied, 'No, don't leave, Idei needs you, you have to stay.'"

Schulhof remained on the job for eight months. In the autumn of 1995, he received a visit from Hashimoto, who delivered him the black spot in the form of a letter from himself admonishing him to be a better team player. Schulhof took this in the spirit in which he assumed it was intended: "I believed Hashimoto was being a messenger from Idei, Iba, the whole top group, and certainly Ohga had to have known

* In fact, Schulhof first heard the news when a staffer who had been on the phone to Tokyo walked into his office and told him. Schulhof blanched, and muttered, "Impossible!"

about it. I was viewed as this aggressive independent, but they couched it in terms of being a team player."

By this time, Schulhof was the focus of growing enmity among top executives at Sony Japan. Teruo "Terry" Tokunaka, then president and CEO of Sony Computer Entertainment, home of Sony's PlayStation, still flushes with anger when he recalls what he claims was Schulhof's effort to block the launch of PlayStation in the United States. The game computer was the brainchild of an engineer named Ken Kutaragi, who was obsessed with developing a machine with work-station power and 3-D graphics but was unable to find a champion anywhere in the audiovisual group. For two years, Kutaragi was homeless, moving himself and his research from one lab to another, until his friend, Tokunaka, took him to Ohga to pitch his idea. Ohga authorized Kutaragi to develop a prototype, installed him in a corner of Sony Music Entertainment (Japan), and allowed him to fund his work with revenue to Sony from a sound processor that he had developed earlier for Nintendo's Super 16 Bit Cartridge. In May 1993, the Executive Committee heard Tokunaka and Kutaragi's presentation and, following Ohga's lead, approved a $50 million investment to develop the computer chip at the heart of the machine, despite the fact that prospects for the new business were at best uncertain. In Japan, Nintendo had a monopoly, dominating the wholesalers and, even more critically, the software providers. Ohga met repeatedly with the largest software companies to overcome their resistance to creating games for Sony's new platform and, in the fall of 1993, as a demonstration of Sony's commitment, he ordained a new subsidiary, Sony Computer Entertainment, and placed Tokunaka in charge. Tokunaka recalls that his hand trembled as he wrote a purchase order for 1.3 million computer chips at a cost of $50 million. On December 3, PlayStation was launched in Japan with eight game titles, and sold three hundred thousand units in the first month, three times the number Sony had prepared. The product went on to become the world's best-selling game computer. By the close of 1998, fifty million units had been sold.

According to Tokunaka, when he attempted to coordinate the U.S. and European launch with Sony Software in New York, he encountered a stone wall. He claims that Schulhof attempted to use his influence with Ohga to halt Sony's entrance into the computer game business altogether, and suggests that he was motivated by a desire to maintain his relationships with Sega and Nintendo, and that he may even have been manipulated by those organizations. From Olaf Olafsson's group, which reported to Schulhof, Tokunaka encountered opposition to the name PlayStation because, in their view, it would convey to native English speakers the image of an infant playing in its crib. Tokunaka wanted one name for the product worldwide and refused to yield, reminding New York that Morita had personally vetoed similar attempts to find better English names for the Walkman. When the product was launched in the United States in the fall of 1995 in spite of problems that Tokunaka insists Schulhof created, it was an immediate success.

Schulhof dismisses Tokunaka's claims as "history rewritten": "There was a brief period, maybe two months, just before the start button was pushed in Japan, when I said we should take one more careful look at the marketplace to be sure we weren't going to get killed, and Ohga agreed. Once the decision was made, I supported it entirely. I certainly would never have objected to the name, because Tokunaka was right to insist on a single name everywhere."

Whatever the facts may have been, and they are beyond recovery, there were hard feelings throughout the organization. On December 4, 1995, Ohga informed Schulhof that his time was up. Officially, his departure was styled as a resignation "to pursue new business interests in the new technologies and entertainment arena." When the news broke in Tokyo, there was applause in the halls of Sony.

It seems likely that Ohga was indeed deciding privately, or possibly had decided, even before Idei and his senior counselors began applying pressure, that, to use his own phrase, endlessly repeated, "Mickey's time had come and gone." Nevertheless, it cannot have been a painless matter to dismiss the man he had treated as his kid brother for twenty years. On the day that Schulhof's "resignation" was announced in New York, there happened to be a party in the Sony Club on the top floor of the building. Both Schulhof and Ohga attended, and others present remember Ohga wandering the room as though in a daze, assuring people that Schulhof was after all "his son" and would definitely be coming back one day. "It was as if Ohga was in some kind of denial about what had happened." To this day, Ohga never discusses Schulhof without finding an opportunity to communicate his abiding affection and esteem: "Have you met Mickey? You should. You'll have to look hard to find someone that smart, and he's really quite a fine fellow, even though at the end I had begun to think he wasn't our man any longer." Nevertheless, as close as he was, as Schulhof himself points out, "Ohga is a friend and loyal, but above all he is a creature of his company. . . . He was getting all this criticism from Idei and others, and he liked me, and he could have said, 'I don't care what you all think, Mickey's staying there and he's doing what he wants.' But that's not his way to run a company: at a certain point, the pressure builds on him and he tells himself, He's my friend, but this is not personal."

Ohga's own explanation seems to confirm Schulhof's reading: "It's not a matter of being cold or warm. When you're running a business, you have no choice. And you can't afford to show all the cards in your hand, either . . . Mickey's era was over. It was very hard because I had treated him like my younger brother. It hurt me. But if it was in Sony's interest, and I was certain it was, I had no choice."

Three years later, Schulhof reflects on his twenty-one-year career at Sony somewhat defensively but without rancor: "The thing to remember is that I was created by Morita and Ohga. If there was a sense of disconnect about me because I went off on my own, it was because that was the way Morita and Ohga wanted it. Morita was an owner, and he could fire anyone in the blink of an eye without thinking twice about it. I was doing what I was doing in the manner in which I was doing it because that's what he wanted . . . and I believe that I was faithful and loyal to both of them. I may be criticized for a lot of things, but in the end, I never did anything that surprised them. I never let anybody say something to them that would be a shock. If you asked me, would I do it all over again knowing the end consequence? In a minute. Would I want to go back there now? Never!"

Schulhof's departure left the U.S. organization headless, but Idei chose not to replace him for the time being. Instead, he assumed direct control of the electronics business himself. To manage the business day-to-day, he left two of Schulhof's lieutenants in place, Carl Yankowski, president of Sony Electronics, and Jeff Sagansky, head of Sony Software. Schulhof had recruited Yankowski in November 1993 to fill the office that had been vacated by Ron Sommer, and, more recently, Ken Iwaki. A Ph.D. in electrical engineering from MIT and a pilot, Yankowski fit the mold for success in Sony's American operation and had been well liked by Ohga, with whom he talked aviation, and by Morita himself, who had approved his hiring after an interview in Tokyo just two weeks before he had been stricken.* Yankowski and Schulhof were friends: when Ohga informed him, "sounding very conflicted," that Schulhof was leaving because of differences with Idei, he was uneasy about his own future in the company without a sponsor. In November 1997, nearly three years down the line, he was sounding cautiously optimistic about his standing with his new boss: "Idei's style is such that it's hard for anybody to get a lot of time with him, but I think basically we're on the same page." Two months later, he was gone. Idei declines to discuss his reasons, citing the "no-attack" clause in Yankowski's severance contract. Almost immediately, Idei replaced Yankowski with a Sony man from Tokyo, Teruo Aoki, an applied physicist who had worked in semiconductors and applied media recording products. Coming as it did in the middle of his effort to build a high-powered American management team, this seemed a curious choice, but Idei shrugs it off as inevitable: "It's very hard to attract top American talent to this particular job. It's as if the Mercedes-Benz distributor in Tokyo showed up to recruit me: I'd tell them to go away."

Jeffrey Sagansky was a former studio executive who had left his job as head of TriStar just after the acquisition and had moved to CBS, which he was credited with having restored to health. Late in the summer of 1994, shortly before Guber left, Schulhof had appointed

---

* Morita received Yankowski in the penthouse apartment in the Ark building. In the course of their conversation, Morita expressed concern that Sony executives were arrogant about the company's success in the seventies and eighties, forgetting that Sony was built on the back of American invention. At the end of an hour and a half, he extended his hand to Yankowski and told him, "I would like you to become a member of the Sony family." "What could I do but say yes?" he said to me. "It was an awesome experience." A few days after Morita was felled on November 30, Yankowski received a letter welcoming him to Sony.

him president of Sony Software, giving rise to rumors that he would soon be replacing Guber as head of Sony Pictures. When Schulhof left, it was again widely assumed that Sagansky would be returned to Hollywood to take charge of the studio, but this was not to be. In the autumn of 1996, Idei moved him to Hollywood to manage Columbia's television business, but in February 1998, shortly after Yankowski had left, he was also terminated. Observers had been wondering how long Idei would suffer Schulhof's colleagues to remain in his organization. Peter Peterson was prophetic: "The turnover is beyond anything by a factor of five that I've ever seen in corporate America: Schulhof, Guber, Peters, Frank Price. Obviously something is very wrong with the way we've been picking people. And before your book is written, the last two choices that Mickey made will also be out of here!"

With Schulhof gone and the electronics business temporarily stabilized under his direct control, Idei turned a dispassionate eye on Sony Pictures Entertainment. From early on, before he was president, he had been one of a growing number of Sony's top managers who were appalled at what Peter Peterson called the company's "funny farm" in Hollywood. But Idei was less reticent than others above him in the organization about speaking his mind publicly. As early as April 1992, the *Los Angeles Times* quoted "Nobuyuki Idei, a company director," on the company's eagerness to squeeze out the "invisible benefits" of owning a Hollywood studio. "We have been hesitant to interfere in CBS Records," Idei was quoted as saying, "but we will be less patient with Sony Pictures."

Following the general exeunt from the studio, which had culminated in Guber's departure in October 1994, Alan Levine had been appointed president and COO of Sony Pictures. Mark Canton, as chairman of Columbia, reported to Levine, who in turn was accountable to Schulhof. When Schulhof left, Ohga once again assumed the title of chairman of Sony Music and Sony Pictures, but Idei was in charge. His first instruction to Alan Levine was to renegotiate Peter Guber's production settlement. Guber bristles at the suggestion that Idei came after him: "I met him only once under the most cordial circumstances," he declared to me. "We met for about thirty minutes in my office at Mandalay, on the Sony lot, and he wished me good luck with all of my productions for Sony." In fact, in a negotiation conducted by Bob Wynne, currently copresident of the studio with John Calley, the term of Sony's production agreement with Mandalay was cut in half. In the spring of 1998, Guber moved his company to Paramount.

In fiscal 1995, the first year of Idei's term, performance at Sony Pictures improved. A major portion of the $200 million in earnings was contributed by television syndication sales, but the twin studios also managed to release a number of box-office successes, the Simpson-Bruckheimer production *Bad Boys*, *Sense and Sensibility*, and *Legends of the Fall*, starring Brad Pitt. At Christmas, TriStar released *Jumanji* with Robin Williams, which became the hit of the holiday season. But the following summer was ruinous; while 20th Century–Fox reveled in *Independence Day*, which Mark Canton had turned away, Columbia offered moviegoers what amounted to a pair of psychopath studies, *The Cable Guy*, a deeply disturbing "comedy" that earned Jim Carrey $20 million and nearly ruined his career, and *The Fan*, a misbegotten thriller starring Robert De Niro as a sports-loving psychotic. *Multiplicity* and *Striptease* also crashed at the box office. By the fall, Sony's market share had dropped once again to sixth place, the bottom of the list, and Canton and Levine were known to be at war.

Since the beginning of the summer, Idei had been quietly looking around for someone to replace Levine as head of Sony Pictures. As the new man in town, he had access to the A-list of Hollywood players, and spent time consulting about the business with Michael Eisner, Michael Ovitz, David Geffen, and Barry Diller. By midsummer there were reports of a short list, which included Stanley Jaffe, the former chairman of Paramount, and Jeff Sagansky. The front-runner was said to be Michael Fuchs, former chairman of Home Box Office.

The winner turned out to be a dark horse named John Calley, a veteran producer and studio chief who was brought to Idei's attention by Peter Peterson. Calley had virtually run Warner Brothers during the 1970s, producing on average a film a month, including the box-office successes *Superman* and *The Exorcist*, and had left Hollywood in 1980 a wealthy man and gone into retirement, first on Fishers Island in Long Island Sound and subsequently on a farm in rural Connecticut.

In 1991, he had surfaced to coproduce two films at Columbia with his best friend, Mike Nichols, *Postcards from the Edge* and *The Remains of the Day*, which had received an Oscar nomination. In 1993, Calley had taken over as president of a foundering United Artists in order to prepare it for a sale by MGM—"putting rouge on the corpse," as he put it—and was contemplating, as Kirk Kerkorian prepared to acquire the studio, a return to independent producing of the smallish, classy films he most enjoyed. Peterson, through his wife, Joan Ganz Cooney, creator of "Sesame Street" and the Children's Television Workshop, was also a good friend of Mike Nichols and his wife, Diane Sawyer, and had heard them describe Calley as a rational and sophisticated man entirely outside the Hollywood mold yet widely respected in the community.

On Peterson's recommendation, Idei requested a meeting, which took place on August 7 in a private room at the Los Angeles International Airport. Asked what he thought had gone wrong at Sony Pictures, Calley cautiously allowed that the management had done more delegating than hands-on management, and that in particular, Peter Guber's insistence on intense control and inaction simultaneously may have been counterproductive. In Calley's view, Sony had mistakenly accepted Hollywood's assertion that no one outside the movie business could hope to fathom it and should "put up the money and keep out of the way." Idei was relieved and impressed. Calley for his part was taken with Idei personally and by what he glimpsed of his vision for Sony: "I was sixty-six. I thought, Enough of this already. I'd been running studios for twenty-odd years and had been in the business for forty-five years. On the other hand, I'd just come from a place [United Artists] that didn't have any money to make movies. The idea of being involved with a company that had from my point of view an unlimited horizon was thrilling. I'd never worked at a Hollywood company that had the same capacity and the same vast potential to connect the dots. And this guy seemed to me to want to change the world, and that's always exciting, unless it happens to be Hitler."

At the end of a three-hour conversation, Calley said he was definitely interested. He was not in need of a job, and his reputation in Hollywood was already secure. Idei's success at intriguing him during a first meeting must be counted as evidence of his persuasiveness, reminiscent of Morita's own. "You hear lots of horror stories about working with the Japanese," Calley told me, "but after meeting Idei, I didn't believe it. I had no sense I was talking to a different species. I had a sense I was talking to a really smart guy. I was astonished by his grasp of things."

With Idei persuaded, there remained the pilgrimage to Tokyo to meet with Ohga. Calley recalls that Ohga was warm, charming, and avuncular, though Calley was slightly older, and that his emphasis was entirely on intuitive creativity. Ohga called on him at his hotel suite, and they chatted for an hour about, among other things, the importance of franchise pictures and the worrisome matter of health as one approached the age of seventy. Peter Guber was not mentioned. Idei showed up and left almost at once to escort Ohga to his car. Calley wondered if he would ever see either of them again. Idei returned in fifteen minutes, having heard Ohga's appraisal, and, over lunch at the Italian restaurant in the hotel, offered Calley the job. Calley accepted.

The days of occasional visits from Tokyo for special screenings or luncheon on the set were now at an end. On the same day he dismissed Alan Levine, Idei installed a man of his own in the Thalberg Building as executive vice president reporting directly to Calley. Masayuki "Yuki" Nozoe had worked for Idei in Tokyo on the team that launched the CD, and had been living in the United States for nine years, working out of Park Ridge, New Jersey, as head of electronics consumer marketing. In the course of promoting DVD, Nozoe had developed relationships in Hollywood in the home video business, and was quick to adapt. His English is fluent enough, his business style Americanized—"let's cut the crap" he remarked during our interview—and with his longish wavy hair and his modish suits, he appears entirely at home on Rodeo Drive.

Shortly after Calley moved in to his office in the Thalberg Building, a friend telephoned to warn him that "someone" in the office next door was passing on to Tokyo everything Calley said in his presence. Calley represents, convincingly, that he expected Nozoe to serve as Idei's eyes and ears in the studio, and that he himself had advised Idei to install a Japanese executive with home-office credibility. In fairness, Nozoe has been more like a cultural mediator than a mole; he understands the kind of business planning Sony expects, and, according to Calley, has enabled the studio for the first time to communicate business plans to Tokyo effectively by coaching him and the other American executives in the kinds of language and statistics to use. In the process, he seems to have earned the Americans' trust and genuine appreciation; in fact, he may now appear to be more closely identified with his Hollywood colleagues than is good for him in Tokyo. Recently, the studio bought him a house in Pacific Palisades which was deemed excessively luxurious and reminiscent of the Guber days.

The stabilization of Sony Pictures today must be considered one of Idei's major achievements, and one that seems to confirm Ohga's prescience about his unique capacity to understand and manage the entertainment businesses. Under Idei, Hollywood's historical disdain for the outsider who dares to meddle in its arcane business has largely disappeared: as a business, the studio is acutely conscious of itself as part of the Sony Group. A business strategy that intersects with Idei's evolving master plan and that is taken very seriously by the studio management is represented in its new slogan, "Lighting up screens around the world."

Performance under Calley and his new team has improved. In his first full year as studio chief, 1997, the studio climbed from fifth to first place in market share and broke Hollywood records with earnings above $1 billion with a succession of smash hits, including *Jerry Maguire*, *My Best Friend's Wedding*, *Air Force One*, and, most spectacularly, *Men in Black*. In a standard Hollywood irony, every one of these films was loaded into the development pipeline by Mark Canton before he was fired, leading John Calley to quip: "In this business, it's important to choose your predecessors carefully."

In 1998, performance cycled downward. *Godzilla*, a project in limbo when Calley arrived, was a $100 million extravaganza that failed significantly to live up to its overwrought marketing campaign and performed poorly in the U.S. market, but will earn money in the long run from foreign sales. Moving into the Christmas season 1998, Sony was in third place in market share, but Calley emphasizes that while big hits have eluded him so far, the studio is now making money on every picture. "It's dangerous to fixate on market share," he told me. "One year Eisner gave Katzenberg a billion dollars and he made hundreds of movies and returned six hundred and fifty million of the billion. Unlimited resources will give you market share, but you can lose your ass."

In the spring of 1997, to fill the vacancy that had been created by Schulhof's departure, Idei hired Howard Stringer, a television broadcast executive with a thirty-year career at CBS, as president of the Sony Corporation of America. In view of his sharpening focus on networks and new media, Stringer was a logical choice: he had produced the "CBS Evening News with Dan Rather" and "CBS Reports" on his way to becoming the president of CBS News, and, from 1988 to 1995, president of the entire CBS Broadcast Group, and was widely connected in the broadcast industry. In 1995, Michael Ovitz had lured him away to head a start-up funded by three Baby Bells. TELE-TV, a "futuristic" experiment in using telephone lines to deliver video into the home, eventually folded, but Stringer had acquired experience in new approaches to content distribution which Idei judged valuable.

At the end of his first meeting with Idei in the summer of 1996, over lunch at Matthew's in New York City, Stringer was confused. Idei had picked his brain about Hollywood and the broadcasting industry, charmingly enough, but had not mentioned a job. At the time, the John Calley appointment had not been announced, and Stringer found himself wondering, hopefully, whether he was being tapped to head the studio. When they met again several months later, at a breakfast meeting in Idei's hotel suite at the St. Regis, around the corner from Sony headquarters in New York, Idei offered him a job, which he defined as strategic coordination between Sony's electronics and entertainment businesses, integrating Sony's new value chain. This position was reminiscent of Schulhof's function, but Stringer knew he was not being offered Mickey Schulhof's job: "In six months of conversations, Idei never mentioned Schulhof, and I sensed I was not about to receive anything like the level of autonomy and independence it sounds like he had enjoyed. My biggest concern was finding myself shut out by the three operating companies. I would in a sense have to take the job on faith."

Idei asked Stringer to think about it, and invited him to Japan to meet Ohga, who reassured him. They had a common ground in music—Stringer was a classical trumpet player—and in CBS, whose gradual decline over the years they lamented together. Ohga assured Stringer that Sony was committed to preservation rather than fragmentation, and would never under any circumstances compromise the value of its brand. A history major at Oxford, Stringer was impressed to have encountered an aesthete at the head of a corporation, and impressed as well by the quality of the Bordeaux Idei poured at dinner at a Japanese restaurant.

The following morning, Idei handed Stringer a deal memo. Having been led to expect "Japanese" vagueness, Stringer was surprised at the cut-and-dried numbers, and further surprised that the offer never changed during the subsequent negotiation. "I had to take it or leave it. I realize now that I was going to symbolize a new way of dealing with American executives. There would be no more taking Sony to the cleaners."

Before he accepted the Sony job, Stringer flew to California to meet Calley for the first time. "People had warned me the music company would be prickly, but I wasn't worried about music and I haven't had much to do with them. Calley was connected to the part where my soul is, and I wouldn't have come to Sony without a relationship to him. Mike Nichols and Diane and Peterson had told him I was OK, and we went out in secret and got along fine. He's older than I am, but we're older than everyone else in a way. We read the same books, and we talk about the past more than we ought to, but if you don't have a shared past, screw it! That's where your identity gets formed."

Stringer has no operational responsibility. In the early nineties, he would doubtless have referred to "synergy" in describing his role, but Idei has declared the "S-word" taboo, perhaps because it was the banner of Mickey Schulhof's expansionism. In the context of the current business, Stringer serves as a coordinator among the three operating companies, a "diplomat" as he puts it, "seducing and persuading them that we have to work together in different kinds of ways to amortize our assets and build digital distribution." In the domain of new business, his energy has been directed at identifying strategic partners and using his wide connections in the industry to facilitate Idei's access to them. Currently, Idei is focused on the software providers that will enable interactive linkage between the home portal and network services. With Stringer as intermediary and along for support, he maintains ongoing dialogues with Microsoft, Sun Microsystems, Cisco, and other members of "the community."

The dialogue has led to a series of licensing agreements for software applications. In March 1998, Sony licensed "personaljava" technology from Sun Microsystems for use in the home entertainment network environment in development at its new Software Platform Development Center. In April, Microsoft and Sony announced plans to cross-license Microsoft's Windows CE, an operating system for digital consumer electronics products, and Sony's Home Networking Module, a "middleware," which allows digital "appliances" to be interconnected and interoperated. "The time has come," Idei declared at a press conference with Bill Gates, "for the PC industry and the AV industry to shake hands." In July, Sony licensed its Home Networking Module to General Instrument, Inc., for use in its digital "set-top devices," the transformed cable boxes that are expected to function as a primary home gateway to digital interactive services. So far, there have been no major deals. "Americans are obsessed with deals," Idei snaps, "but before you get to a deal you have to discover the logic of a deal, the direction that makes sense to both partners. I'm interested in building relationships that make sense."

In the process of expanding Sony's "semantic borders," Idei has established himself as a presence on the international scene. He befriended a number of Europe's leading CEOs, Percy Barnevik of ABB and Peter Brabeck-Letmathe of Nestlé among them, and has accepted an invitation to sit on the board of the Swedish manufacturer Electrolux. In January 1998, he was a focus of attention when he attended the World Economic Forum in Davos, the elite convocation of global business and political leaders which Morita had used as a theater for promoting Sony and his thoughts on global competition. In 1999, cochairman of the Forum for the year, he was the moderator on two panels, "The Media Revolution" and "Corporate Governance in a Global Context."

In 1997, with Stringer working behind the scenes to make it happen, Idei became the first Asian to be invited to Herbert Allen's annual summer retreat for entertainment and communications moguls at Allen's resort in Sun Valley, Idaho. He is now a regular member of a core group, which includes Eisner, Geffen, Katzenberg, Semel, Edgar Bronfman, Jr., and Barry Diller from Hollywood, Warren Buffet and Rupert Murdoch, John Malone, Ralph Roberts of Comcast, Time-Warner chairman Gerald Levin and president Richard Parsons—the only African American—Andy Grove and Bill Gates, and, the only woman, Katharine Graham. Stringer also attends, and has observed Idei in action with surprise and what is clearly genuine admiration: "No foreigner I've ever seen moves in and out among the big guns the way he does. He exudes power in a very Western way. This is no Panasonic or Toyota executive: he has that French look, he carries himself with a lot of authority and style, he manages to be the essence of corporate power, and they like him and take him very seriously. It's remarkable to watch."

The Idaho retreat has been a treasure house of useful connections. When Bill Gates visits Japan, he makes time for golf with Nobuyuki Idei, who flies him in a Sony helicopter to the Glen Oaks Country Club forty-five miles east of Tokyo, among the most exclusive and certainly the most expensive golf clubs in the world. "They would never have let me or my father in there," Idei smiles, "but now that I'm head of Sony, they sort of have no choice." Clearly, though he doesn't say so, Idei enjoys hobnobbing with this crowd and is proud to have been accepted as a peer. But his eye remains on the prize: "When I came back from the first summer," he boasts, "I had five new deals in my briefcase." One of them was a preliminary agreement with John Malone to purchase a 5 percent share of General Instrument; Idei intends to take Sony into the set-top business.

Ohga is proud of Idei's growing visibility in the business world, an international prominence he never achieved, and points to it as a ratification of his choice of successor: "He is doing an extraordinary job of presenting Sony to the world. He has a brilliant touch with the media, and he's managed to create an awareness worldwide that Sony has Idei! No one else could have done that, and it's critical, because as president, you must promote and sell yourself if you are going to promote and sell the company. Mr. Morita was a genius at that, and Idei is following in his footsteps."

Observing his protégé in action for nearly four years, Ohga is not always so admiring. He maintains emphatically, in public and in private, that he was seeing true when he beheld in Idei the "extraordinary something" he would need to lead Sony into the next century. At the same time, he has continued to question Idei's business judgment, his "nose for profit." For the first three years of his term as president, fiscal 1995, 1996, and 1997, despite the darkening economic environment in Asia and a general flattening in Japan, Sony's worldwide sales and profits climbed. But Ohga declines to credit Idei with the favorable results. On the contrary, he asserts that he prepared the company for its new leader in a manner that guaranteed success in the short term: "I cleared our books so he wouldn't be burdened with any negative assets from our long history. And I pushed start buttons before he took over. The success of Sony Pictures today is due to projects that were greenlighted three or four years ago. And I approved PlayStation a year before I chose him—in 1998, PlayStation profits alone will exceed profits of the rest of our electronics businesses combined! I shouldn't say this, but I arranged things so that he would be profitable automatically, with no effort on his part, the minute the yen weakened against the dollar, and that began to happen when he took over. On top of everything else I had prepared for him, fortune presented him a marvelous gift!"

In the second stanza of the same refrain, Ohga emphasizes that Idei has yet to initiate a profitable business during his term as president. As recently as October 1998, he was harping on the same theme: "Idei is doing all kinds of things right now, and working at them very hard, and I've told him I'll support him one hundred percent in everything he does. The fact remains, nothing we've started since he took over has earned any money for us yet. . . . Look where our profit comes from today—PlayStation, Minidisc, Trinitron, even VAIO laptops—

all businesses we started in my day. Idei has gone into the broadcast business only, into SKY PerfecTV!, and there's no real money in broadcasting . . . I'm continually pushing him to develop a nose for profit. There is no one at Sony, not today at any rate, as good as Idei—that's why I made him president. But the biggest problem with him is that he doesn't have as sharp a sense for making money as I do: he has to become more finely attuned to profit."

Ohga had chosen Idei despite misgivings about his business nose. During the first year, he met with him in his office once a week, on Tuesday mornings at 8:00 A.M., for a consultation over Japanese breakfast. More than once, their heated arguments carried over, or were reignited, at the Tuesday-morning meetings of the Executive Committee, which followed. Early on, Idei's determination to reenter the home computer business was the subject of contention. Ohga was reluctant, but willing to approve so long as the business was limited to laptops. Sony had years of experience designing and manufacturing laptops for both Apple and Dell, and had developed the technology; in Ohga's emphatic view, moving into desktops, a new and unfamiliar domain, was asking to lose money. With the same temerity in the face of Ohga's ponderous certainty that he had demonstrated as a younger man, Idei refused to yield: admittedly, computers were risky, but they were in his view an essential step toward the network services of the future. In the end, it was Ohga who stepped aside. Recounting the story eighteen months later, he observes with a smile, matter-of-factly, that while VAIO laptops have been profitable, the desktop machines, as he predicted, are still "deeply in the red."

Regular members of the Executive Committee have observed the shifting balance of power between Ohga and Idei with admiration for the restraint they perceive being exercised on both sides. According to Yoshiyuki Kaneda, one of five executive deputy presidents, Idei has become a master at conveying the deference due Ohga as his mentor even as he opposes him unflinchingly. In Kaneda's view, Ohga's "supreme confidence" allows him to defer to Idei without fear of losing face: "Mr. Ohga has been a superman in Sony history, but he knows that the time has come for radical change, and that Idei is the right man to lead the revolution. He is able increasingly to cede his control to Idei even when his decisions don't feel entirely right to him. And because he's so sure of himself, he is able to defer without feeling any rivalry."

Idei might not agree. He acknowledges Ohga's self-restraint: "One of the amazing things about him is that he won't interfere—even if he disagrees, he'll stand by and let me push the green-light button because he's decided that's my job." At the same time, Idei conveys the sense that managing the relationship can be stressful and exhausting: "He's a formal man. When he enters a room for a meeting, even when he's late, he expects you to stand. And he always talks down. He doesn't communicate, he announces. . . . He confronts me once a week, every time we meet, with the same challenge: 'I created the CD, the Minidisc, the this and that: *What have you done? What are you going to do?*' And he's speaking as a rival, as though we were competing. Lately he complains about my speech at the management conference in May [1998]. Too conceptual! I say that it's my job as president to be conceptual, but you can't argue with him outside his context. So I promise to behave the way he expects the president of Sony to behave. After all, he's the daddy."

While Ohga continues to grouse, his grip is relaxing. Initially, he appointed Idei president and chief operating officer and retained the title of chief executive officer for himself. Idei was nettled, and refused to use the COO title in America, where he felt it conveyed a qualification of his authority. In May 1998, when Ohga was named chairman of the committee for revitalizing commerce and industry at the Keidanren, he promoted Idei to co-CEO. The timing mirrored Morita's relinquishment of the chief executive position to Ohga in 1990 when he threw himself into his campaign for Keidanren chairman. Since May, Ohga's responsibilities as official representative of the Japanese business community have occupied him increasingly and there has been no time for private meetings with Idei.

At board meetings, he remains the paterfamilias. In the summer of 1998, the U.S. boards convened in Kona, Hawaii, and someone had ordained that business attire would be worn during the working day.

Ohga appeared at the Mauna Lani Resort dressed comfortably for the tropics and ordered the directors on the Sony Music board to remove their coats and ties at once.

To the American executives, observing what Stringer describes as the "creative tension" between Ohga and Idei is like watching live theater. Ohga's stance continues to be mildly admonitory; he interrupts Idei frequently to chide him, and his corrections are designed to clarify for all present, in case there was ever any doubt, who the master is and who the journeyman. In the beginning, Idei forbore. Recently, he has appeared less willing, or less able, to mask his impatience at Ohga's posturing, and indulges in grandstanding of his own, rolling his eyes or slumping glumly in his chair when Ohga begins to expound. The Americans observe that Ohga's role is becoming ceremonial. "We're in the middle of a meeting," I was told, "and there's a pause and Ohga's assistant jumps up and passes out Sony CDs: Ravel's *Bolero* and a Mozart piano concerto, and Ohga announces that we should always strive to use Sony classical music on our movie tracks. He's giving us these titles so we can understand what he means by good classical music. It's as if there's a programmed moment when he fires up and shows us that he's still the fist of fury." In private, Ohga continues to insist that Idei has yet to prove himself: "People inside Sony are not going to feel that Idei was a good choice until he does something on his own that works, that generates profit. Until then, to Sony people, all the talking will just be talk."

Meanwhile, Idei continues to ponder Sony's future. As 1998 drew to a close, his postulations became more radical. Perversely, or possibly in hopes of achieving maximum impact, he chose the occasion of the Technology Fair in October 1998, a two-day display and celebration of Sony's latest innovations in consumer electronics, to assert in his keynote speech that the electronics hardware business in its current form was obsolescent. As evidence, he pointed to the domestic karaoke industry and the cellular phone. Early in the year, the Japanese market was introduced to transmission over phone lines of the sing-along music videos used in karaoke. Subscribers can now phone in selections and receive them on a monitor at home or in a karaoke bar. In a matter of months, the new media distribution network destroyed Sony's CD changer and laser disc business. Similarly, cellular phones are now included free of charge as an incentive to subscribe to cellular service. Idei predicts that the next target will be television and set-top boxes. In the "Age of Networks," hardware will be deprived of its stand-alone value: "It won't matter whether TV screens are bright or have beautiful resolution," he told a stunned audience that included the developers of Trinitron. "What matters will be the content: who creates it, and who controls the network that distributes it."

If Sony's hardware business stands in jeopardy, its entertainment businesses are also imperiled: in Idei's view, the age of packaged media—music on CDs, celluloid-based film in theaters or on videotape—is in its final days. Whether Sony's entertainment businesses can survive as one link in a value chain organized around network services remains to be seen. What is clear to Idei is that Sony must prepare to create and control distribution of "new media," sports, financial transactions, shopping, even digital avatars who interact on the subscriber's behalf in a virtual network society. Idei asserts that failure to evolve an entirely new business model will result in Sony's devolution into a supplier of electronic components to the network operators.

Even the Sony brand will have to undergo a transformation. Ohga staked his life's work on Idei understanding the value of "the four letters, S-O-N-Y," and preserving it at any cost. Idei, who is a Sony man when all is said and done, accepts the responsibility. Nonetheless, he expects that the perception of the brand will have to change: "In the past, Sony stood for product excellence. In the future, if we are going to flourish in the network age, it will have to stand for different things, security, privacy, trust. When people see 'SONY,' they will have to feel safe about conducting their electronic commerce with us. If we get stuck in product excellence, we run the risk of turning into a company like Leica, which still manufactures the world's greatest camera but has almost no business!" Lying awake at night, Idei wonders uneasily whether the question "Do you dream in Sony?" points to the future, as he intends, or harkens to the past. "I worry it could mean 'I remember Sony, they used to make wonderful TV sets.' The issue is, does 'dream' have the proper time frame, does it have any time frame at all?"

Idei does not share arcane concerns of this nature with his organization: "Ninety percent of the people here are fully absorbed in executing the current business, and that is as it should be. It makes no sense to confuse and frustrate those people with talk about the future that will be incomprehensible." In the United States, where the success or failure of Sony's migration to a networked world will be determined, Howard Stringer seems to be in step with Idei. And John Calley comprehends the logic of the vision as it applies to his business: "We spent fourteen million on release prints for *Godzilla* alone. There would be a vast savings if we created pictures in a digital format other than film and then transmitted them to satellite and from there back into theaters—and we would make all the hardware." But emotionally, Calley feels disconnected from the new formulations, in his phrase, less digital or even analog than abacus: "I'm just not interested in creating a distribution universe that accommodates the capacity to leverage our assets into a piece of the cable system in Tierra del Fuego." Tommy Mottola remains immovably focused on hit albums, a reality Idei recognizes and, for the time being, accepts: "I don't talk to Calley so much about these things, or to Mottola—music is his life. Whether Sony Music becomes baggage to Sony in the future is not something he needs to concern himself about."

Sony's senior managers in Tokyo, the men who built and manage the core business, are also having trouble tracking Idei's orbit out of Sony's traditional atmosphere. In early October 1998, Idei took the inner circle to an off-site retreat in the resort town of Gotenba for two days of brainstorming about future business strategy and a new corporate architecture. Iba, Morio, and Ohsone were there, and so was Yuki Nozoe from Hollywood and Ohga himself. Stringer flew from New York to explain his proposal to then chairman of CBS Michael Jordan for a possible merger between Sony Pictures and the network, a connection between content controlled by Sony and access to distribution which fits perfectly into Idei's evolving business model. At the conclusion of his presentation, there was an awkward silence. Stringer asked for questions and received none. "I would have thought the old guard would be delighted to sell off Sony Pictures," Stringer told me, "but they seemed to have nothing to say." Idei acknowledges that one of his "weak points" is converting the people he calls "Sony mainstreamers" to apostles of his vision. "I've always felt compelled to place myself in opposition to the establishment. Now I'm the establishment and that's a problem for me. The people around here who do a fine job of managing our current businesses, Iba, Ohsone, Morio, all the senior managers, are basically conservative, and I'm just not very good at converting them to change agents." The consequence is a gap that widens as Idei grows messianic about his vision and impatient with the burden of explanation. There are times when he seems alone with his thoughts and possibly troubled by his isolation.

Idei will reach retirement age in 2002. Once again, the necessity of choosing a successor looms. Ohga will have a voice in the process; he has said he will not retire until Idei is ready to move into the chairman's office. He also represents that the burden of the decision will be Idei's. Sony people already wonder aloud, "Is Masao Morita being groomed for the job? Are the younger presidents of the ten companies, people like Kunitake Ando, in the running?"

Such speculation may be idle. Idei continues to declare his intention to choose not one but several leaders to succeed him: "No one," he told me, "can understand music, pictures, electronics, and the dynamics of the network age. But if we can divide the business into smaller playing fields, there are any number of people who could manage a portion of it."

The divisions Idei has in mind are electronics on the one hand, old and new media on the other. He is also feeling that it may be time to separate the management of businesses based in Japan and in the United States: "American governance is fundamentally different from the way we run business in Japan, and we have to acknowledge that. We speak about a global Sony, an international Sony, but we are actually a Japan-based company with a large American organization, which we tell ourselves we manage from here in our way, but it's simply too different."

There are signs that Idei is already moving to place the American operation on its own foundation. In December 1998, he elevated Stringer to chairman and CEO of the Sony Corporation of America (he had been chairman of Sony Electronics and CEO of Sony Canada). Stringer continues to report to Idei, but Thomas Mottola and John Calley, who were named chairman and CEO of their respective businesses, will now report to Stringer instead of Idei. The proliferation of Stringer's titles is reminiscent of the Schulhof era and conjures at least the possibility that Idei may be preparing to augment significantly Stringer's autonomy.

An independently managed U.S. business has a place in Idei's larger vision of a "Sony World Strategic Holding Company," in which the Sony family of businesses reside as separate entities. In March 1999, after mulling over the idea privately for a year, Idei took a first step in this direction by realigning the electronics business in preparation for what he now called "the network-centric era." Effective April 1, 1999, Sony's ten divisional electronics companies were grouped inside three new business units: the Home Network Company, the Personal IT Network Company, and the Core Technology and Network Company. Each unit was given its own R and D laboratory, relocated from corporate headquarters, and an independent board and management committee. Idei also announced the establishment of a strategic Group Headquarters—an incipient holding company—to oversee group operations and to allocate resources swiftly. In the new Sony Group organization chart, Group Headquarters is positioned alongside (rather than above, as previously) the three pillars of Sony's business: electronics, entertainment, and insurance and finance.

In Idei's view, the common ground that will enable what he calls a "unified dispersed" business model will be provided by numbers, the unemotional arithmetic of sales, profit and loss, and measurement: "Think of me and Mr. Ohga. If you look for them, there are many differences between us. But there are also similarities. Business is logical: if we can reduce our business to numbers as bankers do, we won't have to concern ourselves with the psychologies and emotions which lie behind the numbers. We are not going to pretend that cultural difference doesn't exist, but we must come up with a mechanism for controlling the business as if it didn't. In a strategic holding company, you don't need culture."

The changes Idei foresees as necessary to Sony's future, in core competencies and fundamental structure, will transform the company unrecognizably. By his own account, "The Sony we know today may have to disappear." A question that occurs is whether the Sony magic can survive in a company that has been released from the constraints of the founders' logic at the heart of its culture?

The value of the Sony brand is the product of fifty years of innovation in electronics products for the home consumer. Personal relationships have been the wellspring of that innovation: specifically, it has been driven by the commitment felt by Morita and Iwama and Ohga and by Sony's engineers, Kihara the wizard, Morizono, Ohsone, Yoshida, Miyaoka, Doi, and all the others, to making Masaru Ibuka's dreams come true.

In the process of preparing Sony to meet the future, Nobuyuki Idei has engineered a discontinuity in the company's fifty-year history. If Idei's promotion early in 1995 marked the beginning of a new era, Mickey Schulhof's termination in December of that year signified the uprooting of a fifty-year tradition planted when Ibuka founded the company. Hereafter, personal relationships are not likely again to figure decisively in the process of business decisions that determine Sony's performance. Nor is any individual inside Sony likely to exercise the authority or enjoy the protection granted to Schulhof as a result of his privileged connection to the Sony founders. That is perhaps as it should be. But what will take the place of the personal connections that have always been the source of Sony's incandescence? Norio Ohga declares that Sony must always shine with a blinding light: what remains to be seen is whether the four letters S-O-N-Y will continue to dazzle on the other side of the divide that begins to separate the company from its sentimental past.

INDEX

A & M Records, 143
ABB (Europe), 290, 315
ABC (American Broadcasting Company), 185
Academy Awards (Oscars), 185, 193, 198, 310
Aérospatiale (helicopter), 166n
Agrod company, 55–56, 58–59, 60
Aichi Bank, 23
Air Force One (film), 312
Aiwa America (Sony subsidiary), 105, 278
Aji-no-Moto (food seasoning company), 87, 284
Akiyama, Chieko, 272
Allen, Herbert Jr., 186–90 *passim*, 192, 196, 201–2, 204, 316
Allen, Robert, 251
Allen and Company, 186, 209
Allien, Bernard, 289
All the King's Men (film), 185
American depositary receipts (ADRs), 64, 66
American Express, 267
American Werewolf in London, An (film), 193
Amnesty International, 242
Ampex tape recorders, 27
Ando, Kunitake, 37–38, 76, 99, 162, 292, 323; on Schulhof, 165, 166
Aoki, Teruo, 293, 307
Aoyama Gakuin (private school), 24
Apple Computer, 253–54, 286, 318
Ara, Masahito, 12–13
Arai, Yoshitami, 83
Asada, Tsunesaburo, 20
Asahi Shimbun (newspaper), 8, 14, 20
Ashizawa, Sato, 38–39n
Aso, Tadakichi, 90
Atlantic, The, 268
AT&T ("Chippendale") building (New York), 250–53
Augusta National golf course, 73, 78
Autometric Laboratory (Paramount Pictures), 43
AV (audiovisual products), 289
AVCO Development Company, 103

Baby Bells, 302, 313
Bad Boys (film), 254, 309
Banc de Paris takeover, 194
Bank of Tokyo, 65, 209

Barnevik, Percy, 290, 298, 315
BASF (Germany), 124
Basic Instinct (film), 225
Batman (film), 193, 202, 220
Battle, Kathleen, 290
Beatty, Warren, 225
Begelman, David, 194
Being Digital (Negroponte), 290
Bell and Howell, 76
Bell Laboratories, 31, 32, 34, 137, 148
Bennett, Tony, 175n
Bentley, Helen, 213
Berlin Philharmonic Orchestra, 142, 154
Bernbach, William, 74
Bernstein, Leonard, 19, 71, 80
Bertelsmann (German media group), 248
Betacam, 190
Betamax (VCR), 27, 48, 147, 161; failure of, 106–11, 141–46 *passim*, 150, 155, 175, 184; tuners for, 113
"Bewitched" (TV program), 185
Billboard (magazine), 142, 146, 195, 242–43
"Birth of the Transistor Radio" (Japanese radio program), 34n
Bisset, Jacqueline, 194
Blackstone Group, 73, 239, 301, 302; and Sony acquisitions, 76, 173, 187, 188, 203, 210, 213
Blass, Bill, 71
Bluhdorn, Charles, 64
Blumenthal, Michael, 73
Boeing aircraft, 103
Bogart, Neil, 194
Bolton, Michael, 245
Boston Consulting Group, 291
Boyz 'n the Hood (film), 224, 228
Brabeck-Letmathe, Peter, 315
Bram Stoker's Dracula (film), 224, 229
Breest, Guenther, 136
Bridge on the River Kwai, The (film), 185
Bridgestone Tire (Japan), 90n
Briesch, John, 250
Brilliant Fetus, The (Ibuka), 49
Bronfman, Edgar, Jr., 316
Brooks, James L., 224, 230
Bruckheimer, Jerry, 254, 309
Buffet, Warren, 316
Bugsy (film), 225, 228
Bulova watch company, 54–55, 164
Bungei Shunju publishing company, 90
Burak, Paul, 61, 131n, 173, 213, 298, 299, 302; and Guber-Peters, 177, 202, 210; and Yetnikoff, 246, 247
Burroughs Corporation, 73, 103
Burton, Tim, 224, 235
Bush, George, 108
Business Week, 258
Byrds, the (music group), 175n

CAA (Creative Artists Agency), 191, 192
Cable Guy, The (film), 309
Cabot, Cabot and Lodge, 104
Caine Mutiny, The (film), 185
Calder, Alexander, 159
Calley, John, 195, 299, 309–13, 314, 322, 324
Camp Winona (Maine), 70
Candies, the (music group), 132
Canon Camera, 53
Canton, Mark, 227–28, 230, 233–35 *passim*, 308, 309, 312
Capra, Frank, 185
Carey, Frank, 73
Carey, Mariah, 245, 246
Carolco films, 228
Carrey, Jim, 309
Carter, Jimmy, 73
Cash, Johnny, 175n

CBS (Columbia Broadcasting System), 96, 135, 169, 307, 314, 322; TV programs, 107, 186
CBS Broadcast Group, 313
"CBS Evening News with Dan Rather" (TV program), 313
CBS News, 313
CBS Records, Inc., 130–31, 137, 184n, 195, 243; International Records Division, 97, 98, 136, 160–61; opposes CDs, 169; Sony acquires, 74, 76, 97, 187, 242, 244, (Blackstone Group and) 76, 187, (Idei and) 308, (profits from) 132–33, 209, (renamed) 248, (Schulhof and) 158, 168, 171–78, (U.S. management promised) 216–17
"CBS Reports" (TV program), 313
CBS/Sony Records, 97–98, 141, 172, 243; boycotted, 132; formed, 131; Ohga at head of, *see* Ohga, Norio; releases first CDs, 144; stock sales, 259n
CCD (charged couple device), 148, 156
CD players: developed, 141–44, 285–86; sales drop, 258, 321
CD-ROMs, 249–50, 292
CDs (compact discs) 19, 150, 155, 289; first release of, 144, 211; MiniDisc, 175, 317; opposition to, 143, 169; overtake LPs in sales, 145, 146; Sony manufacture of, 143–44, 169; Sony-Philips joint development of, 49, 97, 137–46
Chaplin (film), 229
Charles, prince of Wales, 78
Chase Manhattan Bank, 73
Children's Television Workshop, 71, 310
Chinese market for Sony, 76
Chirac, Jacques, 267
Christensen, Terry, 202–3, 204, 210, 211
Chromatron (color television), 3, 43–45, 64, 68
Cineplex Odeon, merger with, 300n
Cisco, 315
City Slickers (film), 228
City Slickers II (film), 236
Cliffhanger (film), 230
Club Amphi (Tokyo), 91
Close Encounters of the Third Kind (film), 185
Coca-Cola, 186, 191, 209
Cohen, Albert, 88
Cohen, Harry and Jack, 185, 220
Color Purple, The (film), 193
Color television. *See* Television
Columbia Law Review, 247n
Columbia Pictures Entertainment, 77, 115, 131, 173, 180–219, 221, 255, 309; film library of, 185; Guber and Peters take over, 220–25, 235; Morita's desire for, 190–91, 232, 240; purchase price, 205; renamed Sony Pictures Entertainment, 219, 227; Sony's behavior during negotiations, 157, 175–78 *passim*, 181–83, 188–91, 198–202 *passim*, 205–8, 212–17, 241–42; Sony writes off investment, 238–40, 268–69, 274
Columbia Record Club, 253
Columbia Records, 97, 130–33 *passim*, 175, 243, 248, 254, 308; drop in sales, 242
Columbia Tristar Pictures, 97, 158–59, 221, 225, 235, 253, 307; movies produced by, 224, 229–30, 309
Comcast, 316
Compact Disc Digital Audio System, 141. *See also* CDs (compact discs)
Compaq computers, 292

Computer market, 253–54; Sony in, 286, 292–93, 295, 300, 304–5, 318
Coney Island (New York), 51
Cooney, Joan Gantz, 71, 310
Coppola, Francis Ford, 224
Copyright problem, 107
Costner, Kevin, 187
Cronkite, Walter, 108
Cruise, Tom, 193

Daly, Robert A. "Bob," 196, 204, 215, 216, 228
"Dancing neon," 7
Dangerous (album), 255
Danone Group (France), 285
DAT (digital audiotape), 175; DAT Walkman, 260
Data Discman (CD-ROM reader), 249–50, 257
Davis, Clive, 98, 160
Davis, Geena, 224
Davis, Martin, 184
Davis, Marvin, 208n
Davis, Miles, 175n
"Days of Our Lives" (TV program), 185
Dazai, Osamu, 12
Deep, The (film), 194
Dekker, Wisse, 129, 130, 139
de Klerk, , 78
Dell computers, 292, 318
Delmonico International, 55, 57–59
Deng Xiaopeng, 76
De Niro, Robert, 309
"Dennis the Menace" (TV program), 185
"Densuke" tape recorder, 31
"Designing Women" (TV program), 186
Deutsche Telekom, 61, 259, 262n
Devito, Danny, 224, 235–36
Didion, Joan, 195
Digital Audio Disc (DAD) Conference, 139, 141
Digital broadcasting, 293–94
Digital cameras, 292
"Digital dream kids," 289–90
Digital vs. analog recording, 137–46
Digital video disc (DVD), 141, 278, 311
Digital videotape, 224
Diller, Barry, 309, 316
Dillon, Robert "Bob," 164
DiscoVision (video playback system), 107
Disneyland, 69, 70
Disney Productions, 108, 175, 184, 193, 234, 290
Dmytryk, Edward, 185
Doi, Toshitada, 138–41 *passim*, 145–46, 169, 206, 265, 325
Dolgen, Jon, 223, 228, 229, 233, 236
Doman, Glen, 49
Donner, Lauren and Richard, 228
Doyle Dane Bernbach (DDB), 72, 106–7, 108
"Dumping," 76, 111
Dunaway, Faye, 195
Dunne, John Gregory, 195
Dylan, Bob, 175n

Eastman Kodak, 148
Eastwood, Clint, 230
Edison, Thomas, 144
Eisner, Michael, 184, 290, 309, 313, 316
Electrolux, 315
Elizabeth II, queen of England, 267
Emerson television, 76
England, Walkman in, 154, 155
Environmental Protection Agency, 102
Epic Records, 133, 175, 255
Evans, David Mickey, 224, 228

Exorcist, The (film), 309
Eyes of Laura Mars (film), 195

Falcon airplane, 166–67, 169, 173
Fan, The (film), 309
Farr, Charlie, 58
Fender Guitar, 98
Fiddler on the Roof (musical comedy), 62n
Fink, Morton, 153
Finn, Kevin, 165
Firestone Tire Company, 181
Fischer-Diskau, Mr. and Mrs. Dietrich, 81
Fisher, George, 267
Flashdance (film), 193
Flatliners (film), 187
Forbes magazine, 258
Foreman, John, 195
Fortune 500 companies, 205
Fortune (magazine), 72
Fox television network, 186
Friedman, Albert, 55, 58
From Here to Eternity (film), 185
Fuchs, Michael, 309
Fuji Bank, 209
Fuji Television Network, 293
Fuji-Xerox, Ltd., 265
Furst, Anton, 220

Galiana-Mingot, Michel, 79
Ganis, Syd, 223, 228, 233
Gates, Bill, 290, 315, 316
GATT talks, 267
Geffen, David, 243, 309, 316
General Electric Company, 45, 110, 290
General Instrument, Inc., 315, 316
General Magic (Apple computer spinoff), 253–54
General Motors Corporation, 74, 238
General Radio Corporation, 6
Genryu (official Sony history), 54, 55
Gere, Richard, 225
Germany, 51, 74; tape and tape recorders from, 27, 28, 35, 124
Geronimo (film), 230
Gerstner, Lou, 290, 299
Ghostbusters (film), 185
"Gidget" (TV program), 185
GI Project (computers), 292
Godzilla (film), 236, 292, 312, 322
Goldman Sachs, Japan, 298, 299
Goto, Noboru, 82–83
Gould, Glenn, 175n
Graham, Katharine, 316
Griffin, Nancy, 203n, 228n
Gross, Adolf, 55–61 *passim*, 134, 285
Gross, Dorothy (Mrs. Adolf), 56
Group de Suez Banks, 285
Grove, Andy, 290, 292, 316
Grubman, Allen, 246
Grundig tape recorders and cassettes, 27, 129
Guber, Lynda (Mrs. Peter), 196, 197
Guber, Peter, 180–83 *passim*, 195–97, 202–5, 242–43, 253, 309; contract with Warner, 193–94, 208–19 *passim;* Morita and Ohga interview, 198–201, 233; Sony hires, 115, 302, (in Hollywood) 220–30, 233–34, 310, (leaves Sony), 235–38, 307–8; teaches film class, 200–201
Guber-Peters Entertainment Company (GPEC), 180, 183, 197–99, 202–5, 209, 210; Sony purchases, 215
Gulf and Western, 64, 184
Gwathmey, Charles, 251, 252

Haggerty, Pat, 73, 74
Hancock, Herbie, 146
Handycam (video movies), 258, 292

Hanks, Tom, 224
Hara, Wakiko "Mother Hara," 38, 40, 41
Harvard Business School, 249, 275
Hasegawa, Junichi, 28, 30
Hashimoto, Tsunao, 190, 274, 275, 279, 280, 298, 299; and Schulhof, 302, 303
Hattori, Gakujun, 20, 21
Hazelden Institute (Minnesota), 192, 193
Hearing aids, 32–33, 154
Heston, Charlton, 208
Hewlett-Packard, 103
Higashikuni, Naruhiko, 9
Higuchi, Akira, 26, 38, 52, 121, 284, 296n; joins Ibuka, 9–10, 14–15, 16; visits U.S., 6, 53
Hiraiwa, Gaishi, 267, 268
Hiramatsu, Kozo, 289
Hirohito, emperor of Japan, 5, 9, 16–17, 20, 217
Hitachi, 31, 65, 68, 110
Hit and Run: How Jon Peters and Peter Guber Took Sony for a Ride in Hollywood (Griffin and Masters), 203n, 228n
Hitler, Adolf, 142, 159
Hitotsubashi University, 298
Hoffman, Dustin, 193, 232
Hokkaido University, 6
Hollywood acquisitions, 61, 74, 178, 184, 189–92, 240; and anti-Japanese sentiments, 181, 216–17, 231–32. *See also* Columbia Pictures Entertainment
Home Box Office, 309
Home Furnishing Daily, 109
Home videotape recorder (VCR). *See* Betamax
Honda, Soichiro, and Honda Motors, 91
Honshu Paper Company, 29
Hook (film), 225, 228, 232, 255
Horowitz, Vladimir, 175n
Hoshikawa, Ken, 198, 199
Hudson Hawk (film), 228
Hüsch, Gerhard, 123, 274

Iba, Tamotsu, 274, 278, 294–98 *passim*, 302, 303, 322, 323
IBM (International Business Machines), 73, 96, 161–62, 238, 290; IBM Japan, 41, 42
Ibuka, Makoto (son), 1, 40, 41–42, 117, 269
Ibuka, Masaru, 76, 91, 325; as author, 49; early life and early career, 6–8, 24, 25n, (postwar return to Tokyo) 9–10; family life, 39–42; "Founding Prospectus," 10–11, 15, 23; illness and death, 268, 269–72, 281; at Japan Measuring Instruments, 8, 9; and Ohga, 121, 274; patents held by, 7; personal traits, (appearance) 4, (fascination with toys) 1–2, 37, (idealism and optimism) 11, 13, (inventiveness) 7–8, 13–15, 25, 43, 289, (single-mindedness) 37–38, (tactlessness) 35–36, 99n; relationship with Kihara, 25–27; relationship with Morita, *see* Morita, Akio; and Schulhof, 165–66

AND SONY, 128, 149, 156, 206, 262, 284, 287; as chairman, 95, 147; early days, 13–16, 53, 120, 123; ignorance of technology, 49–50, 138, 148; incorporates, 23; and Morita's move to New York, 68–69; retires, 38; and Sony products, 106, 110, (color TV) 43–48, 68, (requests portable player) 150, 151, 152, (tape recorders) 27–28,

30–31, 35, 120–21, (transistors) 31–36, 137; U.S. subsidiary, 59, 64, 66; visits U.S., 31–34, 62n; widowed mother (Sawa), 39, 40
Ibuka, Motoi (grandfather), 6
Ibuka, Sekiko Maeda (Mrs. Masaru), 39–41
Ibuka, Shizuko (daughter), 40
Ibuka, Taeko (daughter), 40, 49
Idei, Nobuyuki, 117, 272; early life and career, 282–86; as international figure, 315–17; as president of Sony, 86, 88, 122, 192, 253, 265, (refuses stock sales) 235n, (and Schulhof) 241, 299–303, 305, 307, 308, 325, (and videodisc project) 137, 286; as successor to Ohga, 280–82, 286–88, 316–25, ("regeneration" plans) 289–94, (structural changes) 294–303, 305, 306–15, 324–25; successor to, 323–24
Idei, Teruyo (Mrs. Nobuyuki), 117, 282
"I Dream of Jeannie" (TV program), 185
IFF (identification of friend or foe) device, 43
I'll Do Anything (film), 230
Imadato, Hiroki, 90, 91
Imperial Bank, 24
Inagaki, Shigeru, 51, 62–63, 94, 95, 105, 111
Independence Day (film), 309
Ingersoll, Mr. and Mrs. Robert, 80
Institute of Electronics and Electrical Engineers (New York trade show), 43
Institutes for the Achievement of Human Potential, 49
Integrated circuits, 49, 74, 148
Intel, 290, 292
International Brotherhood of Electronic Workers, 103n
International Labor Organization, 282
International Music Industry Conference, 142
Internet, the, 253, 292
In the Line of Fire (film), 230
Intuition at Work (Van Deckter), 289
IRS (Internal Revenue Service), 55
Ishibashi, Kanichiro, 90n
Ishihara, Hideo, 298
Ishihara, Shintaro, 217, 218
Ishihara, Shun, 90
Isuzu Motors, 74
IT (information technology), 289
It Happened One Night (film), 185
Ito, Kyoko, 126
Itochu Trading Company, 59, 293
Iwai, Ichiro, 119, 274
Iwaki, Ken, 101, 149–50, 151, 190–91, 264, 278, 296, 300, 307; as possible successor to Ohga, 149, 275–77, 288, 295
Iwama, Kazuo, 33–34, 104, 162, 166, 274, 286; heads Sony America, 95–96, 98, 101, 102, 111–12, 275; Schein and, 111–13, 114, 276; as Sony president, 33, 49–50, 147–50, 156, 163, 285, 325 (death of) 155–56, 263
Iwama, Kikuko Morita (Mrs. Kazuo), 33, 96, 148
Iwama Report, The, 34

Jackson, Michael, 171, 175n, 218, 242–46 *passim*, 254–55
Jaffe, Stanley, 309
Japan: automobile industry in, 74, 90; color television in, 42; Commercial Code of, 147, 205, (U.S.

Japan (*cont.*)
Occupation Revised Code) 205–6; corporate taxes in, 104; first transistor radio in, 35; GNP, growth of, 57; Imperial Army, 21, 85; Imperial Household Agency, 53; Imperial Navy, 4, 17, 21, 90n; Institute of Commerce and Industry, 283; lawsuits rare in, 107–8; Liberal-Democratic party, 267; linguistic insularity of, 81n; "made in Japan" stigma, 52, 57, 72; Ministry of Communications, 6; Ministry of Education, 21, 30, 120; Ministry of Finance, 59, 64, 65; Ministry of Trade and Industry (MITI), 31–32, 33, 36, 56, 74, 97; Office of Aviation Technology, 20; postwar, 11–17, 21, 28, 30, 34, 51, 60, 119; private clubs in, 90–91; semiconductors developed in, 31–32; Supreme Court, 30; symphony orchestra, 117; -U.S. relations, 9, 64, 73–74, 266, 268, (anti-Japanese sentiments) 181, 216–17, 231–32; War Office, 6; wartime, 4–6, 17, 85, 90n, 119
Japan Association of Microtrains, 1
Japan Bankers' Guild, 24
Japan Center for International Exchange, 268
Japan Development Bank, 46
Japan House (New York), 9
Japan IDG (publishing company), 289
Japan Measuring Instruments, 8, 9
Japan Musical Instrument Company (later Yamaha), 125
Japan Optical Sound, 9
Japan Record Dealers Association, 132
Japan Sky Broadcasting (JSkyB) Company, Ltd., 293
Japan Steel, 90
Japan That Can Say No, The (Morita), 89n, 217–18, 266
Japan Victor Corporation (JVC), 109, 110, 141
Jaws (film), 193
"Jeopardy!" (TV program), 185
Jerry Maguire (film), 312
Jobs, Steve, 286
Joel, Billy, 144
Johns, Jasper, 159
Johnson, Philip, 251
Jones, Quincy, 218
Jordan, Michael, 322
Jumanji (film), 309
Jurassic Park (film), 230
Justice Department, , 213
Juvilier, Adolph, 55

Kabat, Herb, 55
Kaji, Ryuzo, 14, 20
Kaloflex camera company, 63
Kaneda, Yoshiyuki, 295–96, 298, 318–19
Karuizawa Golf Club, 283
Katzenberg, Jeffrey, 184, 193, 313, 316
Kaufman, Victor, 186, 191
Kawai Piano Company, 98
Kazan, Elia, 185
Keidanren (Federation of Economic Organizations), 89, 232, 266, 267, 319; Council for Better Corporate Citizenship in the United States, 266
Keio University, 17
Kendall, Donald, 73
Kerkorian, Kirk, 185, 202, 204, 310
Kihara, Nobutoshi, 28–31, 35, 43, 106, 110, 206, 325; relationship with Ibuka, 25–27
Kihara-Sony Research Institute, 30
Kikuchi, Makoto, 33, 36, 49, 50, 148, 166

Kindergarten Is Too Late (Ibuka), 49
Kissinger, Henry, 75–76, 77, 78, 272
Kobayashi, Kenzo, 9
Kobayashi, Yotaro "Tony," 265, 272
Kobunsha publishing company, 217
Kodera, Junichi "Steve," 103–4, 278
Kogan, Leonid, 37
Korea proposed as base of operations, 103
Korean War, 31
Kowa, Viktor de, 124n
Kurahashi, Masao, 120
Kurata, Hiroko, 38–39, 47, 281
Kurosawa, Yoshiko, 40–41
Kutaragi, Ken, 304
Kyuzaemon Morita the Tenth, 23n, 52
Kyuzaemon Morita the Fourteenth (father), 17–18, 19–20, 21–22, 72, 149; and fledgling company, 22, 23n, 28; as largest shareholder, 24, 54
Kyuzaemon Morita the Fifteenth. *See* Morita, Akio
Kyuzaemon Morita the Sixteenth, 87. *See also* Morita, Hideo

Labor unions, 103–4
Lagore, Joe, 163
Lambsdorff, Otto Graf, 74, 75
Land, Edwin, 73, 76, 297
Landau, Jon, 242, 243
Landro, Laura, 301, 302
Last Action Hero (film), 230, 233
Lauper, Cyndi, 175n
Lawrence, E.O., 43, 44
Lawrence of Arabia (film), 185
League of Nations, 283
League of Their Own, A (film), 224, 229
Lean, David, 185
Legends of the Fall (film), 309
Lehman Brothers, 76
Levin, Gerald, 316
Levine, Alan, 202, 210, 211, 221, 222, 225–31 *passim*, 233–36 *passim;* at Sony Pictures, 308, 309, 311
Levine, "Jimmy," 71
Levinson, Barry, 225
Lévi-Strauss, Claude, 291
Lieberson, Goddard, 74, 132, 175n
"Light telephone," 7
Liman, Arthur, 209
Lion Toothpaste Company, 8, 9
Li Peng, 78
Loews Theater chain, 172, 180, 253; becomes Loews Cineplex Entertainment Corporation, 300n
Longshoremen's union, 103n
Los Angeles Times, 222, 225, 231, 308
Lowy, Frank, 197
Lugosi, Bela, 106

Maazel, Lorin, 19, 117
MacArthur, General Douglas, 21, 119, 206n
MacNaughton, Donald, 74–75
Macy's of San Francisco, 102
MAD ("Musicians Against Digital"), 145
Made in Japan (Morita), 20, 199
Maeda, Tamon, 8–9, 14, 23, 30, 54, 206; Ibuka marries daughter of, 39–40
Magic Johnson Theatres, 300n
Magnavox, 110
Magnetic tape and tape recorder, 26, 27; "Densuke" recorder, 31; Ibuka and, 27–28, 30–31, 35, 270n, (Tape-corder) 120–21; Japanese vs. German, 124; Kihara's production of, 28–31; Ohga and, 128
Mainichi (newspaper), 31
Malone, John, 316

Mandai, Junshiro, 24, 65, 206
Mandalay Entertainment, 237, 308–9
Mandela, Nelson, 78
Mansfield, Mike, 75
"Marque" (heat-seeking missile), 4
"Married . . . with Children" (TV program), 186
Masaki, Teruo "Ted," 182, 191, 202, 207, 210, 212–13, 214
Mascarich, Frank, 32, 33
Masters, Kim, 203n, 228n
Masutani, Rin, 24
Mathis, Johnny, 175n
Matsubara, Midori. *See* Ohga, Midori Matsubara
Matsushita, Konosuke, 110–11, 263
Matsushita Electric, 31, 139, 239, 260; as Sony competitor, 68, 109–11, 141, 250, 258
Matsuzono, Hisami, 91
"Mavica" (video still camera), 27
Maxim's (Tokyo branch), 39, 80
MCA (Music Corporation of America)/Universal, 106–7, 108, 184
Medavoy, Mike, 221, 222, 224–25, 228–31 *passim*, 235
Meinl, Jurius, 124n
Men in Black (film), 236, 312
Meriton (Sony subsidiary), 105
Merv Griffin Enterprises, 185
Metropolitan Opera Orchestra, 170
Mettler, Rube, 73
MGM (Metro-Goldwyn-Mayer), 183, 184–85, 204, 209; MGM/UA, 185, 310
Microsoft, 286, 290, 292, 315
Midnight Express (film), 194
Mifune, Toshiro, 283
Milstein, Nathan, 68
Minidisc. *See* CDs (compact discs)
Minnesota Mining and Manufacturing (3M), magnetic tape made by, 27, 29
Mishima, Yukio, 71n
Missouri, USS, 10
MIT (Massachusetts Institute of Technology) Lincoln Labs, 137
MITI (Ministry of Trade and Industry). *See* Japan
Mitsubishi Real Estate Company, 181, 239
Mitsubishi Trading Company, 65, 110
Mitsui Bank, 24, 59, 65, 124, 209
Mitsui Trading Company, 59, 65, 293
Miyake, Issey, 39
Miyaoka, Senri, 43–48 *passim*, 138, 166, 325
Miyatake, Kazuya, 60
Miyazawa, Kiichi, 74
MJJ Music, 255
Monet, Claude and Philippe, 171
Monetary Control Association, 23
"Monkees, The" (TV program), 185
Morgan, , 239, 302
Morgan Guaranty Trust, 73, 74
Morimoto, Masayoshi "Mike," 103
Morio, Minoru, 265, 278, 286, 288, 294–300 *passim*, 322, 323
Morita, Akio (Kyuzaemon Morita the Fifteenth), 35, 129, 227, 243–44, 307; in dealing with Western world, 63–64, 88–89, 91–92, 93–96, 99, (and unions) 103, (visits U.S. and Europe) 32, 51–52, 53–67; education, 18–20; and family business, 22, 298; in family hierarchy, 69, 82, 84–88, (as eldest son) 3, 17–20 *passim*, 22, 63, 72–73; and Idei, 284–85; as international

figure, 267–68, 285, 315, 317; personal characteristics, (appearance) 4, (fascination with toys and mechanics) 1–2, 18, 151, 290, (friendships) 61–62, 299, (social presence) 73–83, 88, (as teetotaler) 18n; in postwar Japan, 16–17; and private clubs, 90–91, 133; relationship with Ibuka, 1–4, 5, 20–21, 24, 38, 39, 45, 54, 157, (in Ibuka's company) 15–16, 23, (Ibuka's son) 42

AND SONY, 287; acquisitions, 97–98, 130–31, 147, 175–78, 183–219, 232, 240; as chairman, 147–48, 262; commitment to, 325; and Iwaki, 275, 276–77; vs. Matsushita, 109–11; and Ohga, 98, 116, 121–37 *passim*, 146–50 *passim*, 156–57, 158, 263, 273–74; and Ohga's successor, 280–82; Walkman, 137, 150–55, 289

AND SONY AMERICA: American protégés, *see* Schulhof, Michael P. "Mickey"; Yetnikoff, Walter; in crisis, 163–64; early days, 51–67, 68–92; and music business, 97–98; opens California plant, 103–4; as president, 95; recruits Iwama, 33; resigns, 23n, 268; Schein and, 96–111, 113–15, 158; Sony America products, 28–32 *passim*, 43–48 *passim*, 142; and stock sales, 302; suffers stroke, 23n, 51n, 77, 79, 82, 265–74 *passim*, 279, 300, 307n; writings, 89n, 217–18, (memoir) 20, 54, 100, 199

Morita, Hideo "Joe" (son), 2, 3, 23n, 63, 70, 88, 133, 270, 282; and family business, 22, 86–87; relationship with father, 69, 82, 85, 86, 272–73

Morita, Kazuaki (brother), 2, 16- 17, 18, 22, 71, 87, 117

Morita, Keizo (uncle), 51n

Morita, Kikuko (sister). *See* Iwama, Mrs. Kazuo

Morita, Masaaki "M. M." (brother), 16–17, 18, 21n, 113, 117, 149, 206, 265, 296; in family hierarchy, 18, 156, 262; as Sony America chairman, 189–90, 214, 262–64

Morita, Masao (son), 1, 2, 70, 83, 114, 270, 272; relationship with father, 51n, 84–85, 273; at Sony, 85–86, 87–88, 124n, 277, 323

Morita, Naoko (daughter), 70, 281

Morita, Yoshiko "Yoshi" (Mrs. Akio), 23n, 41, 66–69 *passim*, 83, 91, 99, 125, 131, 142, 198, 267n, 269; delivers eulogy at Ibuka's funeral, 269–72; family publishing house, 217; and Idei family, 282; social life, 71, 80–81, 96, 199, 232, 268; and Walkman, 152

Morita Company Ltd. (family business), 17, 20, 22, 86, 87

Morita Foundation, 22–23

Morizono, Masahiko, 123, 149, 190, 206, 296n, 325

Moss, Jerry, 143

Motion Picture Association of America, 108–9

Motion Picture Group, 229

Motorola, 253, 267, 286

Mottola, Thomas "Tommy," 244–47, 248, 322, 324

Mr. Jones (film), 225

Mr. Saturday Night (film), 229

MSX Computer, 286, 289

Multiplicity (film), 309

Murdoch, Rupert, 192, 234, 293, 316

Museum of Modern Art (New York), 71
My Best Friend's Wedding (film), 312
My Fair Lady (musical comedy), 56, 175n

Nakajima, Heitaro, 137, 138, 140–41
Nakasone, Yasuhiro, 267
Nakatani, Iwao, 298
Nakayama, Teiichi, 119, 122
Nanao Wireless, 9
Narita, Mitsuzo, 149–50
National Music Publishers' Association, 260
Nation Records, 255
Nawrocki, Eileen, 246
NBC (National Broadcasting Company), 107
Negroponte, Nicholas, 290
Nestlé, 315
Nevin, John, 111
New Kids on the Block (music group), 245
Newsweek, 72, 181
New York Philharmonic Orchestra, 154
New York Stock Exchange, 66
New York Times, 128, 145n, 217, 225, 229, 232, 260, 261, 264
New York Times Magazine, 174
NHK (Japan national broadcasting service), 27, 34n
Nicholas, Nick, Jr., 209
Nichols, Mike, 236, 310, 314
Nicholson, Jack, 193, 236
Nightingale-Bamford School (New York), 70
Nintendo video game, 250, 304, 305
Nippon Columbia, 97, 248
Nippon Electric Company (NEC), 93
Nishida, Kahei, 119–20
Nishikawa Electric, 123
Nishiyama, Sen, 99
Nissan Motors, 90
Nissho Iwai, 293
Nixon, Richard, 73, 75, 76, 187
Nobel Prize, 43, 283
Nolan Sisters, 86
Nomura, Kodo, 39–40
Nomura Securities, 65, 66, 90n
Nozoe, Masayuki "Yuki," 311–12, 322

Ochanomizu Women's College, 283
O'Donnell, John, 288–89
Oe, Kenzaburo, 12, 283
Ohga, Midori Matsubara (Mrs. Norio), 123, 125, 135, 269
Ohga, Norio, 163, 227, 251–52; early life and career, 35, 118–26; friendships with Jews, 62; injuries and illnesses, 139, 150–51, 189, 277; as musician/artist, 116–18, 122–26, 132–33, 144, 170, 290, 314, 320; officiates at Ibuka's funeral, 269–70
AND SONY: acquisitions, 176–77, 184–90 *passim*, 192, 199–202, 207, 211–16 *passim*, 223; arrives at (1959), 122, 126–30; commitment to, 325; in company hierarchy, 262, 286; decision-making process, 150, 189, 206–8; and digital recording/CDs, 138, 145n, 146, 150, 155, 169, 286, 289; heads CBS/Sony, 86, 98, 130–37, 138, 144–51 *passim*, 173; information technology, 292; and Masao Morita, 88, 124n, 263; Morita's chosen successor, 148, 156, 240, 263, 274, 281; movie business, 221, 229, 231–35, 236–37, 239, 274, 302, 311; protégés of, *see* Schulhof, Michael P. "Mickey"; Yetnikoff, Walter; and Sony im-

age, 127–29, 287, 300, 321, 325; and Stein, 99, 113–14, 115; and successor to, 78–79, 149, 274–79, 280–82, 286–88, 294–95, 297, 312, 317–20 (*see also* Idei, Nobuyuki); and technology, 121, 138–39, 155; writes off loss, 239, 274
Ohgoshi, Akio, 47
Ohira, Masayoshi, 73, 267
Ohsone, Kozo, 150–55 *passim*, 206, 274, 278, 294–98 *passim*, 322, 323, 325
Okazaki, Yuka, 86
Okochi, Hiroshi, 60
Okumura, Tsunao, 90n
Olafsson, Olaf, 249, 250, 255, 305
Onoyama, Hiroko, 39n, 148, 163, 268, 269
On the Waterfront (film), 185
Opel, John, 73
Optic laser technology, 49, 155
Orion Pictures, 221
Osaka Imperial University, 4
Ottens, L. F., 139, 169
Overseas Debt Service Company, 289
Ovitz, Michael, 309, 313; and Hollywood acquisitions, 184, 187, 188, 191–92, 201, 239
Ozawa, Seiji, 19, 283

Paley, William, 97, 133, 173, 174, 176
Palm Pilot, 254
Pan American, 73
Pan American International Advisory Committee, 76
Panasonic, 45, 76, 162
Papilio Cosmetics, 29
Paramount Pictures, 43, 64, 184, 223, 309
Paris World's Fair (1933), 7
Parker, Charlie, 145
Parsons, Richard, 316
Pavarotti, Luciano, 200
PCL (Photo-chemical Laboratories), 7–8, 24; later Sony-PCL, 42
Pei, I. M., 192
Peltz, Nelson, 172, 175
Pepsi-Cola, 73
PerfecTV Corporation, 293
Peters, Jon, 180–83 *passim*, 194–97 *passim*, 202, 203–5, 253, 255; contract with Warner, 193, 208–19 *passim;* Sony hires, 115, (arrives in Hollywood) 220, 222, 224, (fired) 225–27, 233, 308
Peterson, Peter, 73, 76–77, 81, 173, 231, 272, 308, 309, 310; and acquisitions, 184, 191, 197, 201, 202, 210–15 *passim;* Schulhof as viewed by, 215, 252, 254, 299; on Sony Board, 77, 187–88, 207, 298, 301
Philips Electronics , 36, 52, 169, 193; CD player of, 142, 144; Sony relationship with, 49, 97, 129–30, 137, 139–46, 261, 286
"Phosphor doping" process, 34
Pierson, Frank, 195
Pitt, Brad, 309
PlayStation (video game). *See* Sony PlayStation
Plaza Agreement (1985), 258
Poehl, Karl Otto, 78
Polaroid Corporation, 73; Polaroid Japan, 297
Pollack, Sidney, 185
Polygram, 193, 198
Porter, Michael, 291
Postcards from the Edge (film), 310
Potsdam Declaration (1945), 5
Pressman (tape recorder), 150, 152
Price, Frank, 193, 221, 222, 224, 227–31 *passim*, 308

Price Waterhouse, 65
Prince of Tides, The (film), 224, 225, 228
Prudential Life Insurance Company, 74
Psygnosis software, 250
Puttnam, David, 204

Qintex Group (Australia), 185
Quilted Giraffe (restaurant), 252

Radio Engineering (Turman), 6
Radio Flyer (film), 224, 228–29
Radios. *See* Transistors
Rain Man (film), 193, 202
RayKay Inc. (holding company), 33, 87
RCA (Radio Corporation of America), 32, 42, 45, 96, 256; competes with Sony Trinitron, 104; Selectavision VHS, 110, 111
Regency radio company, 35
Reischauer, Edwin, 99n
Reitman, Ivan, 185
Remains of the Day, The (film), 310
Reproducta, Inc., 159, 168
Return to the Blue Lagoon (film), 224, 228
Revenge (film project), 187
Riboud, Antoine, 285
Rikkyo University, 41
River Runs Through It, A (film), 229
Roberts, Julia, 232
Roberts, Ralph, 316
Robinson, James, 267
Roche, James, 74
Rockefeller, David, 71, 73–74
Rockefeller, Mrs. David, 71
Rockefeller Center, 181, 239
Roepke, Theodore, 284
Rogers, Fred ("Mr. Rogers"), 109
Rosenman, Colin, Freund, Lewis, and Cohen, 131n, 173
Rosenman and Stern, 61
Rosing, Wayne, 286
Rosiny, Edward, 94–95, 98, 162, 164, 173; Morita's relationship with, 58, 61, 62, 71, 103n, 131n
Ross, Courtney (Mrs. Steven), 208, 218
Ross, Steven, 114, 164, 184, 241; and Columbia Record Club, 253; Sony vs., 115, 208–11 *passim*, 215, 216, 218, 246
Rossen, Robert, 185
Ruthless People (film), 184

Sagansky, Jeffrey, 307–8, 309
Sagor, Irving, 55–64 *passim*, 67, 96, 102, 105, 112
St. Bernard's School (New York), 69–70, 71
Saito, Eishiro, 90
Saji, Keizo, 90n, 117
Sakura Bank, 298
Sanger, David, 128
Santa Fe Computer Institute, 290
Sanyo, 76, 110, 250
Sato, Eisaku, 267
Sawyer, Diane, 310, 314
Schaeffer, Paul, 197, 199
Schein, Harvey, 61, 62, 158, 160, 165, 278; and CBS/Sony, 130–36 *passim*, 172, 177; heads Sony America, 96–97, 98–109, 111–15, 161–66 *passim*, 275–76
Schlessinger, Stanley, 246
Schmuckli, Jack, 187, 258, 261–62, 297
Schoenberg, Arnold, 175n
Schulhof, Hannalore (Mrs. Rudolph), 159
Schulhof, Michael P. "Mickey": early life and career, 159–66; Morita's and Ohga's relationship with, 62, 142, 143, 161–71 *passim*,

173–79 *passim*, 233, 241, 261–62, 279, 301, (adopted into Sony family) 158, 248–49, 305–6, 325, (fired) 241, 247, 302–4, 305–8, 325, (positions of power) 61, 115, 159, 241–42, 264–65, 300
AND SONY AMERICA: acquisitions, 158, 168, 171–78, 183–97 *passim*, 201–2, 208–18 *passim*, 224, 241–42, 250; creates new businesses, 253–55, 258; Guber and, 223–27, 233, 235–37; Idei and, 241, 299–303, 307, 308, 325; ignores Masaaki Morita, 190, 263–64; relocates headquarters, 250–53; severance agreement, 182; on Sony Board, 77, 168; and Sony software and electronics (Sony USA), 233, 249–50, 255–64 *passim*, 277, 302–8 *passim;* and Yetnikoff, 244, 245–47
Schulhof, Paola (Mrs. Michael), 143, 160, 208
Schulhof, Rudolph, 159n
Schwartzenbach, Ernest, 61, 65, 66, 94, 98
Schwarzenegger, Arnold, 230
Schwarzman, Stephen, 173, 177, 187, 197, 202, 210, 301, 302
Scully, John, 253–54
SEC (Securities and Exchange Commission), 65, 107, 238n
Seeger, Pete, 175n
Sega video game, 250, 305
"Seinfeld" (TV program), 186n
Sekimoto, Tadahiro, 93
Semel, Terry, 196, 204, 215, 216, 228, 316
Semiconductors, Japanese production of, 31–32, 95, 148
Sense and Sensibility (film), 309
"Sesame Street" (TV program), 310
Shadow mask system. *See* Television (color)
Sharp television, 76, 110, 141
Sheinberg, Sidney, 106–8
Shibaura Electric Company, 7, 95. *See also* Toshiba
Shidehara, Kijuro, 9
Shimoda conference, 73
Shiroyama, Saburo, 88–89
Shore, Paul, 161
Showa Bank, 23
Siegel, Bugsy, 225
Simon and Garfunkel, 132, 133, 175n
Simon and Schuster, 218
Simpson, Don, 254, 309
Single White Female (film), 229
Sinjou, J. P., 140
Skadden Arps, 202, 210
SKY PerfecTV!, 293–94, 318
Sleepless in Seattle (film), 230
Smith, Will, 254
Smith Barney, 65, 68, 69, 94
Snapple company, 175
SOBAX electronic calculator, 161
Softbank software (Japan), 293, 307
Sommer, Ron, 61, 258–62, 277, 307
Son, Masayoshi, 293
Sony Aviation (New York), 243
Sony of Canada, Ltd., 88, 324
Sony Club, 251–52
Sony Computer Entertainment, 295, 304
Sony Corporation: acquisitions, 74, 180–219 (*see also* CBS Records, Inc.; Columbia Pictures Entertainment); advertising, 128–29, 153–55, 287, 289; aircraft owned by, 166–68, 169, 173, 243–44; American branch, *see* Sony Corporation of America (Son-Am); Audio Division, 86, 145; audiotape rights repurchased, 105;

Sony Corporation (*cont.*)
Betamax debacle, *see* Betamax (VCR); board of directors, 156, (approves Columbia purchase) 205–8, (first American on) 77, 207, ("outsiders" invited on) 207, (Schulhof on) 77, 168; Bulova offer declined, 54–55, 164; Business Machines Division, 161; CDs of, *see* CDs (compact discs); Central Research Laboratory, 36, 138, 148; Chinese market for, 76; and color television, *see* Chromatron; Trinitron; company image, 127–29, 287, 290, 300, 321–22, 325; and computer market, 286, 292–93, 295, 300, 304–5, 318; Consumer A&V Products Company, 278; Consumer Products Division, 164, 250; Core Technology and Network Company, 324; Corporate Communications, 286; Corporate Planning Division, 276; early history, 13–16, 51–67, (first development loan) 46, (first European interest in) 125, (investment in) 22, 24, 38, 40; early success, 8; Executive Committee, 147, 176, 215, 262, 275–76, 304, (balance of power) 318, (decision-making) 149–50, 151, 189, 199, 206–8; fiftieth anniversary, 127; Group Headquarters, 324; growth of, 26, 28, 35, 38, 54, 68, 122, (earnings drop) 143, 150, 181, 238, 258–59, (new strategy) 291; hierarchy in, 28, 262, 264, 286 (*see also* Morita family in, *below*); Home Network Company, 324; Home Video Department, 286; Ibuka's son joins, 42; Idei as head, *see* Idei, Nobuyuki; innovative products from, 26–27, 43, 148 (*see also* tape recorders, *below*); international trade division, 63, 68; *Iwama Report* on file at, 34 (*see also* Iwama, Kazuo); joint ventures, 64, 74, 75, 114, 130, 286, 292, 293 (*see also* Philips Electronics N.V.); lawsuits against, 76, 108, 111, 182, 247, 260, (Warner Brothers) 180, 183, 211, 215–19; location of worldwide headquarters, 16, 97; Micro-Computer Office Automation Group (MCOA), 286; Morita family in, 18, 22–23, 85–88, 262, (adopted members) 156–57, 158–59, 240, 297 (*see also* Morita, Akio); in movie business, 228–34; and music, 19, 117–18; music business, 61, 97–98; naming of, 52–53; as OEM (original equipment manufacturer), 165; official history (*Genryu*), 54, 55; Ogha at, *see* Ogha, Norio; opens American manufacturing plant, 103–4; Personal and Mobile Communication Company, 88; personal commitment to, 325; Personal IT Network Company, 37–38, 324; Product Planning Center, 128; Professional and Broadcast Products Division, 164; Recording Media and Energy Company, 278; "regenerated," 289–94; Research Laboratory, 152; scholarship established by, 58; severance agreements, 182; stock options, 98–99; stock sales, public, 239, 259n, 301–2, (first Japanese sales in U.S.) 64–67, (refused) 235n; subsidiaries (Aiwa America, Meriton), 105; Systems Business Company, 278; tape recorders and tape, 27–31, 128, 175, 270n, (Tape-corder)

120–21, (Walkman) 137, 150–55, 183; transistor radios, 35, 53–59 *passim*, 124–25, (Delmonico returns inventory) 58–59; transistor television, 27, 57, (micro-) 67, 69, 71–72 (*see also* Chromatron; Trinitron); video recorders, 96, 143, 148; women in Organization, 39n; writes off loss, 238–40, 268–69, 274; Yoshiko (Mrs. Morita) and, 270

Sony Corporation of America (Son-Am): advertising by, 106, 109; American management of, 93–94, 256, 313, 324 (*see also* Schein, Harvey; Schulhof, Michael P. "Mickey"; Steiner, Ray); Compensation Committee, 77, 131n, 299; emotional distance from Japan, 109, 114, 115, 162; established, 59–67; growth of, 96–97; Iwaki at, 264, 277; lack of profitability, 104–5, 113; lawsuits against, 108, 111; Masaaki Morita heads, 189–90, 214, 262–64; moves to new building, 100, (relocates) 250–53; opens California plant, 103–4; as parent subsidiary, renamed, reverts to original name, 256n, 264; and stock sales, 301–2; Tamiya at, 278

Sony Electronic Publishing (now Sony Electronics, Inc. [SEL]), 61, 255, 256n, 258, 293, 324; renamed, 256n, 264; Schulhof and, 249, 264, 300, 307

Sony Europe, 61, 187, 258, 262n, 297

Sony France, 79, 285

Sony Germany, 258, 262n

Sony Group, 293, 312, 324

Sony Hawaii, 100

Sony Industries, created, 164

Sony Life Insurance Company, Ltd., 75, 264n, 277, 295, 296

Sony Music Entertainment (SME), Inc., 178, 218, 245–49 *passim*, 253–56 *passim*, 293, 301, 308, 320; CBS Records renamed, 248; and DAT format, 260; future of, 322; SME Japan, 88, 304

Sony Music Foundation, 117

Sony Open golf tournament, 272

Sony Overseas, 284, 285

Sony-PCL, 42. *See also* PCL (Photochemical Laboratories

Sony Pictures Entertainment, 249, 292, 293; as American company, 231–32; costs, 229, 232; Idei and, 308–14, 322–23; name adopted, 219, 227; public offering refused, considered, 235n, 301–2

Sony PlayStation (video game), 250, 304–5, 317

Sony Plaza, 252

Sony-Prudential Life Insurance Company, 75. *See also* Sony Life Insurance Company, Ltd.

Sony Retail Entertainment, 253, 300n

Sony Signatures, 253

Sony Software, Inc., 233, 249–50, 255–58, 302, 304, 307, 308; Home Networking Module, 315; Platform Development Center, 315

Sony Theatres, 300n

Sony USA (SUSA), 256n. *See also* Sony Corporation of America (Son-Am)

"Sounds of Silence, The" (hit single), 132

Spielberg, Steven, 185, 193, 225, 228–32 *passim*, 255

Springsteen, Bruce, 171, 175n, 242–43, 244

Stallone, Sylvester, 230
Stancil and Hoffeman Company, 31
Stanton, Frank, 96, 98
Star Is Born, A (film), 195
Stark, Ray, 186
Star Theatres, 300n
Steiner, Ray, 94–95, 101, 102, 109, 111, 112; as president of Sony America, 113, 164, (death of) 162–63
Steinway, Henry, and Steinway Piano, 98
Stern, Isaac, 267
Stewart, Potter, 200
Sting, The (film), 114
Strauss, Robert, 266
Streisand, Barbra, 171, 175n, 195, 224
Stringer, Howard, 299, 313–15, 316, 320, 322–23, 324
Striptease (film), 309
Suematsu, Kenichi, 298
Sumitomo, 293
Sun Microsystems, 286, 315
Suntory Group, 90n, 117
Superman (film), 309
Suprascope company, 105, 162
Sushi-ko (Tokyo sushi bar), 91
Suzuki, Dr. Akio, 265
Suzuki, Masayoshi "Big," 59, 62
Suzuki, Shinichi, and Suzuki music method, 49
SW (Sony and Warner) Networks, 253
Sweden, Walkman in, 154
Sylvania, 95, 110, 163

Tachikawa, Shozaburo, 23, 28, 41–42, 263
Tajima, Michiharu, 23–24, 38, 53, 54, 206
Tajima Company, 104
Tamiya, Kenji, 162–63; heads Sony America, 164, 165, 168, 178, 256, 278; and Schein, 100–102, 109, 112; on Sony Board, 206
Tanaka, Michiko, 124–25
Tape recorder. *See* Magnetic tape and tape recorder
Taylor, Arthur, 135–36
Teamsters Union, 103n
Technology Fair (1998), 320
Techtronix, Inc., 103
Telefunken, 27, 35, 124, 129, 141
TELE-TV experiment, 313
Television, 27, 57; color, (in Japan and U.S.) 42, (shadow mask system) 42, 43, 45, 47; micro-, 67, 69, 71–72. *See also* Chromatron; Trinitron
Terasawa, Yoshio "Terry," 65, 66
Terminator 2 (film), 228
Texas Instruments, 35, 73; Texas Instruments Japan, 74
Thalberg, Milton, 59, 60
Thriller (hit record), 243
Tiffany building, 181
Time (magazine), 72, 106
Time-Warner, 209–10, 215, 218, 316. *See also* Warner Brothers/Warner Communications
Timmer, Hans, 143
Tisch, Lawrence "Larry," 146, 172, 173, 174–75, 176–77
Toho film studios, 8, 236
Tojo, Hideki, 9, 85
Tokugawa family, 120
Tokunaka, Teruo "Terry," 188, 304–5
Tokyo Broadcasting System, 293
Tokyo Electric Power Company, 267
Tokyo Imperial University, 23
Tokyo Institute of Technology, 20, 21
Tokyo Medical and Dental University Hospital, 265
Tokyo Symphony Orchestra, 117

Tokyo Telecommunications Research Institute, 10
Tokyo Telecommunications Engineering Co. Ltd. "Totsuko," 263; formally incorporated, 23, 33; investment in, 24; name changed to Sony, 52–53; Ohga and, 120–25 *passim;* and tape recorders, 27, 30, 124; and transistor radios, 35. *See also* Sony Corporation
Tokyo University of the Arts, 119, 120
Tootsie (movie), 185
Toshiba (formerly Shibaura), 95, 110; Ibuka fails at, 7; as Sony competitor, 31, 45, 65, 68, 141
Totsuko. *See* Tokyo Telecommunications Engineering Co. Ltd.
Transistors: germanium contact point, 32; Sony's development of, 31–35; transistor radios, 27, 33–35, 53–59 *passim,* (Japanese program on "Birth of") 34n; transistor television, *see* Television
Transistor Technology (Bell Laboratories), 32, 33
Trenchant, Georges, 284
Trilateral Commission, 73–74, 147
Trinitron (color television), 38, 102, 106, 112, 128, 321; development of, 42, 46–48, 103–4, 113, 278, (secrecy about) 166; profit from, 317
Tristar, 180, 191, 209, 221. *See also* Columbia TriStar Pictures
TRW, 73, 165
Tsukamoto, Koichi, 91
Tsunoda, Koichi " Mike," 164
Tsurushima, Katsuaki, 37
Turman, F. E., 6
Turner, Ted, 185, 234
Tushinsky brothers, 105
20th-Century-Fox, 192, 223, 309
Twombley, Cy, 159

Uemura, Taiji, 7–8, 38
Union Carbide, 64
United Artists, 221, 310; MGM/UA, 185
United States: broadcast industry standards, 258; color television in, 42; Ibuka and Iwama visit, 31–34, 62n; Japanese acquisitions in, 181; -Japan relations, 9, 64, 73–74, 266, 268, (anti-Japanese sentiments) 181, 216–17, 231–32; Morita moves to, 22, 32, 51, 53–67, 68–73; Sony stock sales in, 64–67; transistor radio sales in, 57, (first) 35; Walkman in, 154. *See also* Sony Corporation of America (Son-Am)
United States, USS, 125
Universal Studios, 221, 230, 239. *See also* MCA (Music Corporation of America) Universal
Unoki, Hajime "Jimmy," 60
Updike, John, 193
U.S.-Japan Economic Relations Group, 73
U.S. Occupation Army, 15, 16
U.S. Office of Civil Information and Education, 27
U.S. Robotics, 254

VAIO (video audio integration operation) laptops, 292–93, 317, 318
Valenti, Jack, 108–9
Van Deckter, Gigi, 289
Vander Dussen, Neil, 178, 256–57, 258, 259
Van Tilberg, Johan, 139
VCR. *See* Betamax
Verhoeven, Paul, 225

VHS (Video Home System), 109–11, 141, 150, 175
Viacom, 236
Victor Records, 57
Video games, 250, 255, 304–5
Video home systems. *See* VHS
Video laser disc project, 137, 286
Video movies (8-millimeter camcorder), 27, 96, 143, 148, 150, 183, 278; Handycam, 258, 292
Video printer (color), 27
Video still camera ("Mavica"), 27
Videotape, digital, 224
Videotape recorder (VTR), 27, 96
Village Voice, 195
VIVA (German TV-music channel), 253
Volcker, Paul, 78
Volkswagen (VW), 51, 71–72
Voltmeter, 15
Von Karajan, Herbert, 19, 137, 142–43, 146, 274, 286
Vreeland, Diana, 71

Wacoal company, 91
Walkman, 137, 150–55, 183, 255, 278, 289, 296; DAT, 260; other names suggested, 154–55, 305; profits from, 256
Wallace, George, 106, 162
Wall Street Journal, 246, 250, 301
Walter, Bruno, 175n
Warner Brothers/Warner Communications, 184, 193, 195, 309; Guber-Peters contract with, 180, 183, 194, 196, 202–10 *passim*, 215; Peters returns to, 227; sues Sony, accepts settlement, 211, 215–19. *See also* Time-Warner
Warner Records, 242, 253
Warren, Lesley Ann, 195
Waseda Journal, 284
Waseda University, 7, 17, 282, 283; Ibuka lectures at, 25
Wasserman, Lew, 107, 184
Watson, James, Jr., 73
Watts, André, 80
Weatherstone, Dennis, 302
Webster Chicago, 27
Welch, Jack, 267, 290, 298, 299
Western Electric, 96; Sony's license from, 31–33, 51
Westgate, R. I.W., 70
Westinghouse, 34
"Wheel of Fortune" (TV program), 185
"Who's the Boss?" (TV program), 185
Williams, Andy, 81, 133
Williams, John, 200
Williams, Robin, 232, 309
Willis, Bruce, 228
Wine, Barry, 252
Wisemen's Council, 73
Witches of Eastwick, The (film), 193
Wolf (film), 236
Wonder, Stevie, and Wonderland Studio, 145–46
World Economic Forum (Switzerland), 78, 315–16
World War II, 23, 90n; attack on Pearl Harbor, 8; cost to Japan, 11–13; Japan during and postwar, *see* Japan; last days of, 4–6, 8–9, 10, 16–17, 20
Wyman, Thomas, 144, 146, 169
Wynne, Bob, 309

Yamada, Shido, 55
Yamaguchi, Hiroshi, 82–84, 89, 90n, 91, 272
Yamaguchi, Momoe, 132
Yamaguchi, Takahiro, 289
Yamaha, 125
Yamamoto, Tadashi, 268

Yanase, Jiro, and Yanase Motors, 90
"Yankee Alley," 15
Yankowski, Carl, 300, 307, 308
Yawata Steel, 65
Yetnikoff, Cynthia Slamar (Mrs. Walter), 173
Yetnikoff, June (Mrs. Walter), 135
Yetnikoff, Walter, 132, 209, 221; and Hollywood acquisitions, 61, 131, 171–79 *passim*, 183–89 *passim*, 192–99 *passim*, 204, 210–18 *passim*, 241–42; Morita's and Ohga's relationship with, 62, 131, 134–37, 169, 178, 241–49 *passim*, (fired, severance agreement) 182, 245–48; and Peters, 226–27; relationship with artists, 242–44; and videodisc roject, 137, 286
Yokogawa Electric, 16
Yoshida, Shigeru, 90
Yoshida, Susumu, 42n, 45–48 *passim*, 104–5, 123, 149, 325
Young, Neil, 145

Zenith, 111
Zero Years Old (Ibuka), 49
Zinnemann, Fred, 185